AGRICULTURE ISSUES AND POLICIES

COVID 19

AGRICULTURAL AND FOOD ISSUES

AGRICULTURE ISSUES AND POLICIES

Additional books and e-books in this series can be found on Nova's website under the Series tab.

AGRICULTURE ISSUES AND POLICIES

COVID 19

AGRICULTURAL AND FOOD ISSUES

KRISTIAN GOLDSCHMIDT
EDITOR

snova
New York

Library of Congress Cataloging-in-Publication Data

ISBN: 978-1-53618-561-4

Published by Nova Science Publishers, Inc. † New York

CONTENTS

Preface

As COVID-19 has spread throughout the United States, it has reduced domestic economic activity and disrupted domestic and international supply chains for goods and services, including food and agricultural products. These disruptions have produced an immediate and very strong demand shock on the U.S. food supply chain. This book looks at current agricultural and food issues including: appropriations, trade, disaster assistance, food safety and security and loan programs.

Chapter 1 - The Agriculture appropriations bill funds the U.S. Department of Agriculture (USDA) except for the U.S. Forest Service. It also funds the Food and Drug Administration (FDA) and—in even-numbered fiscal years—the Commodity Futures Trading Commission (CFTC). Agriculture appropriations include both mandatory and discretionary spending. Discretionary amounts, though, are the primary focus during the bill's development. The largest discretionary spending items are the Special Supplemental Nutrition Program for Women, Infants, and Children (WIC); agricultural research; rural development; FDA; foreign food assistance and trade; farm assistance loans and salaries; food safety inspection; animal and plant health programs; and technical assistance for conservation programs. For FY2021, the Trump Administration requests $19.9 billion for discretionary-funded accounts within the jurisdiction of Agriculture appropriations subcommittees. The request would be a reduction of $3.6 billion from FY2020 (-15%). The primary changes in the Administration's request includes reductions to foreign food assistance (-$1.97 billion); domestic nutrition assistance (-$1.06 billion); and agricultural research, rural development, regulatory, and farm production and conservation agencies (-$530 million). The overall nondefense spending cap for FY2021 that Congress may use to develop the FY2021 appropriations is about 1% higher than in FY2020. Separate from the regular FY2020 annual appropriation, Congress passed four supplemental appropriations acts in response to the COVID-19 pandemic, all designated as emergency spending. These acts provide $53 billion during FY2020 to programs in the jurisdiction of the Agriculture appropriations subcommittees. Of this, $36 billion are specified supplemental amounts, $3 billion is from an indefinite authorization, and $14 billion is reimbursement for past obligations that is comparable to the annual appropriation.

Chapter 2 - Sales of U.S. agricultural products to foreign markets absorb about one-fifth of U.S. agricultural production, thus contributing significantly to the health of the farm economy. Farm product exports, which totaled $136 billion in FY2019 (see chart), make up about 8% of total U.S. exports and contribute positively to the U.S. balance of trade. The economic benefits of agricultural exports also extend across rural communities, while overseas farm sales help to buoy a wide array of industries linked to agriculture, including transportation, processing, and farm input suppliers. A major area of interest for the 116th Congress during its first session was the loss of export demand for agricultural products in the wake of tariff increases imposed by the Trump Administration on U.S. imports of steel and aluminum from certain countries and other imported products from China. Some of the affected countries levied retaliatory tariffs on U.S. agricultural products, contributing to a 53% decline in value of U.S. agricultural exports to China in 2018 and a broader decline in exports across countries imposing retaliatory tariffs in 2019. To help mitigate the economic impact from export losses, the U.S. Department of Agriculture (USDA) authorized two short-term assistance ("trade aid") programs to producers of affected agricultural commodities, valued at up to $12 billion in 2019 and $16 billion in 2019. Other major agricultural trade developments in 2019 included efforts to ratify the U.S.-Mexico-Canada Agreement (USMCA), trade negotiations with China, Japan, and the European Union, and continued review of U.S. participation in the World Trade Organization (WTO). The USMCA was ratified by Mexico and the U.S. Congress, and awaits ratification by Canada before it can enter into force. The United States and Japan signed an agreement increasing market access for many U.S. agricultural exports to Japan. This agreement, which does not require congressional approval, excludes provisions pertaining to non-tariff measures that could become future trade barriers for U.S. agricultural exporters. A second-stage negotiation toward a more comprehensive pact could commence in 2020. In January 2020, President Trump signed a "Phase One" executive agreement (that also does not require congressional approval) with the Chinese government on trade and investment issues, including agriculture. Under the agreement, China is not required to repeal any tariffs, but it has reduced certain retaliatory tariffs and is granting tariff exclusions for various agricultural products in order to reach a target level of U.S. imports—$32 billion (relative to a 2017 base of $24 billion) over a two-year period. The coronavirus outbreak since January 2020 may affect China's ability to meet these commitments. In addition to further negotiations with Japan and China, the Administration has stated its intent to pursue trade agreements with the European Union, India, Kenya, the United Kingdom, and possibly other countries. The Trump Administration has also indicated that reforming the WTO is a priority for 2020. The WTO Ministerial Conference in June 2020 presents an opportunity to address pressing concerns over agricultural reform efforts. Among other agricultural trade issues that may arise in the 116th Congress are proposed changes to U.S. trade remedy laws to address imports of seasonal produce affecting growers in the Southeast, the establishment of a common international framework for approval, trade, and marketing of the products of agricultural biotechnology, and foreign restrictions on U.S. exports of meat that are inconsistent with international trade protocols. Additionally, U.S. beef and pork face trade barriers in several markets because of U.S. producers' use of growth promotants and the feed additive ractopamine.

Chapter 3 - The U.S. Department of Agriculture (USDA) offers several programs to help farmers recover financially from natural disasters, including drought and floods. All the programs have permanent authorization, and one requires a federal disaster designation (the emergency loan program). Most programs receive mandatory funding amounts that are "such sums as necessary" and are not subject to annual discretionary appropriations. The federal crop insurance program offers subsidized policies designed to protect crop producers from risks associated with adverse weather, as well as weather-related plant diseases and insect infestations and declines in commodity prices. Policies must be purchased prior to the planting season. Eligible commodities include most major crops and many specialty crops (including fruit, tree nut, vegetable, and nursery crops), as well as forage and pastureland for livestock producers. Producers who grow a crop that is currently ineligible for the federal crop insurance program may apply for the Noninsured Crop Disaster Assistance Program (NAP). NAP provides a basic level of coverage (when loss exceeds 50% of expected production) as well as options to purchase additional coverage. Similar to crop insurance, policies must be purchased prior to the planting season. There are four permanently authorized disaster programs for livestock and trees. Producers do not pay a fee to participate, and advanced sign-up is not required. The programs are:

1. the Livestock Indemnity Program (LIP), which provides payments to eligible livestock owners and contract growers at a rate of 75% of market value for livestock deaths in excess of normal mortality or sold at a reduced sale price caused by adverse weather, attacks by reintroduced wild animals, and disease;
2. the Livestock Forage Disaster Program (LFP), which provides payments to eligible livestock producers who have suffered grazing losses on drought-affected pasture or

grazing land or on rangeland managed by a federal agency due to a qualifying fire;

3. the Emergency Assistance for Livestock, Honey Bees, and Farm-Raised Fish Program (ELAP), which provides payments to producers of livestock, honey bees, and farm-raised fish as compensation for losses due to animal disease, adverse weather, and feed or water shortages; and
4. the Tree Assistance Program (TAP), which provides payments to orchardists/nursery tree growers for losses in excess of 15% to replant trees, bushes, and vines damaged by natural disasters.

Separately, for all types of farms and ranches, when a county has been declared a disaster area by either the President or the Secretary of Agriculture, producers in that county may become eligible for low-interest emergency disaster loans from USDA. USDA has several permanent disaster assistance programs designed to help producers repair damaged land following natural disasters. It also has authority to issue disaster payments to farmers with funds from the Commodity Credit Corporation (CCC) or "Section 32." Finally, USDA can use a variety of existing programs to address disaster issues as they arise, such as allowing emergency grazing on land enrolled in the Conservation Reserve Program. Congressional actions through the Agriculture Improvement Act of 2018 (P.L. 115-334) and the FY2019 supplemental appropriation (P.L. 116-20) have made a number of amendments to the agricultural disaster assistance programs and authorized supplemental appropriations covering crop and livestock losses from natural disasters in recent years. USDA has implemented this ad hoc assistance through two versions of the Wildfires and Hurricanes Indemnity Program (WHIP). Losses occurring in 2017 were eligible for 2017 WHIP. An expanded set of losses occurring in 2018 and 2019 are eligible for WHIP Plus (WHIP+). In addition to WHIP+, USDA has implemented two subprograms—the On-Farm Storage Loss Program and the Milk Loss Program—as well as block grants with states and payments for acres that were prevented from being planted. Economic and market losses related to the COVID-19 pandemic are not considered eligible losses under the USDA disaster assistance programs. While USDA has provided some flexibilities through existing programs, most disaster assistance programs require some level of production or physical loss from an unavoidable natural cause in order to trigger eligibility.

Chapter 4 - Coronavirus Disease 2019 (COVID-19) is thought to spread mainly from person to person through respiratory droplets produced when an infected person sneezes, coughs, or speaks. Currently there is no evidence indicating the transmission of COVID-19 through food, according to U.S. and international public health agencies. However, China—a major U.S. agricultural export market—has reportedly started to institute requirements to address COVID-19-related concerns about the safety of imported food, which could impact global agricultural trade.

Chapter 5 - Some Members of Congress have demonstrated an ongoing interest in global food security, including how the Coronavirus Disease 2019 (COVID-19) pandemic may affect global food security needs in 2020. Some policymakers are examining whether existing U.S. global food security programs are sufficient to meet potentially increased needs. They are also considering what legislative changes, if any, might help address the emerging global food assistance challenges caused by COVID-19.

Chapter 6 - During the Coronavirus Disease 2019 (COVID-19) pandemic, rising unemployment rates and other factors led an increasing number of Americans to seek food assistance. Food banks, food pantries, and other emergency feeding organizations—sometimes referred to as the *charitable food assistance network*—are often the first place people go when they become food insecure, and these organizations have experienced an influx in demand during the pandemic. At the same time, they have had to address safety concerns resulting from COVID-19, often operating with fewer staff or utilizing the National Guard to distribute food, and setting up drive-through and mobile food distributions. While the charitable food assistance network relies heavily on private donations, federal programs also provide a significant source of food and revenue for emergency feeding organizations. The primary federal programs that support food banks and other emergency feeding organizations are administered by the U.S. Department of Agriculture (USDA) and the Department of Homeland Security's (DHS's) Federal Emergency Management Agency (FEMA). Ongoing programs, including The Emergency Food Assistance Program (TEFAP) and Emergency Food and Shelter Program (ESFP), have received supplemental appropriations during the pandemic. Other programs have been activated (USDA's Disaster Household Distribution program and FEMA's Public Assistance (PA) program) or created (USDA's Farmers to Families Food Box program) in response to the pandemic. Recent legislation has proposed additional funding for existing programs as well as the creation of new programs and funding streams for emergency feeding organizations.

Chapter 7 - As COVID-19 has spread throughout the United States, it has reduced domestic economic activity and disrupted domestic and international supply chains for goods and services, including food and agricultural products. These disruptions have produced an immediate and very strong demand shock on the U.S. food supply chain that has sent many commodity prices sharply lower. The food supply chain refers to the path that raw agricultural commodities take from the farm where they are produced, through the food processing and distribution network to the consumer where they are used. Supply chain disruptions have been primarily due to two factors: widespread shutdowns of all but essential businesses; and uncertainty about the availability of labor for the food distribution network—whether from illness, fear of illness, or immigration status. The food supply chain has been deemed essential; however, many institutional purchasers (including restaurants, hotels, schools, and entertainment venues) have been closed. According to the U.S. Department of Agriculture (USDA), U.S. consumers normally spend 54% of their food and drink dollars on away-from-home food purchases. Thus, prior to the COVID-19 pandemic a large share of U.S. food products traveling through the food supply chain was going to institutional buyers, often in bulk or vendor-ready form, for away-from-home consumption. In order to redirect this food product flow towards retail outlets and at-home consumption, much of this food would require processing into more consumer-usable forms, repackaging, and relabeling. This requires some level of retooling by food packagers and processors. In addition, several plants in the food processing industry, including meat processing plants, have experienced severe COVID infection outbreaks among workers and been forced to shut (at least temporarily). Several industry groups from the U.S. agricultural sector have released estimates of the economic damage experienced by producers and ranchers. Most of these early assessments are limited to evaluating the effect of the price decline on any unsold production of crops or livestock remaining under farmer's control, and the unexpected marketing costs of unsold products due to the shutdown of most institutional buying of agricultural products. Cumulatively, industry estimates of COVID-related losses approach $40 billion (this would represent over 10% of annual cash receipts). The effect on farm net income is expected to be similarly negative. In response to the COVID-19 pandemic, Congress has passed and the President has signed four supplemental appropriations acts (P.L. 116-123, P.L. 116-127, P.L. 116-136, and P.L. 116-139) that have included both direct and indirect funding for the U.S. agricultural sector. On April 17, 2020, Secretary of Agriculture Sonny Perdue announced the Coronavirus Food Assistance Program (CFAP), valued at $19 billion, to provide immediate financial relief to farmers, ranchers, and consumers in response to the COVID-19 national emergency. According to Secretary Perdue, CFAP will include $16 billion in direct payments to producers that have been impacted by the sudden drop in commodity prices and the disruption in food supply chains that has occurred since the outbreak, and $3 billion in commodity purchases for distribution through food banks, faith- based organizations, and other nonprofit organizations. CFAP direct payments are intended to partially offset the loss of market revenue from the price decline, and the unexpected carry-cost of unsold commodities for producers and ranchers of products that have been negatively affected by COVID-19. USDA has released limited information on the specifics of how CFAP's direct payment program will be implemented. Many of the program specifics are expected to be delineated in an expedited rulemaking process. CFAP direct payments are expected to go out to producers by the end of May or early

June. CFAP's commodity purchase and distribution program serves the dual roles of supporting commodity market prices by temporarily increasing demand, and of expanding the availability of food distribution to consumers that have lost their jobs or have limited resources. Expectations are that it will be operated differently than, and separate from, existing USDA commodity purchase programs such as Section 32 or Food and Nutrition Service (FNS) food distribution programs in two major ways. First, USDA plans to purchase about $100 million per month of fresh produce, $100 million per month of dairy products, and $100 million per month of meat products (chicken and pork). Second, the CFAP purchase program intends for vendors to deliver household-ready boxes—potentially a diversified mix of the previously mentioned produce, meat and dairy products—which are ready for more convenient and immediate distribution ("off the truck and into the trunk") that is consistent with social distancing. Initial program delivery may be as early as May 15. The program is expected to operate through the end of 2020. Potential congressional concerns include how to channel assistance to industries affected by COVID-19, long-term effects of the pandemic, and the capacity of rural banks to help with recovery. Several issues related to CFAP and the U.S. agricultural sector in a post-COVID economy that could be of potential interest to Congress are presented at the end of this chapter.

Chapter 8 - On April 17, 2020, Secretary of Agriculture Sonny Perdue announced that the U.S. Department of Agriculture (USDA) would initiate the Coronavirus Food Assistance Program (CFAP) to provide immediate financial relief to farmers, ranchers, and consumers in response to the COVID-19 national emergency. CFAP consists of two distinct initiatives: a $16 billion direct payment program for agricultural producers that have been impacted by the decline in commodity prices and the disruption in food supply chains related to COVID-19, and a $3 billion food purchase and distribution program. This chapter focuses on the larger of these two CFAP initiatives—the $16 billion direct payment program. The CFAP direct payment program derives its funding from two sources: $9.5 billion emergency program spending authorized by the Coronavirus Aid, Relief, and Economic Security Act (CARES Act; P.L. 116-136) and an existing $6.5 billion balance from the Commodity Credit Corporation (CCC) Charter Act of 1948 (P.L. 80-806; 15 U.S.C. 714 *et seq.*). On May 19, 2020, USDA released the final rule that details CFAP's $16 billion direct payment program, including the list of eligible commodities, eligibility requirements for producers, payment calculations, and application procedures. In its announcement, USDA stated that it would begin accepting applications on May 26, through the Farm Service Agency (FSA), from agricultural producers who have suffered losses. To be eligible, an application must be submitted to any FSA county office by the close of business on August 28, 2020. Payments to eligible producers are expected to be made soon after each application is processed. USDA's FSA will be the principal agency charged with implementing CFAP's direct payment program. FSA is to be assisted by the Agricultural Marketing Service (AMS) with respect to matters dealing with producers of specialty crops. To be eligible for a payment, a commodity must have suffered a price loss of at least 5% during the mid-January to mid-April period, or was subject to additional significant marketing costs for unexpected supply chain disruptions, including unsold inventories. Specialty crops may also qualify for a payment if product was shipped off the farm, but spoiled prior to receipt of payment due to COVID-19-caused disruption of the food supply chain. For an individual or legal entity to be eligible, they must complete an application to determine the quantities affected and to meet certain other criteria, including conservation compliance, share in the risk of profit and loss from the farm's operation, and an average Adjusted Gross Income (AGI) for 2016, 2017, and 2018 that is less than $900,000, unless at least 75% of AGI is from farming and ranching activities. USDA is to track payments made from the CARES Act and the CCC Charter Act funds separately, to ensure that the payments are consistent with each respective authority. As a result, every eligible commodity may receive a payment based on two different types of losses or costs: (1) income losses due to mid-January-to-mid-April price declines, which is to be partially offset by payments from CARES Act funds; and (2) unexpected marketing costs due to food supply chain disruptions, which is to be partially offset by payments from CCC funds. USDA is to pay an initial tranche equal to 80% of the aggregate payment to each qualifying producer. If funds remain available after the initial payment to all eligible applicants, and depending on whether USDA adds other commodities to the list, USDA is to disburse the remaining funds such that the initial and final payments do not exceed a total of $9.5 billion of CARES Act funds and $6.5 billion of CCC funds. Funds may be prorated if necessary to stay within these limits. There is a payment limitation of $250,000 per person or entity for all commodities combined. Applicants who are corporations, limited liability companies, or limited partnerships may qualify for payments of up to $750,000 if at least three members each actively provide at least 400 hours of personal labor or personal management to the farming operation. Congress may be interested in evaluating whether CFAP payments mitigate the damage caused by the COVID-19 pandemic for individual farming operations. Will additional assistance be necessary later this year? Is additional assistance needed to support certain affected industries that are not eligible under USDA's initial rollout? Another immediate concern may be oversight of the large sums of taxpayer money that are to flow out through the USDA and the CCC. Producer self-certification of losses may create an incentive to over-report losses.

Chapter 9 - U.S. Department of Agriculture (USDA) Food and Nutrition Service (FNS) programs are often part of emergency response efforts, providing program flexibilities, foods for distribution, and benefits for redemption. Emergencies generate different FNS responses, which can vary with states' requests. During the COVID-19 pandemic, access to food—particularly in light of school closures—has been a concern for many. Some also view the Supplemental Nutrition Assistance Program (SNAP) as a force for economic stimulus. This Insight discusses related provisions of the second and third COVID-19 legislative packages, which supplement FNS's prior COVID-19 response with new funds and authorities.

Chapter 10 - The U.S. fruit and vegetable industry entered 2020 following a difficult 2019 characterized by both lower prices and reduced production in some sectors. By broad market category, the U.S. Department of Agriculture (USDA) reported that grower prices for both fruits and tree nuts and vegetables and pulses began 2020 at low levels, with expectations for continued price declines. In mid-March 2020, the foodservice industry, which accounts for a substantial share of the produce industry's sales, was largely shut down as most states closed all but essential businesses in response to the coronavirus (COVID-19) public health emergency. This situation resulted in major supply chain disruptions for the U.S. produce industry. Reportedly, up to one-half of the produce industry sells its products to foodservice distributors (e.g., Sysco Corporation, U.S. Foods), according to former California Agricultural Commissioner A.G. Kawamura, which in turn supply schools, restaurants, and other institutional customers. The sudden decrease in the foodservice market left many produce growers with nowhere to sell their crops. Given the highly perishable (non-storable) nature of most fresh fruits and vegetables, many growers were unable to pivot quickly to other marketing channels. Some growers reportedly resorted to plowing under their fields to avoid incurring the added costs of harvesting, cooling, and packaging. Some sectors are reporting plans to reduce acreage due to loss of foodservice demand. Another up to one-half of the produce industry reportedly relies on sales to larger retail chain stores (e.g., Walmart or membership clubs such as Costco) with some produce growers selling nearly all of their production to some combination of the foodservice market and larger retail chain stores. Fruits and vegetables also are sold to wholesale distributors and produce markets that serve independent grocery stores and restaurants. Direct-to-consumer sales of fruits and vegetables comprise a small portion of total U.S. produce sales. Available data from the 2017 *Census of Agriculture* indicate that the value of all food sold directly to consumers accounted for an estimated 3% of the value of total U.S. agricultural production. *Figure 1* broadly depicts the U.S. value chain and the various produce marketing channels. The U.S. produce sector is highly streamlined and specialized. Many perishable produce items are pre-sold and packaged in bulk to certain specifications, and producers cannot readily shift such items from one marketing channel to another.

Most produce is planted, harvested, packed, and shipped to a precise schedule to replenish stores' inventories "just in time," according to a University of California report. This limits the industry's flexibility in responding to market disruptions. Cash flow concerns pose an added challenge to produce growers, according to A.G. Kawamura, especially if a business's supplier or buyer defaults on a commitment, or is unable to deliver supplies or move outgoing products due to the current pandemic. Challenges for the U.S. produce industry are further compounded by ongoing agricultural labor concerns. Fruit and vegetable production is labor-intensive and few automation/mechanization options are available for handling perishable fruits and vegetables. On April 15, 2020, in an effort to lessen labor disruption impacts on the industry, the Department of Homeland Security and USDA announced a temporary final rule to change certain H-2A visa requirements. Related concerns involve the potential effects if farmworkers and other produce industry workers become ill from the virus, as happened in some U.S. meatpacking plants. COVID-19 outbreaks have disrupted meat processing and caused some shortages of certain meat products in grocery stores.

Chapter 11 - U.S. livestock and poultry producers entered 2020 with an optimistic outlook for prices and income. Then in mid-March the food service sector, which accounts for a substantial share of meat consumption, was largely shut down as most states closed all but essential businesses. A temporary surge in retail meat purchases offset some of the reduction in food service demand. However, in early April, the situation worsened for producers as COVID-19 outbreaks began spreading in meatpacking plants around the United States, disrupting meat processing and leading to some shortages of meat products in grocery stores. The reduction in slaughter is weighing on livestock prices, reflecting shrinking meatpacker demand for slaughter-ready livestock and poultry. On April 12, 2020, Smithfield Foods, the largest pork processor in the United States, indefinitely closed its Sioux Falls, SD, pork processing plant because of the spread of COVID-19 among plant workers. With an estimated daily slaughter capacity of 19,500 head, the Sioux Falls plant accounts for 4% of estimated daily U.S. hog slaughter. More pork slaughter capacity has been lost since then because of COVID-19 outbreaks in plants operated by other processors. On April 13, JBS USA announced it was closing its Greeley, CO, beef plant until April 24 because of an outbreak of COVID-19 among plant workers. *Cattle Buyers Weekly* estimates daily slaughter capacity of 6,000 head at the Greeley plant, which would be 4%- 5% of daily national cattle slaughter capacity. Other beef plants have had to close due to COVID-19 outbreaks among workers. Outbreaks of COVID-19 have hit beef, pork, and poultry plants across the country. As of April 27, at least 15 meatpacking plants were reported closed. Some plants have shut down temporarily to clean and reconfigure work stations; others have not set reopening dates. Reportedly, some plants have implemented onsite temperature monitoring of employees, spread out workers within plants, set up plexiglass dividers between workers who normally operate in tight spaces, and spread the work of one shift over two shifts, thus slowing operating speed. Even so, some meat plant employees have asserted that packers have not implemented adequate protective measures in their plants. On April 26, 2020, the Occupational Safety and Health Administration updated guidelines to address workplace conditions for meat plant workers. On April 28, 2020, President Trump announced that he plans to sign an executive order that would use the Defense Production Act to classify meat processing as critical infrastructure to keep production plants open.

Chapter 12 - Milk price prospects for U.S. dairy producers in 2020 have weakened as the U.S. economy deteriorates under the expanding Coronavirus 2019 (COVID-19) pandemic. The year began with a positive outlook for the dairy industry that included higher milk prices and increased income for dairy producers in 2020, according to the U.S. Department of Agriculture (USDA). The dairy industry has gone through several years of relatively low milk prices, leading some dairy farms to exit the industry and two large dairy processors to file for bankruptcy. In 2019, according to USDA, the number of licensed dairy herds declined from 37,468 to 34,187, a nearly 9% drop. Almost every state lost dairies last year. As a safety net for dairy producers, Congress enacted Section 1401 of the 2018 farm bill (P.L. 115-334) establishing the Dairy Margin Coverage (DMC) program. The program guarantees dairy producers could receive payments when the margin or difference narrows between the USDA-reported all milk price and a calculated feed ration value based on the prices of corn, soybean meal, and alfalfa hay. The DMC replaced the 2014 farm bill's (P.L. 113-79's) Margin Protection Program (MPP). The MPP program made very few payments to dairy producers during the 2015-2017 period ($12 million), and many producers considered the program ineffective as a safety net during years of relatively low milk prices. Congress amended MPP in the Bipartisan Budget Act of 2018 (BBA; P.L. 115-123) which resulted in higher MPP payments in 2018 ($253 million).

Chapter 13 - On April 28, 2020, President Trump issued an executive order (EO) invoking the Defense Production Act (DPA) to address the food supply chain for meat and poultry products during the national emergency caused by the COVID-19 outbreak. The release of the EO was preceded by some state and congressional calls for such an order, and significant media coverage about the potential contents of such an order. This Sidebar addresses some frequently asked questions about the EO's contents, legal bases, and potential effects.

Chapter 14 - On May 21, 2020, the U.S. Department of Agriculture (USDA) temporarily expanded the Disaster Set- Aside (DSA) provision to allow flexibility for farm loan repayment due to the economic effects of the Coronavirus Disease 2019 (COVID-19) pandemic. The set-aside provision allows a borrower to move a loan payment owed to USDA's Farm Service Agency (FSA) to the end of the loan or, in the case of an annual operating loan, to extend the payment by a year. Interest continues to accrue on the deferred principal; neither the interest nor the principal is forgiven. The set-aside (deferment) is meant to provide financial relief and cash flow flexibility during a crisis. Earlier during the pandemic, on March 26, 2020, USDA announced certain flexibilities in its loan-making and servicing procedures. These included relaxing deadlines, accommodating social distancing, and temporarily suspending loan accelerations and new foreclosures.

Chapter 15 - The Coronavirus Aid, Relief, and Economic Security Act (CARES Act; P.L. 116-136) created the Small Business Administration's (SBA's) Paycheck Protection Program (PPP) and Emergency Economic

Injury Disaster Loan (EIDL) grants to provide short-term, economic relief to certain small businesses and nonprofits. Important note: The SBA started accepting PPP and Emergency EIDL grant applications on April 3, 2020. The SBA stopped accepting new PPP and Emergency EIDL grant applications on April 16, because funding provided for these programs by the CARES Act was either exhausted (PPP) or nearing exhaustion (Emergency EIDL grants). The SBA resumed accepting PPP applications on April 27, following enactment of the Paycheck Protection Program and Health Care Enhancement Act (P.L. 116- 139). The act provided an additional $321.335 billion for the PPP and $10 billion for Emergency EIDL grants. The SBA continued to process Emergency EIDL grant applications submitted prior to April 16. However, because of the large number of Emergency EIDL grant applications submitted, the SBA did not accept new Emergency EIDL grant applications until May 4, 2020, and then only on a limited basis to accommodate agricultural enterprises. P.L. 116-139 included a provision (Division A, §101) to temporarily permit agricultural enterprises, as defined by Section 18(b) of the Small Business Act (15 U.S.C. §647(b)), with not more than 500 employees to receive Emergency EIDL grants and EIDLs through December 31, 2020. This chapter will be updated to reflect any legislative changes affecting these programs.

Chapter 16 - The U.S. Department of Agriculture (USDA) Rural Development agency (RD) administers loan, grant, and technical assistance programs that support infrastructure, housing, and business development in rural areas. In response to the COVID-19 pandemic, Congress provided supplemental funding to certain RD programs. In addition, RD has taken a number of actions to provide relief for current program participants and assist potential applicants in applying for program funding.

Chapter 17 - In March and April 2020, Congress passed and the President signed four supplemental appropriations acts in response to the COVID-19 pandemic (P.L. 116-123, P.L. 116-127, P.L. 116-136, and P.L. 116-139). This In Focus summarizes over $36 billion of appropriations and policy changes in the jurisdiction of the Agriculture appropriations subcommittees. These supplemental appropriations acts are referred to here by order of passage.

In: COVID 19: Agricultural and Food Issues ISBN: 978-1-53618-561-4
Editor: Kristian Goldschmidt

Chapter 1

AGRICULTURE AND RELATED AGENCIES: FY2021 APPROPRIATIONS*

Jim Monke

ABSTRACT

The Agriculture appropriations bill funds the U.S. Department of Agriculture (USDA) except for the U.S. Forest Service. It also funds the Food and Drug Administration (FDA) and—in even-numbered fiscal years—the Commodity Futures Trading Commission (CFTC).

Agriculture appropriations include both mandatory and discretionary spending. Discretionary amounts, though, are the primary focus during the bill's development. The largest discretionary spending items are the Special Supplemental Nutrition Program for Women, Infants, and Children (WIC); agricultural research; rural development; FDA; foreign food assistance and trade; farm assistance loans and salaries; food safety inspection; animal and plant health programs; and technical assistance for conservation programs.

For FY2021, the Trump Administration requests $19.9 billion for discretionary-funded accounts within the jurisdiction of Agriculture

---

* This is an edited, reformatted and augmented version of Congressional Research Service, Publication No. R46437, dated July 1, 2020.

appropriations subcommittees. The request would be a reduction of $3.6 billion from FY2020 (-15%). The primary changes in the Administration's request includes reductions to foreign food assistance (-$1.97 billion); domestic nutrition assistance (-$1.06 billion); and agricultural research, rural development, regulatory, and farm production and conservation agencies (-$530 million). The overall nondefense spending cap for FY2021 that Congress may use to develop the FY2021 appropriations is about 1% higher than in FY2020.

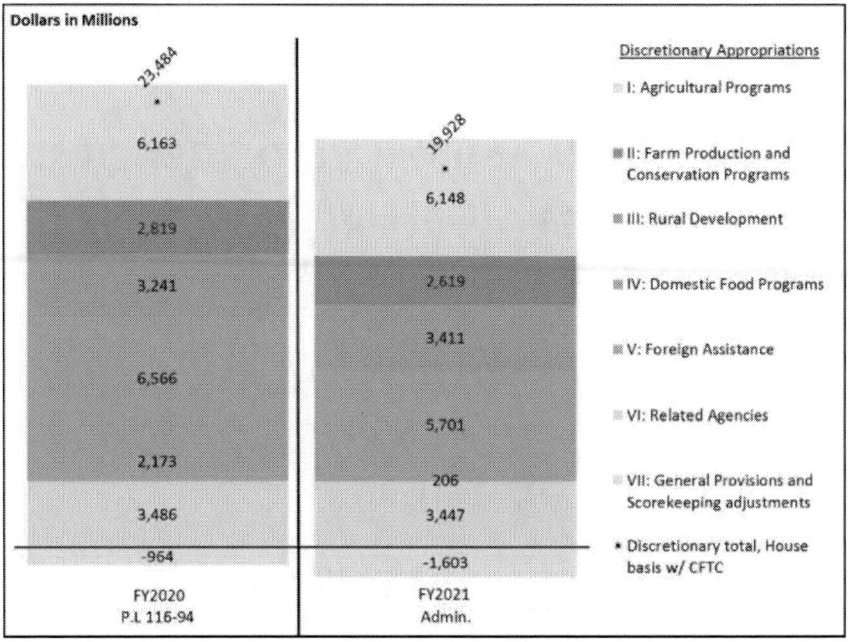

Source: CRS, using P.L. 116-94, House Cmte. Print 38-679, unpublished CBO tables, OMB *FY2021 Budget of the U.S. Government*, Appendix, and USDA *FY2020 Budget Explanatory Notes.*

Note: Admin. = Trump Administration Budget Request. CFTC = Commodity Futures Trading Commission. For comparability, includes CFTC in Related Agencies in all columns regardless of jurisdiction.

Discretionary Agriculture Appropriations, by Title, FY2020-FY2021.

Separate from the regular FY2020 annual appropriation, Congress passed four supplemental appropriations acts in response to the COVID-19 pandemic, all designated as emergency spending. These acts provide $53 billion during FY2020 to programs in the jurisdiction of the Agriculture appropriations subcommittees. Of this, $36 billion are specified supplemental amounts, $3 billion is from an indefinite authorization, and $14 billion is reimbursement for past obligations that is comparable to the annual appropriation.

STATUS OF FY2021 AGRICULTURE APPROPRIATIONS

On February 10, 2020, the Trump Administration released its FY2021 budget request to Congress (Table 1). As of June 29, 2020, neither the House nor the Senate has yet marked up an FY2021 Agriculture appropriations bill. See Figure 1 for a comparison of timelines and Appendix D for more details.

For accounts in the jurisdiction of the Agriculture appropriations bill, the Administration's budget requests $19.9 billion in discretionary spending, a $3.6 billion reduction from FY2020 (-15%, Table 2). This is based on the jurisdiction of the House Agriculture subcommittee, including the Commodity Futures Trading Commission (CFTC).[1]

Table 1. Status of FY2021 Agriculture Appropriations

Administration Request	**House Action**				**Senate Action**				**Final Appropriation**
	Subcmte.	Cmte.		Floor	Subcmte.	Cmte.	Floor		
2/10/2020 OMB Budget Appendix USDA FDA CFTC FCA	—	—	—	—	—	—	—	—	—

Source: CRS, compiled from Congress.gov, OMB, and agency websites.

Note: OMB = Office of Management and Budget, USDA = U.S. Department of Agriculture, FDA = Food and Drug Administration, CFTC = Commodity Futures Trading Commission, and FCA = Farm Credit Administration.

[1] Jurisdiction for CFTC appropriations differs between the chambers. Since FY2008, CFTC is marked up in the Agriculture Subcommittee of the House Appropriations Committee and in the Financial Services and General Government Subcommittee of the Senate Appropriations Committee. The enacted CFTC appropriation is carried in the Agriculture bill in even-numbered fiscal years and in the Financial Services bill in odd-numbered fiscal years.

Table 2. Agriculture and Related Agencies Appropriations, by Title, FY2020-FY2021, Budget authority in millions of dollars

	FY2020		FY2021				Change
Title of Agriculture Appropriations Act	P.L. 116-94	COVID-19 Actsa	Admin. requestb	House	Senate	Enacted	FY2020 to Admin.
I. Agricultural Programs: Discretionary	6,163.1	133.8	6,148.0	—	—	—	-15.1
Mandatory	1,404.0	—	1,431.0	—	—	—	+27.0
Subtotal	7,567.1	—	7,579.0	—	—	—	+11.9
II. Farm Production and Conservation Programs	2,819.0	3.0	2,618.7	—	—	—	-200.2
Mandatory	36,268.5	14,000.0	34,269.6	—	—	—	-1,998.9
Subtotal	39,087.5	14,003.0	36,888.3	—	—	—	-2,199.2
III. Rural Development	3,240.6c	45.5	3,410.5	—	—	—	+169.9
IV. Domestic Food Programs: Discretionary	6,566.0	1,350.0	5,700.7	—	—	—	-865.3
Mandatory	91,436.3	24,710.0	93,309.8	—	—	—	+1,873.5
Subtotal	98,002.3	26,060.0	99,010.5	—	—	—	+1,008.2
V. Foreign Assistance	2,172.7	4.0	205.9	—	—	—	-1,966.8
VI. Food and Drug Administration	3,171.5c	163.0	3,220.4	—	—	—	+48.9
Commodity Futures Trading Commission	315.0	—	226.5	—	—	—	-88.5
VII. General Provisions:							
CHIMPS and rescissionsd	-1,006.1	—	-1,200.0	—	—	—	-193.9
Emergency appropriations	535.0	—	0.0	—	—	—	-535.0
Other appropriations	440.3	12,600.0	0.0	—	—	—	-440.3
Scorekeeping adjustmentse	-398.0	—	-403.0	—	—	—	-5.0
Emergency designation in this bill	-535.0	—a	0.0	—	—	—	+535.0
Discretionary: Senate basis w/o CFTC	[23,169.1]	14,299.3	19,701.3	—	—	—	-3,467.9
Discretionary: House basis w/ CFTC	23,484.1	14,299.3	19,927.8	—	—	—	-3,556.4
Mandatory	129,108.8	38,710.0	129,010.4	—	—	—	-98.4

	FY2020		FY2021				Change
Total: Senate basis w/o CFTC	152,278.0	53,009.3	148,711.7	—	—	—	-3,566.3
Total: House basis w/ CFTC	152,593.0	53,009.3	148,938.2	—	—	—	-3,654.8

Source: CRS, using appropriations text and reports; unpublished CBO tables; OMB *FY202 1 Budget of the U.S. Government*, Appendix; and USDA *FY2020 Budget Explanatory Notes*.

Notes: Amounts are nominal discretionary budget authority unless labeled otherwise. Bracketed amounts are not in the official totals due to differing House-Senate jurisdiction for CFTC but are shown for comparison.

a. Total from four coronavirus supplemental appropriations acts (P.L. 116-123, P.L. 116-127, P.L. 116-136, P.L. 116-139), all of which was emergency spending. Although amounts were not designated mandatory, they are grouped here for comparison to accounts in the annual appropriations. See detail in Table 3.

b. The request is based on current law and does not reflect the Administration's various legislative proposals.

c. Excludes amounts that are provided separately in General Provisions (see Table B-3).

d. Changes in Mandatory Program Spending (CHIMPS) revise mandatory programs via appropriations (Table B-1). Rescissions are permanent cancellations of previous authorizations (see Table B-2).

e. "Scorekeeping adjustments" are part of the official accounting for budget enforcement (Table B-4).

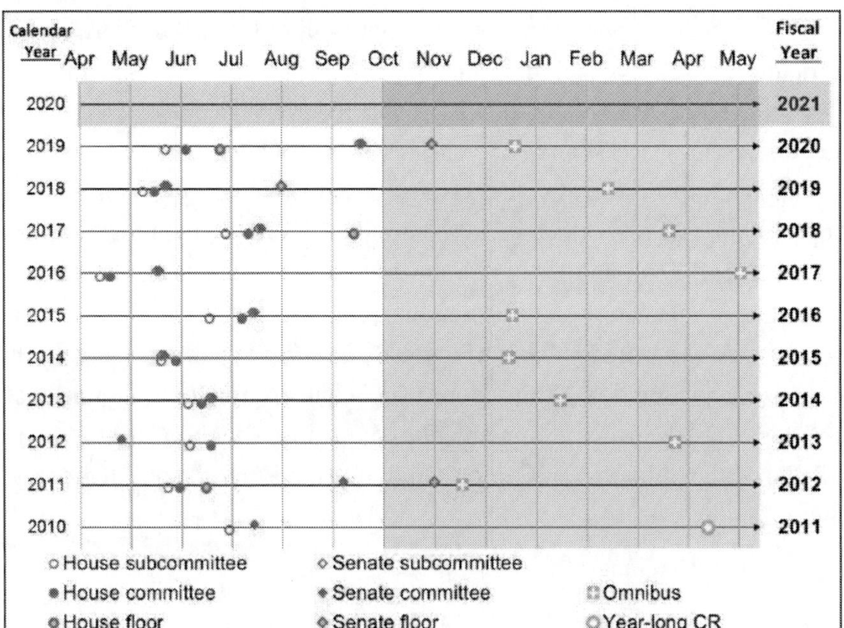

Source: CRS. For specific dates and links to bills, see Appendix D.

Notes: Blue shading shows the beginning of each fiscal year. Yellow shading is the current appropriations cycle.

Figure 1. Timeline of Action on Agriculture Appropriations, FY2011-FY2021.

The primary changes in the Administration's FY2021 request compared to FY2020 include reductions to foreign assistance (-$1.97 billion) and domestic nutrition assistance (-$1.06 billion) and reductions to agricultural research, rural development, regulatory, and farm production and conservation agencies (-$530 million).

As Congress develops appropriations bills, the overall nondefense spending cap for FY2021 that was set in the Bipartisan Budget Act of 2019 (P.L. 116-37) is about 1% higher than in FY2020.

SCOPE OF AGRICULTURE APPROPRIATIONS

The Agriculture appropriations bill—formally known as the Agriculture, Rural Development, Food and Drug Administration, and Related Agencies Appropriations Act—funds all of the U.S. Department of Agriculture (USDA), excluding the U.S. Forest Service.[2] It also funds the Food and Drug Administration (FDA) in the Department of Health and Human Services (HHS) and, in even-numbered fiscal years, CFTC.

Jurisdiction is with the House and Senate Committees on Appropriations and their Subcommittees on Agriculture, Rural Development, Food and Drug Administration, and Related Agencies. The bill includes mandatory and discretionary spending, but the discretionary amounts are the primary focus (Figure 2). Some programs are not in the authorizing jurisdiction of the House or Senate Agriculture Committees, such as FDA; Special Supplemental Nutrition Program for Women, Infants, and Children (WIC); or child nutrition (checkered regions in Figure 2).

---

[2] Although the Forest Service is part of USDA, its appropriations jurisdiction is in the Interior Appropriations Act.

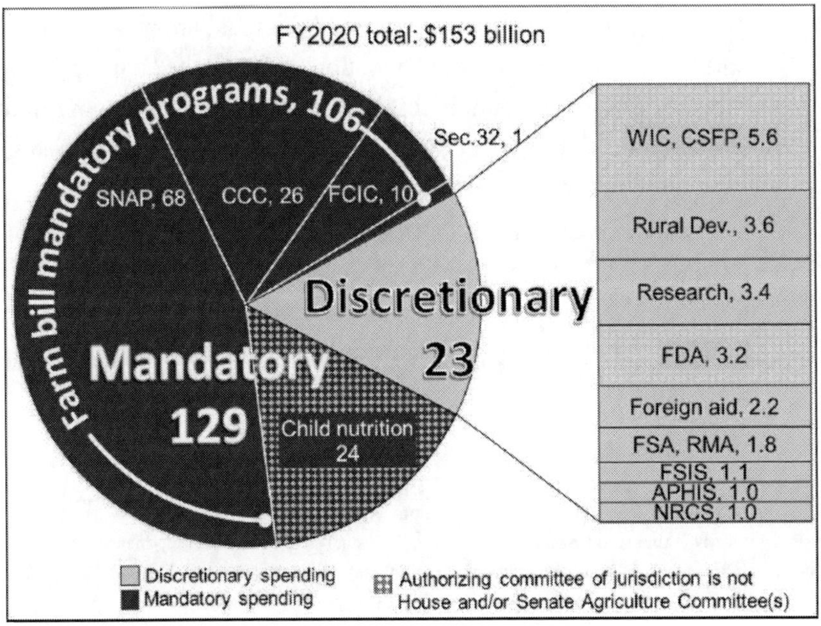

Source: CRS.

Notes: SNAP = Supplemental Nutrition Assistance Program; CCC = Commodity Credit Corporation; FCIC = Federal Crop Insurance Corporation; Section 32 = Funds for Strengthening Markets, Income and Supply; WIC = Special Supplemental Nutrition Program for Women, Infants, and Children; CSFP = Commodity Supplemental Food Program; FDA = Food and Drug Administration; FSA = Farm Service Agency, incl. Farm Production and Conservation Business Center; RMA = Risk Management Agency; FSIS = Food Safety and Inspection Service; APHIS = Animal and Plant Health Inspection Service; NRCS = Natural Resources Conservation Service. Does not show agencies under $0.5 billion. Spending in the General Provisions are presented with the agencies.

Figure 2. Scope of Agriculture and Related Agencies Appropriations. Budget authority in billions of dollars.

The federal budget process treats discretionary and mandatory spending differently:[3]

[3] See CRS In Focus IF11032, *Budgetary Decisionmaking in Congress*; and CRS Report R44582, *Overview of Funding Mechanisms in the Federal Budget Process, and Selected Examples*.

- *Discretionary spending* is controlled by annual appropriations acts and receives most of the attention during the appropriations process. The annual budget resolution[4] process sets spending limits for discretionary appropriations. Agency operations (salaries and expenses) and many grant programs are discretionary.
- *Mandatory spending*[5] is carried in the appropriation and usually advanced unchanged, since it is controlled by budget rules during the authorization process.[6] Spending for so-called entitlement programs is determined in laws such as the 2018 farm bill[7] and 2010 child nutrition reauthorization.[8]

Key Budget Terms

Budget authority is the main purpose of an appropriations act or a law authorizing mandatory spending. It provides the legal basis from which to obligate funds. It expires at the end of a period, usually after one year unless specified otherwise (e.g., two years or indefinite). Most funding amounts in this chapter are budget authority. Obligations are contractual agreements between a government agency and its clients or employees. These occur when an agency agrees to spend money from its budget authority. The Antideficiency Act prohibits agencies from obligating more budget authority than is provided in law, such as during a government shutdown. Outlays are the payments (cash disbursements) that satisfy a valid obligation. Timing of outlays may differ from budget authority or obligations because payments from an agency may not occur until services are fulfilled, goods are delivered, or construction is completed, even though an obligation occurred. Program level represents the sum of the activities undertaken by an agency. A program level may be higher than a budget authority if the program (1) receives user fees, (2) includes loans that are leveraged by an expectation of repayment (loan authority may exceed budget authority), or (3) receives transfers from other agencies. Rescissions are actions that reduce budget authority after enactment. They generally score budgetary savings. CHIMPS (Changes in Mandatory Program Spending) are adjustments via an appropriations act that can change available funding for mandatory programs. CHIMPS usually change spending for one year and may score either as an increase or decrease to outlays. They do not change the underlying authority of the program in law.

For more background, see CRS Report 98-721, *Introduction to the Federal Budget Process*.

---

4 See CRS Report R42388, *The Congressional Appropriations Process: An Introduction*.

5 Mandatory spending creates funding stability and consistency compared to annual discretionary appropriations. In agriculture, it originated with farm commodity programs that had uncertain outlays due to weather and markets.

6 See CRS Report 98-560, *Baselines and Scorekeeping in the Federal Budget Process*.

7 P.L. 115-334 (Agricultural Act of 2018). See CRS Report R45425, *Budget Issues That Shaped the 2018 Farm Bill*.

8 P.L. 111-296 (Healthy, Hunger-Free Kids Act of 2010). See CRS In Focus IF10266, *Child Nutrition Reauthorization (CNR): An Overview*.

In the FY2020 appropriation (P.L. 116-94), the discretionary amount is 15% ($23 billion) of the $153 billion total. Mandatory spending carried in the act comprised $129 billion, about 85% of the total, of which about $106 billion is attributable to programs in the 2018 farm bill.

Within the discretionary total, the largest spending items are WIC, agricultural research, rural development, FDA, foreign food aid and trade, farm assistance loans and salaries, food safety inspection, animal and plant health programs, and technical assistance for conservation program.

The main mandatory spending items are the Supplemental Nutrition Assistance Program (SNAP) and other food and nutrition act programs, child nutrition (school lunch and related programs), crop insurance, and farm commodity and conservation programs that are funded through USDA's Commodity Credit Corporation (CCC). SNAP is referred to as an "appropriated entitlement" and requires an annual appropriation.[9] Amounts for the nutrition program are based on projected spending needs. In contrast, the CCC appropriations reimburse spending from a line of credit.[10]

RECENT TRENDS IN AGRICULTURE APPROPRIATIONS

Discretionary Agriculture appropriations reached a 10-year low in FY2013 after decreasing from a high in FY2010. Since FY2013, they have gradually increased 19% over seven years. Changes within titles have generally been proportionate to changes in the overall bill, though some areas have sustained relative increases, such as FDA (related agencies) and rural development.

The stacked bars in Figure 3 represent the discretionary authorization for each appropriations title. The total of the positive stacked bars is the budget authority in Titles I-VI. In FY2018, USDA reorganization affected the placement of some programs between Titles I and II of the bill (most

---

[9] See CRS Report RS20129, *Entitlements and Appropriated Entitlements in the Federal Budget Process*.

[10] See CRS Report R44606, *The Commodity Credit Corporation: In Brief*.

noticeably the Farm Service Agency). In most years, the cumulative appropriation for the agencies is higher than the official discretionary total in the spending allocation (the blue line) because of the budgetary offset from negative amounts in Title VII (general provisions) and other negative scorekeeping adjustments. These negative offsets are mostly due to rescissions of prior-year unobligated funds and, before FY2018, limits placed on mandatory programs.

Historical trends may be tempered by inflation adjustments, as shown in the dotted line. In FY2020 the inflation-adjusted total is 0.2% less than it was in FY2011 and 2.8% less than it was in FY2018.

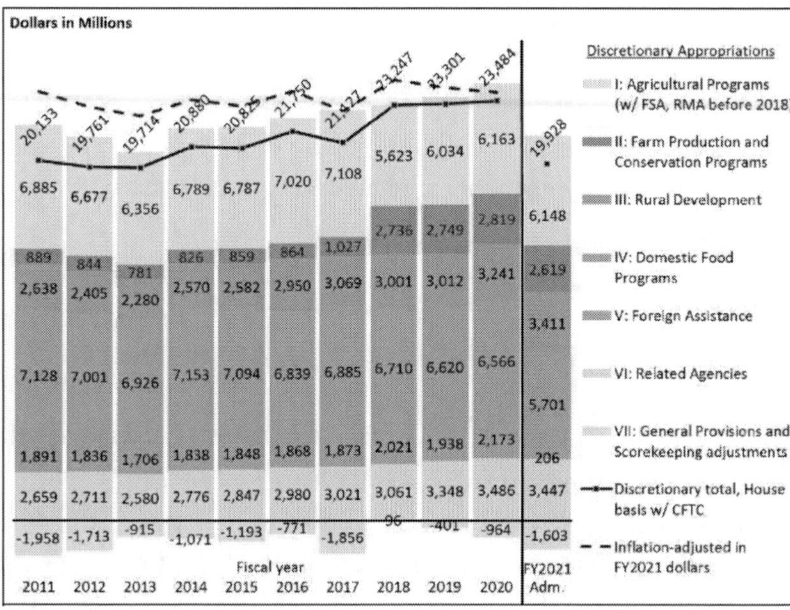

Source: CRS.

Note: For FY2021, Adm. is the Trump Administration's request. For comparability, includes CFTC in Related Agencies in all columns regardless of jurisdiction. The inflation-adjusted line was calculated using the gross domestic product price deflator. Amounts printed diagonally indicate the solid line, which is the subcommittee allocation.

Figure 3. Discretionary Agriculture Appropriations, by Title, FY2011-FY2021.

ACTION ON FY2021 APPROPRIATIONS

Administration's Budget Request

The Trump Administration released its FY2021 budget proposal to Congress on February 10, 2020.[11] USDA concurrently released its more detailed budget summary and justification,[12] as did the FDA,[13] the CFTC,[14] and the Farm Credit Administration.[15] The Administration also highlighted separately some of its proposed reductions and eliminations.[16]

For accounts in the jurisdiction of the Agriculture appropriations bill, the Administration's budget requests $19.9 billion, a $3.6 billion reduction from FY2020 (-15%; Table 3, Figure 3).[17] This is compiled based on the jurisdiction of the House Agriculture subcommittee, including the CFTC.

The primary changes in the Administration's request compared to what Congress enacted in FY2020 include reductions to the following:

- Foreign assistance (-$1.97 billion, -90%), mostly by eliminating funding for Food for Peace Title II grants (-$1.73 billion) and the McGovern-Dole International Food for Education (-$220 million).[18]
- Domestic nutrition assistance programs (-$1.06 billion), mostly from reductions to the WIC program (-$549 million, -9%) and the commodity assistance program (-$264 million, -77%). In addition, the Administration proposes rescinding more from unspent WIC balances than was done by the enacted FY2020 General Provisions

---

[11] OMB, *FY2021 Budget of the U.S. Government*, Appendix.

[12] USDA, *FY2021 USDA Budget Summary*; and USDA, *FY2021 Budget Explanatory Notes for Committee on Appropriations*.

[13] FDA, *FY2021 Justification of Estimates for Appropriations Committees*.

[14] CFTC, *FY2021 President's Budget Request*.

[15] Farm Credit Administration, *FY2021 Proposed Budget and Performance Plan*.

[16] OMB, *FY2021 Budget of the U.S. Government*, Major Savings and Reforms.

[17] This represents the Administration's request based on current law and does not reflect the Administration's various legislative proposals that would change mandatory programs and any related changes to appropriations.

[18] A similar reduction is discussed in CRS Report R46255, *International Food Assistance: FY2020 Appropriations*.

(-$200 million), which contributes to the overall budget effect of the request.

- Farm Production and Conservation mission area (-$200 million, -7%), mostly from defunding watershed programs that are typically not requested by the Administration.
- Agricultural research agencies (-$151 million net, -4%), including from the Agricultural Research Service (ARS) facilities (-$143 million, -74%); ARS salaries and expenses (-$46 million, -3%); the Economic Research Service (-$23 million, -27%); and formula research grants (-$53 million). The request would increase competitive research (+$175 million) and create funding for the Office of Chief Scientist (+$6 million).[19]
- Rural development (-$130 million), from reductions to rural business development programs (-$95 million, -90%); rural housing programs (-$92 million, -5%); rural water and the regular rural broadband and distance learning/telemedicine accounts (-$58 million, -8%); and a net reduction to the rural broadband E-Connect program (-$50 million net,[20] -17%). The Administration would increase rural development salaries and expenses (+$64 million net) and create a new Higher Blends Infrastructure Initiative (+$100 million).
- Regulatory agencies (-$50 million), mostly by reducing the Agricultural Marketing Service (-$40 million, -21%) and the Animal and Plant Health Inspection Service (-$10 million, -1%).
- Commodity Futures Trading Commission (-$89 million, -28%, to be replaced by new user fees).
- Eliminating other appropriations that were in General Provisions (-$140 million).

---

[19] CRS Report R46341, *Federal Research and Development (R&D) Funding: FY2021*.

[20] The FY2021 budget request would move funding for the rural broadband E-Connect program into the regular Rural Utilities Service program account (+$250 million) while eliminating funding that was in the General Provisions title of the FY2020 appropriation (-$300 million). This is separate from the supplemental $100 million for E-Connect that was in the CARES Act coronavirus appropriation.

These reductions are offset by increases to the following:

- USDA administration (+$142 million, +30%), mostly by increasing the Office of Chief Information Officer (+$76 million, +155%); buildings and facilities (+$25 million, +19%); the Office of Budget and Program Analysis (+$11 million, +117%); and Departmental Administration (+$6 million, +27%). The request would create an Office of Safety, Security and Protection (+$25 million) and an Office of Information Affairs (+$2.5 million). It would reduce the Office of Public Partnership and Engagement (-$4 million, -68%).
- Food Safety Inspection Service (+$38 million, +4%).
- Food and Drug Administration (+49 million, +2%).

Budget Caps, Subcommittee Allocations, and Sequestration

Budget enforcement has procedural and statutory elements. The procedural elements relate to a budget resolution and are enforced with points of order. The statutory elements impose discretionary spending limits and are enforced with budget caps and sequestration.[21]

Budget Resolution

Typically, each chamber's Appropriations Committee receives a top-line limit on discretionary budget authority, referred to as a "302(a)" allocation, from the Budget Committee via an annual budget resolution. The Appropriations Committees then in turn subdivide the allocation among their subcommittees, referred to as the "302(b)" allocations.[22]

For FY2021, the House and Senate Budget Committees have not reported a budget resolution.

---

[21] CRS Report R42388, *The Congressional Appropriations Process: An Introduction*.

[22] References to 302(a) and 302(b) are to sections of the Congressional Budget Act of 1974.

Budget Caps

The Budget Control Act of 2011 (BCA, P.L. 112-25) set discretionary budget caps from FY2013 through FY2021 as a way of reducing federal spending.[23] Sequestration is an across-the-board backstop to achieve budget reductions if spending exceeds the budget caps (2 U.S.C. §901(c)).[24]

Subsequent Bipartisan Budget Acts (BBAs) have avoided sequestration on discretionary spending—with the exception of FY2013—by raising those caps four times in two-year increments in 2013, 2015, 2018, and 2019 (Figure 4).[25]

Most recently, the BBA of 2019 (P.L. 116-37) raised the cap applicable to FY2021 nondefense discretionary spending to $627 billion. Agriculture appropriations are part of the nondefense spending category. The BBA provides language to execute (or "deem") the caps for the appropriations process without a budget resolution.[26]

The $627 billion nondefense budget cap for FY2021 is about 1% greater than the cap that was used to write the FY2020 appropriations. This cap is to be shared among many appropriations subcommittees when spending allocations are made.

Discretionary Spending Allocations

In the absence of a budget resolution and without committee meetings during the coronavirus pandemic, subcommittee allocations have not been announced in either chamber.

---

[23] CRS Report R44874, *The Budget Control Act: Frequently Asked Questions*.

[24] CRS Report R42972, *Sequestration as a Budget Enforcement Process: Frequently Asked Questions*.

[25] CRS Insight IN11090, *Increasing the BCA Spending Limits: Characteristics of Previously Enacted Legislation*.

[26] CRS Insight IN11148, *The Bipartisan Budget Act of 2019: Changes to the BCA and Debt Limit*.

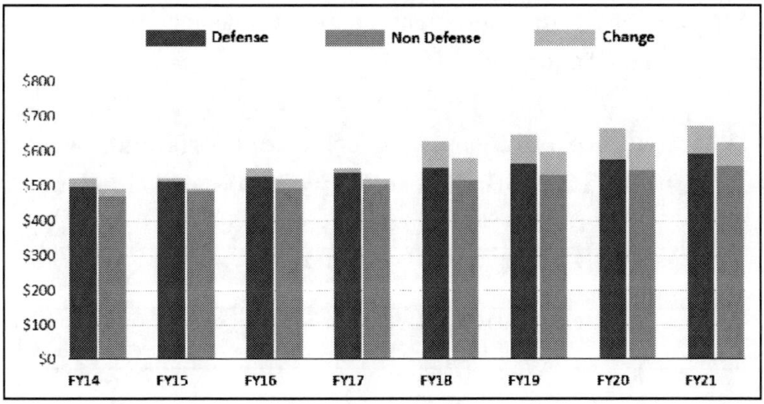

Source: CRS Insight IN11148, The Bipartisan Budget Act of 2019: Changes to the BCA and Debt Limit.

Figure 4. BCA Discretionary Limits, FY2014-FY2021.
Budget authority in billions of nominal dollars.

House Action

The House Agriculture Appropriations Subcommittee held hearings in February and March 2020 on the request for USDA overall, the FDA and FCA budget requests, and the USDA and HHS Offices of Inspector General.[27] Neither the Agriculture subcommittee nor the Appropriations Committee has marked up a FY2021 Agriculture appropriations bill.

Senate Action

The Senate Agriculture Appropriations Subcommittee held a hearing on the USDA budget request in March 2020.[28] Neither the Agriculture

---

[27] House Agriculture Appropriations Subcommittee, hearing videos at https://appropriations.house.gov/subcommittees/agriculture-rural-development-food-and-drug-administration-and-related-agencies-116th/congress_hearing.

[28] Senate Agriculture Appropriations Subcommittee, hearing videos at https://www.appropriations.senate.gov/subcommittees/agriculture-rural-development-food-and-drug-administration-and-related-agencies.

subcommittee nor the Appropriations Committee has marked up a FY2021 Agriculture appropriations bill.

Table 3. Agriculture and Related Agencies Appropriations, by Agency, FY20 1 9-FY202 1 Budget authority in millions of dollars

	FY2019	**FY2020**		**FY2021**				**Change: FY2020 to...**	
Agency or major program	P.L. 116-6	P.L. 116-94	COVID-19 Actsa	Admin. requestb	House	Senate	Enacted	Admin. request	House
Title I. Agricultural Programs									
Departmental Administration (Table A-1)	390.4	472.8	0.8	614.7	—	—	—	+141.9	—
Research, Education and Economics									
Agricultural Research Service	1,684.5	1,607.1	—	1,418.0	—	—	—	-189.1	—
Agency or major program	P.L. 116-6	P.L. 116-94	COVID-19 Actsa	Admin. requestb	House	Senate	Enacted	Admin. request	House
National Institute of Food and Agriculture	1,471.3	1,527.4	—	1,590.8	—	—	—	+63.4	—
National Agricultural Statistics Service	174.5	180.3	—	177.5	—	—	—	-2.8	—
Economic Research Service	86.8	84.8	—	62.1	—	—	—	-22.6	—
Under Secretary	0.8	0.8	—	6.8	—	—	—	+6.0	—
Marketing and Regulatory Programs									
Animal and Plant Health Inspection Service	1,014.3	1,045.9	55.0	1,035.6	—	—	—	-10.3	—
Agricultural Marketing Service	160.3	188.2	45.0	148.4	—	—	—	-39.7	—
Section 32 (M)	1,374.0	1,404.0	—	1,431.0	—	—	—	+27.0	—
Under Secretary	0.9	0.8	—	0.9	—	—	—	+0.1	—
Food Safety									
Food Safety and Inspection Service	1,049.3	1,054.3	33.0	1,092.4	—	—	—	+38.1	—
Under Secretary	0.8	0.8	—	0.8	—	—	—	+0.0	—
Subtotal, Title I									
Discretionary	6,033.9	6,163.1	133.8	6,148.0	—	—	—	-15.1	—
Mandatory (M)	1,374.0	1,404.0	—	1,431.0	—	—	—	+27.0	—

	FY2019	FY2020		FY2021				Change: FY2020 to...	
Subtotal	7,407.9	7,567.1	—	7,579.0	—	—	—	+11.9	—
Title II. Farm Production and Conservation									
Business Center	216.4	203.9	—	243.6	—	—	—	+39.7	—
Farm Service Agencyc	1,494.2	1,541.7	3.0	1,484.9	—	—	—	-56.8	—
FSA Farm Loans: Loan Authorityd	7,987.7	8,431.0	—	8,916.8	—	—	—	+485.8	—
Risk Management Agency	58.4	58.4	—	59.4	—	—	—	+1.1	—
Federal Crop Insurance Corporation (M)	15,410.6	9,959.0	—	8,716.0	—	—	—	-1,243.0	—
Commodity Credit Corporation (M)	15,410.0	26,309.0	14,000.0e	25,553.1	—	—	—	-755.9	—
Conservation Operations	819.5	829.6	—	830.2	—	—	—	+0.6	—
Agency or major program	P.L. 116-6	P.L. 116-94	COVID-19 Actsa	Admin. requestb	House	Senate	Enacted	Admin. request	House
Watershed and Flood Prevention	150.0	175.0	—	0.0	—	—	—	-175.0	—
Watershed Rehabilitation Program	10.0	10.0	—	0.0	—	—	—	-10.0	—
Under Secretary	0.9	0.9	—	1.1	—	—	—	+0.2	—
Subtotal, Title II									
Discretionary	2,748.8	2,819.0	3.0	2,618.7	—	—	—	-200.2	—
Mandatory (M)	30,821.1	36,268.5	14,000.0	34,269.6	—	—	—	-1,998.9	—
Subtotal	33,569.9	39,087.5	14,003.0	36,888.3	—	—	—	-2,199.2	—
Title III. Rural Development									
Salaries and Expenses (including transfers)f	686.8	697.8	—	761.5	—	—	—	+63.6	—
Rural Housing Service	1,606.0	1,686.7	—	1,595.0	—	—	—	-91.7	—
RHS Loan Authorityd	28,293.8	28,646.0	—	27,260.0	—	—	—	-1,386.0	—
Rural Business-Cooperative Service	98.6	105.0	20.5	110.4	—	—	—	+5.4	—
RBCS Loan Authorityd	1,026.4	1,088.9	1,000.0	1,500.0	—	—	—	+411.1	—
Rural Utilities Service	620.2g	750.3g	25.0	942.6h	—	—	—	+192.3	—

Table 3. (Continued)

	FY2019	**FY2020**		**FY2021**				**Change: FY2020 to...**	
RUS Loan Authorityd	8,419.9	8,401.2	—	7,518.0	—	—	—	-883.2	—
Under Secretaryi	—i	0.8	—	1.0	—	—	—	+0.2	—
Subtotal, Title III	3,011.7g	3,240.6g	45.5g	3,410.5	—	—	—	+169.9	—
Subtotal, RD Loan Authorityd	37,740.0	38,136.1	1,000.0	36,278.0	—	—	—	-1,858.1	—
Title IV. Domestic Food Programs									
Child Nutrition Programs (M)	23,140.8	23,615.1	8,800.0	25,040.9	—	—	—	+1,425.8	—
WIC Program	6,075.0	6,000.0	500.0	5,451.5	—	—	—	-548.5	—
SNAP, Food and Nutrition Act Programs (M)	73,476.9	67,886.3	15,910.0j	68,281.4	—	—	—	+395.1	—
Agency or major program	P.L. 116-6	P.L. 116-94	COVID-19 Actsa	Admin. requestb	House	Senate	Enacted	Admin. request	House
Commodity Assistance Programs	322.1	344.2	850.0	80.7	—	—	—	-263.5	—
Nutrition Programs Administration	164.7	155.9	—	155.3	—	—	—	-0.6	—
Under Secretary	0.8	0.8	—	0.8	—	—	—	+0.0	—
Subtotal, Title IV									
Discretionary	6,620.3	6,566.0	1,350.0	5,700.7	—	—	—	-865.3	—
Mandatory (M)	96,560.0	91,436.3	24,710.0	93,309.8	—	—	—	+1,873.5	—
Subtotal	103,180.3	98,002.3	26,060.0	99,010.5	—	—	—	+1,008.2	—
Title V. Foreign Assistance									
Foreign Agricultural Service	213.9	215.5	4.0	193.7	—	—	—	-21.8	—
Food for Peace Title II, and admin. expenses	1,500.1g	1,725.1	—	0.1	—	—	—	-1,725.0	—
McGovern-Dole Food for Education	210.3	220.0	—	0.0	—	—	—	-220.0	—
CCC Export Loan Salaries	8.8	6.4	—	6.4	—	—	—	+0.0	—
Office of Codex Alimentarius	4.0	4.8	—	4.8	—	—	—	+0.0	—
Under Secretary	0.9	0.9	—	0.9	—	—	—	+0.0	—
Subtotal, Title V	1,938.0	2,172.7	4.0	205.9	—	—	—	-1,966.8	—
Title VI. Related Agencies									

	FY2019	**FY2020**		**FY2021**				**Change: FY2020 to...**	
Food and Drug Administration	3,080.5	3,171.5	163.0	3,220.4	—	—	—	+48.9	—
Commodity Futures Trading Commissionk	[268.0]	315.0	—	226.5	—	—	—	-88.5	—
Subtotal, Title VI	[3,348.5]	3,486.5	163.0	3,446.9	—	—	—	-39.6	—
Title VII. General Provisions									
Changes in Mandatory Programs, CHIMPSl									
a. Conservation programs	[-60.2]m	-60.2	—	-60.2	—	—	—	+0.0	—
b. Nutrition programs	+5.0	+9.0	—	—	—	—	—	-9.0	—
c. Farm Production & Conservation Bus. Ctr.	[+60.2]m	+60.2	—	+60.2	—	—	—	+0.0	—
Agency or major program	P.L. 116-6	P.L. 116-94	COVID-19 Actsa	Admin. requestb	House	Senate	Enacted	Admin. request	House
d. Other CHIMPS and mandatory rescissions	+10.0	—	—	—	—	—	—	—	—
Subtotal, CHIMPS (Table B-1)	+15.0	+9.0	—	+0.0	—	—	—	-9.0	—
Rescissions (discretionary)n (Table B-2)	-505.0	-1,015.1	—	-1,200.0	—	—	—	-184.9	—
Other appropriations (Table B-3)									
a. Emergency/ disaster programs	—	535.0o	—	—	—	—	—	-535.0	—
b. Water and Waste Water	75.0	—	—	—	—	—	—	—	—
c. Broadband piloth	125.0	300.0	100.0	—h	—	—	—	-300.0	—
d. Food for Peace	216.0	—	—	—	—	—	—	—	—
e. Coronavirus Food Assistance Program	—	—	12,500.0p	—	—	—	—	—	—
f. Other appropriations	77.5	140.3	—	—	—	—	—	-140.3	—
Subtotal, Other appropriations	493.5	975.3	12,600.0	0.0	—	—	—	-975.3	—
Subtotal, Title VII	3.5	-30.8	12,600.0	-1,200.0	—	—	—	-1,169.2	—
Scorekeeping Adjustmentsq (Table B-4)									
Emergency declaration in this bill	—	-535.0o	—a	—	—	—	—	+535.0	—

Table 3. (Continued)

	FY2019	FY2020		FY2021				Change: FY2020 to...	
Other scorekeeping adjustments	-404.0	-398.0	—	-403.0	—	—	—	-5.0	—
Subtotal, Scorekeeping adjustmentsq	-404.0	-933.0	—	-403.0	—	—	—	+530.0	—
Totals									
Discretionary: Senate basis w/o CFTCk	23,032.7	[23,169.1]	14,299.3	19,701.3	—	—	—	-3,467.9	—
Agency or major program	P.L. 116-6	P.L. 116-94	COVID-19 Actsa	Admin. requestb	House	Senate	Enacted	Admin. request	House
Discretionary: House basis w/ CFTCk	[23,300.7]	23,484.1	14,299.3	19,927.8	—	—	—	-3,556.4	—
Mandatory (M)	128,755.1	129,108.8	38,710.0	129,010.4	—	—	—	-98.4	—
Total: Senate basis w/o CFTC	151,787.8	[152,278.0]	53,009.3	148,711.7	—	—	—	-3,566.3	—
Total: House basis w/ CFTC	[152,055.8]	152,593.0	53,009.3	148,938.2	—	—	—	-3,654.8	—

Source: CRS, using appropriations text and reports; unpublished CBO tables; OMB *FY2021 Budget of the U.S. Government*, Appendix; and USDA *FY2020 Budget Explanatory Notes*.

Notes: Amounts are nominal discretionary budget authority in millions of dollars unless labeled otherwise. "(M)" indicates that the account is mandatory authority (or primarily mandatory authority). Bracketed amounts are not in the official totals due to differing House-Senate jurisdiction for CFTC but are shown for comparison.

a. Total from four coronavirus supplemental appropriations acts (P.L. 116-123, P.L. 116-127, P.L. 116-136, P.L. 116-139), all of which was emergency spending. Although amounts were not designated as mandatory, they are grouped here for comparison to accounts that are in the annual appropriations. See detail in CRS In Focus IF11491, *Supplemental Appropriations for Agriculture and Related Agencies Due to COVID-19*.

b. The request is based on current law and does not reflect the Administration's various legislative proposals.

c. Includes regular FSA salaries and expenses plus transfers for farm loan program salaries and administrative expenses. Also includes farm loan program loan subsidy, State Mediation Grants, Dairy Indemnity Program (mandatory funding), and Grassroots Source Water Protection Program. Does not include appropriations to the Foreign Agricultural Service for export loans and P.L. 480 administration that are transferred to FSA.

d. Loan authority is the amount of loans that can be made or guaranteed with a loan subsidy. This amount is not added in the budget authority subtotals or totals.

e. The CARES Act (P.L. 116-136) provided $14 billion of reimbursement to the CCC after a June 2020 financial statement. Although it is not budget authority for direct coronavirus support and not counted in CARES Act accounting, it is comparable to the CCC reimbursement in regular annual appropriations.

f. Rural Development salaries and expenses include a base amount plus transfers from the three rural development agencies. Amounts presented for the agencies therefore include program funds for loans and grants.

g. Excludes a portion of the other appropriations that are provided separately in General Provisions (see Table B-3).

h. The ReConnect Broadband Pilot Program that was created in the FY2018 appropriation has been funded by separate appropriation in the General Provisions title. In FY2019 and FY2020 it is augmented by a transfer from the Cushion of Credit account that is available outside the appropriations caps. For example, in FY2020, appropriators direct $555 million to ReConnect from $300 million appropriated in General Provisions and $255 million from the Cushion of Cred it account. For FY2021, the Administration requests $250 million in the Rural Utilities Service account rather than through General Provisions.

i. The USDA-initiated reorganization in 2017 created an "Assistant to the Secretary for Rural Development" as part of the Office of the Secretary rather than the previously Senate-confirmed undersecretary position. The 2018 farm bill reinstated the undersecretary position, and the FY2020 appropriation funds it as such.

j. In addition to the specified supplemental amounts for the SNAP account in the CARES Act (P.L. 116-136), the Families First Coronavirus Response Act (P.L. 116- 127, §1101) authorized and appropriated open-ended funding for the Pandemic Electronic Benefit Transfer (P-EBT) program. The amount displayed does not include the P-EBT funding.

k. Jurisdiction for the Commodity Futures Trading Commission (CFTC) is in the House Agriculture Appropriations Subcommittee and the Senate Financial Services Appropriations Subcommittee. After FY2008, CFTC is carried in the enacted Agriculture appropriations in even-numbered fiscal years. It is always carried in House Agriculture subcommittee markup but never in Senate Agriculture subcommittee markup. Bracketed amounts are not in the Agriculture appropriations totals due to differing House-Senate jurisdiction for CFTC but are shown for comparison.

l. Includes reductions (limitations and rescissions) and increases to mandatory programs that are known as CHIMPS.

m. These bracketed amounts were not in the official CBO scoring of CHIMPS. Appropriations acts in FY2018 and FY2019 transferred mandatory conservation funding into the Farm Production and Conservation Business Center, but the official CBO scoring of appropriations at that time did not record it as a CHIMP the way that the FY2020 scoring reflects. For more background, see CRS Report R46011, *FY2020 Appropriations for Agricultural Conservation*.

n. Rescissions are actions that permanently reduce a budget authority subsequent to an enacted appropriation. They score budgetary savings. Any rescissions from mandatory programs are included with the CHIMPS.

o. Emergency appropriations in FY2020 include $535 million for Ebola and $1.5 billion for the Wildfires and Hurricane Indemnity Program, which was offset with $1.5 billion in rescissions (Table B-3).

p. Includes $9.5 billion specified in the CARES Act for payments to farmers and $3 billion from indefinite authority in the Families First Coronavirus Relief Act that USDA used for the Farmers to Families Food Boxes Program.

q. "Scorekeeping adjustments" are not necessarily appropriated items and may not be shown in Appropriations Committee tables but are part of the official CBO score (accounting) of the bill. They predominantly include "negative subsidies" in loan program accounts (mostly from receipt of fees) and adjustments for emergency designations in the bill.

OTHER APPROPRIATIONS ISSUES

Policy-Related Provisions

Besides setting spending authority, appropriations acts are also a vehicle for policy-related provisions that direct executive branch actions.[29] These provisions, limitations, or riders may have the force of law if they are included in the act's text, but their effect is generally limited to the current fiscal year unless they amend the *U.S. Code*, which is rare in appropriations acts.

Report language may also provide policy instructions. Although report language does not carry the force of text in an act, it often explains congressional intent, which the agencies may be expected to follow.

In the past, Congress has said that committee reports and the joint explanatory statement need to be read together to capture all of the congressional intent for a fiscal year. For example, the explanatory statement for the FY2020 Further Consolidated Appropriations act instructs that the earlier House and Senate reports should be read together with the conference agreement:

> Congressional Directives. The statement is silent on provisions that were in both the House Report (H.Rept. 116-107) and Senate Report (S.Rept. 116-110) that remain unchanged by this agreement, except as noted in this statement.
>
> The House and Senate report language that is not changed by the statement is approved and indicates congressional intentions. The statement, while repeating some report language for emphasis, does not intend to negate the language referred to above unless expressly provided herein.[30]

---

[29] See CRS Report RL30240, *Congressional Oversight Manual*.

[30] U.S. Congress, House Committee on Appropriations, *Further Consolidated Appropriations Act, 2020*, committee print, 116th Cong., 2nd Sess., January 2020, 38-679 (Washington, DC: GPO, 2020), p. 305.

COVID-19 Supplemental Appropriations in FY2020

In March and April 2020, Congress passed and the President signed four supplemental appropriations acts in response to the novel coronavirus pandemic—designated Coronavirus Disease 2019, or COVID-19.[31] These acts in FY2020 include support for programs in the jurisdiction of the Agriculture appropriations subcommittees. The four acts are:

1. Coronavirus Preparedness and Response Supplemental Appropriations Act (P.L. 116-123, March 6, 2020);
2. Families First Coronavirus Response Act (P.L. 116-127, March 18, 2020);
3. CARES Act (P.L. 116-136, March 27, 2020); and
4. Paycheck Protection Program and Health Care Enhancement Act (P.L. 116-139, April 27, 2020).

These acts together specify $36 billion of appropriations to programs in the jurisdiction of the Agriculture appropriations subcommittees. In addition, the second act authorized indefinite funding for commodity food assistance, from which the USDA has announced that it would obligate $3 billion to date.[32] The third act provided for $14 billion of reimbursement for the CCC that, while not immediately available for obligation and not counted as COVID-19 relief, is comparable to the regular CCC reimbursements in annual Agriculture appropriations acts and is included in this chapter.

Columns labeled "COVID-19 Acts" in Table 2, Table 3, and other appendix tables summarize these combined $53 billion of appropriations through the coronavirus supplemental acts. All of this spending is designated as emergency spending and not counted against budget caps.

---

[31] CRS In Focus IF11491, *Supplemental Appropriations for Agriculture and Related Agencies Due to COVID-19.*

[32] The Farmers to Families Food Box Program, discussed in CRS Report R46347, *COVID-19, U.S. Agriculture, and USDA's Coronavirus Food Assistance Program (CFAP).*

Budget Sequestration Continues on Mandatory Spending

Sequestration is a process to reduce federal spending through automatic, largely across-the-board reductions that permanently cancel budget authority. Sequestration is triggered as a budget enforcement mechanism when federal spending exceeds statutory budget goals.

Despite the BBA agreements that have raised discretionary spending caps to avoid sequestration on discretionary accounts, sequestration still impacts mandatory spending. Sequestration on mandatory accounts began in FY2013 and continues to the present. Congress has amended the BCA five times to extend sequestration on mandatory spending accounts. The original sunset of sequestration was to be FY2021; it is now scheduled to continue through FY2029. Sequestration has reduced mandatory spending by about 6% per year, amounting to about $1.5 billion of reductions annually from agriculture accounts in recent years. See Appendix C for effects.

APPENDIX A. APPROPRIATIONS IN ADMINISTRATIVE ACCOUNTS

Table A.1. USDA Departmental Administration Appropriations Budget authority in millions of dollars

	FY2019	FY2020		FY2021				Change: FY2020 to...	
Agency or major program	P.L. 116-6	P.L. 116-94	COVID-19 Acts	Admin. request	House	Senate	Enacted	Admin. request	House
Office of the Secretary									
Office of the Secretary	5.1	5.1	—	5.1	—	—	—	+0.1	—
Assistant to the Sec. for Rural Developmenta	0.8	—	—	—	—	—	—	—	—
Office of Homeland Security	1.5	1.5	—	1.3	—	—	—	-0.2	—
Public Partnership and Engagementb	4.7	6.2	—	2.0	—	—	—	-4.3	—

	FY2019	FY2020		FY2021				Change: FY2020 to...	
Agency or major program	P.L. 116-6	P.L. 116-94	COVID-19 Acts	Admin. request	House	Senate	Enacted	Admin. request	House
Assistant Secretary for Admin.	0.9	0.9	—	0.9	—	—	—	+0.0	—
Departmental Administration	22.3	21.4	—	27.2	—	—	—	+5.8	—
Asst. Sec. Congressional Relations	3.9	3.9	—	3.9	—	—	—	+0.1	—
Office of Communications	7.5	7.3	—	7.4	—	—	—	+0.1	—
Subtotal	46.6	46.1	—	47.9	—	—	—	+1.7	—
Executive Operations									
Office of Chief Economist	21.3	24.0	—	20.9	—	—	—	-3.2	—
Office of Hearings and Appeals	15.2	15.2	—	15.5	—	—	—	+0.3	—
Office of Budget, Program Analysis	9.5	9.5	—	20.7	—	—	—	+11.1	—
Subtotal	46.0	48.8	—	57.1	—	—	—	+8.3	—
Other Administration									
Chief Information Officer	55.6	66.6	—	142.8	—	—	—	+76.2	—
Chief Financial Officer	6.0	6.0	—	11.1	—	—	—	+5.1	—
Assistant Secretary for Civil Rights	0.9	0.9	—	0.9	—	—	—	+0.0	—
Office of Civil Rights	24.2	24.2	—	20.6	—	—	—	-3.6	—
Buildings and facilities	60.0	128.2	—	152.8	—	—	—	+24.6	—
Office of Safety, Security, and Protection	—	—	—	25.0	—	—	—	+25.0	—
Hazardous materials management	3.5	4.5	—	3.5	—	—	—	-1.0	—
Office of Inspector General	98.2	98.2	—	100.4	—	—	—	+2.2	—
General Counsel	45.1	45.1	0.8	45.9	—	—	—	+0.7	—
Office of Information Affairs	—	—	—	2.5	—	—	—	+2.5	—
Office of Ethics	4.1	4.1	—	4.2	—	—	—	+0.1	—
Subtotal	297.7	377.9	0.8	509.8	—	—	—	+131.9	—
Total, Departmental Administration	390.4	472.8	0.8	614.7	—	—	—	+141.9	—

Source: CRS, using appropriations text and reports; unpublished CBO tables; OMB *FY2021 Budget of the U.S. Government*, Appendix; and USDA *FY2020 Budget Explanatory Notes*.

a. The USDA-initiated reorganization in 2017 created an "Assistant to the Secretary for Rural Development" as part of the Office of the Secretary rather than the previously Senate-confirmed undersecretary position. The 2018 farm bill reinstated the undersecretary position, and the FY2020 appropriation funds it as such.

b. Formerly, this was the Office of Advocacy and Outreach.

APPENDIX B. APPROPRIATIONS IN GENERAL PROVISIONS

Table B.1. General Provisions: Changes in Mandatory Program Spending (CHIMPS) Budget authority in millions of dollars

	FY2019	FY2020		FY2021				Change: FY2020 to...	
CHIMPS and mandatory rescissions	P.L. 116-6	P.L. 116-94	COVID-19 Acts	Admin. request	House	Senate	Enacted	Admin. request	House
Farm bill programs									
Farm Security Rural Investment Act conservation programs	[-60.2]0	-60.2	—	-60.2	—	—	—	+0.0	—
Farm to School	+5.0	+9.0	—	0.0	—	—	—	-9.0	—
Subtotal, farm bill programs	+5.0	-51.2	—	-60.2	—	—	—	-9.0	—
Other CHIMPS									
Farm Production & Conservation (FPAC) Business Center	[+60.2]0	+60.2	—	+60.2	—	—	—	+0.0	—
Food for Progress	+10.0	—	—	—	—	—	—	—	—
Total CHIMPS	+15.0	+9.0	—	+0.0	—	—	—	-9.0	—

Source: CRS, using appropriations text and reports; unpublished CBO tables; OMB *FY202 1 Budget of the U.S. Government*, Appendix; and USDA *FY2020 Budget Explanatory Notes*.

Note: a. Amounts in brackets were not in the official CBO scoring of CHIMPS. The FY2019 appropriations act transferred mandatory conservation funding into the Farm Production and Conservation Business Center, but the official CBO scoring of appropriations at that time did not record it as a CHIMP the way that the FY2020 scoring reflects. For more background, see CRS Report R460 1 1, *FY2020 Appropriations for Agricultural Conservation*.

Table B.2. General Provisions: Rescissions from Discretionary Accounts Budget authority in millions of dollars

	FY2019	**FY2020**		**FY2021**				**Change: FY2020 to...**	
Rescissions from discretionary accounts	P.L. 116-6	P.L. 116-94	COVID-19 Acts	Admin. request	House	Senate	Enacted	Admin. request	House
Special Supplemental Nutrition Program for Women, Infants, and Children	-500.0	-1,000.0	—	-1,200.0	—	—	—	-200.0	—
Electric loan refinancing	—	-15.1	—	—	—	—	—	+15.1	—
USDA unobligated balances	-5.0	—	—	—	—	—	—	—	—
Total	-505.0	-1,015.1	—	-1,200.0	—	—	—	-184.9	—

Source: CRS, using appropriations text and reports; unpublished CBO tables; OMB *FY2021 Budget of the U.S. Government*, Appendix; and USDA *FY2020 Budget Explanatory Notes*.

Table B.3. General Provisions: Other Appropriations Budget authority in millions of dollars

	FY2019	**FY2020**		**FY2021**				**Change: FY2020 to...**	
Program	P.L. 116-6	P.L. 116-94	COVID-19 Acts	Admin. request	House	Senate	Enacted	Admin. request	House
FDA buildings and facilities	—	20.0	—	—	—	—	—	-20.0	—
Coronavirus Food Assistance Program	—	—	12,500.0a	—	—	—	—	—	—
Agriculture Risk Coverage pilot	5.0	—	—	—	—	—	—	—	—
Dairy Innovation	—	20.0	—	—	—	—	—	-20.0	—
Mitigation banking	—	5.0	—	—	—	—	—	-5.0	—
Conservation Reserve Program pilot	1.0	1.0	—	—	—	—	—	-1.0	—
Citrus greening	8.5	8.5	—	—	—	—	—	-8.5	—
Protecting Animals with Shelter Grants	—	2.0	—	—	—	—	—	-2.0	—
Fruit fly quarantine payments	9.0	—	—	—	—	—	—	—	—

Table B.3. (Continued)

	FY2019	FY2020		FY2021				Change: FY2020 to...	
Program	P.L. 116-6	P.L. 116-94	COVID-19 Acts	Admin. request	House	Senate	Enacted	Admin. request	House
Geographically disadvantaged farmers	2.0	2.0	—	—	—	—	—	-2.0	—
Farming Opportunities Training and Outreach	—	5.0	—	—	—	—	—	-5.0	—
Water Bank	4.0	4.0	—	—	—	—	—	-4.0	—
Rural Energy Savings Program	10.0	12.0	—	—	—	—	—	-12.0	—
Maturing mortgage pilot	1.0	1.0	—	—	—	—	—	-1.0	—
Rural microentrepreneur assistance program	3.0	—	—	—	—	—	—	—	—
Water and Waste Water	75.0	—	—	—	—	—	—	—	—
Waste water pilot	—	5.0	—	—	—	—	—	-5.0	—
Broadband pilotb	125.0	300.0	100.0	—b	—	—	—	-300.0	—
Distance Learning Telemedicine	16.0	—	—	—	—	—	—	—	—
Rural Hospital Technical Assistance	—	1.0	—	—	—	—	—	-1.0	—
Tribal Demonstration Projects	—	3.0	—	—	—	—	—	-3.0	—
Healthy Food Financing Initiative	2.0	5.0	—	—	—	—	—	-5.0	—
Healthy Fluid Milk	—	1.0	—	—	—	—	—	-1.0	—
RISE grants	—	5.0	—	—	—	—	—	-5.0	—
NIFA military veterans grants	5.0	5.0	—	—	—	—	—	-5.0	—
Genome to Phenome	—	1.0	—	—	—	—	—	-1.0	—
NIFA 1890 land grants, Centers of Excellence	5.0	6.0	—	—	—	—	—	-6.0	—
Tribal Student Scholarships	—	5.0	—	—	—	—	—	-5.0	—
International Agriculture Education	—	1.0	—	—	—	—	—	-1.0	—
Pollinator Research Coordinator	—	0.4	—	—	—	—	—	-0.4	—

	FY2019	FY2020		FY2021				Change: FY2020 to...	
Program	P.L. 116-6	P.L. 116-94	COVID-19 Acts	Admin. request	House	Senate	Enacted	Admin. request	House
Urban Agriculture Office	—	5.0	—	—	—	—	—	-5.0	—
Food Loss Liaison	—	0.4	—	—	—	—	—	-0.4	—
Food for Peace	216.0	—	—	—	—	—	—	—	—
Food for Progress	6.0	—	—	—	—	—	—	—	—
Nutrition Assistance Program Study	—	6.0	—	—	—	—	—	-6.0	—
Micro-grants for Food Security	—	5.0	—	—	—	—	—	-5.0	—
School breakfast expansion	—	5.0	—	—	—	—	—	-5.0	—
Subtotal, other spending provisions	493.5	440.3	12,600.0	—	—	—	—	-440.3	—
Disaster/Emergency programs									
Ebola prevention and treatment	—	+535.0	—	—	—	—	—	—	—
Rescission of unobligated emergency funding	—	-1,500.0	—	—	—	—	—	—	—
WHIP (Wildfire, Hurricanes Indemnity Program)	—	+1,500.0	—	—	—	—	—	—	—
Subtotal, disaster/emergency programs	—	+535.0c	—	—	—	—	—	—	—
Total	493.5	975.3c	12,600.0	0.0	—	—	—	-975.3	—

Source: CRS, using appropriations text and reports; unpublished CBO tables; OMB *FY202 1 Budget of the U.S. Government*, Appendix; and USDA *FY2020 Budget Explanatory Notes*.

Note: Excludes supplemental appropriations outside the annual appropriations act.

a. Includes $9.5 billion specified in the CARES Act for payments to farmers and $3 billion from indefinite authority in the Families First Coronavirus Relief Act that USDA used for the Farmers to Families Food Boxes Program.

b. The ReConnect Broadband Pilot Program that was created in the FY2018 appropriation has been funded by separate appropriation in the General Provisions title. In FY2019 and FY2020 it is augmented by a transfer from the Cushion of Credit account that is available outside the appropriations caps. For example, in FY2020, appropriators direct $555 million to ReConnect from $300 million appropriated in General Provisions and $255 million from the Cushion of Credit account. For FY2021, the Administration requests $250 million in the Rural Utilities Service account rather than through General Provision s.

c. Appropriations designated as emergency spending in FY2020 and are not counted against the discretionary spending limit include $535 million for Ebola and $1.5 billion for WHIP, which was offset with $1.5 billion in rescissions (see Table B.4).

Table B.4. General Provisions: Scorekeeping Adjustments Budget authority in millions of dollars

	FY2019	FY2020		FY2021				Change: FY2020 to...	
Scorekeeping adjustment	P.L. 116-6	P.L. 116-94	COVID-19 Acts	Admin. requesta	House	Senate	Enacted	Admin. request	House
Loan program negative subsidies									
Rural housing negative subsidy	-135.0	-125.0	—	-125.0	—	—	—	+0.0	—
Rural community facilities negative subsidy	-72.0	-72.0	—	-72.0	—	—	—	+0.0	—
Rural elec. & tele. loan negative subsidy	-173.0	-176.0	—	-176.0	—	—	—	+0.0	—
Rural water & waste loan negative subsidy	-2.0	-3.0	—	-3.0	—	—	—	+0.0	—
Ag credit loan negative subsidy	-23.0	-23.0	—	-23.0	—	—	—	+0.0	—
Subtotal, negative subsidies	-405.0	-399.0	—	-399.0	—	—	—	+0.0	—
Denali Commission	—	—	—	-4.0	—	—	—	-4.0	—
Child nutrition equipment grants	1.0	1.0	—	—	—	—	—	-1.0	—
Emergency designations not counted in allocation	—	-535.0b	—c	—	—	—	—	+535.0	—
Total	-404.0	-933.0	—	-403.0	—	—	—	+530.0	—

Source: CRS, using appropriations text and reports; unpublished CBO tables; OMB *FY2021 Budget of the U.S. Government*, Appendix; and USDA *FY2020 Budget Explanatory Notes*.

Notes: A negative subsidy for a loan program generally occurs when the interest rate and/or fees charged to the borrowers are more than sufficient to cover the costs of the risk of default (Government Accountability Office, "Credit Reform: Current Method to Estimate Credit Subsidy Costs Is More Appropriate for Budget Estimates Than a Fair Value Approach," GAO-16-41, January 2016, p. 8).

a. The Administration's request does not specify the negative subsidies for the loan programs. This table assumes continuation of amounts in FY2020 pending CBO scorekeeping that will accompany FY2021 House and Senate markup.

b. Appropriations designated as emergency spending in FY2020 include $535 million for Ebola and $1.5 billion for the Wildfires and Hurricane Indemnity Program, which was offset with $1.5 billion in rescissions of prior-year emergency appropriations (see Table B-3).

c. Amounts in the coronavirus supplemental appropriations acts were all designated as emergency spending and are not noted further here because the acts faced no budget constraint.

APPENDIX C. BUDGET SEQUESTRATION

Sequestration is a process to reduce federal spending through automatic, largely across-the-board reductions that permanently cancel mandatory and/or discretionary budget authority.[33] A sequestration rate is the percentage reduction that is subtracted from an appropriated budget authority to achieve an intended budget goal.

Table C.1. Summary of Sequestration on Agriculture Accounts (million dollars)

Fiscal Year	Sequestration Ratea	Budget Authority Subject to Sequestrationb	Sequestered Budget Authorityb
Discretionary Spending			
2013	5.0%	23,064	1,153
Mandatory Spending			
2013	5.1%	13,987	713
2014	7.2%	14,610	1,052
2015	7.3%	15,795	1,153
2016	6.8%	26,748	1,819
2017	6.9%	24,429	1,686
2018	6.6%	19,938	1,316
2019	6.2%	24,682	1,530
2020	5.9%	23,352	1,378
2021	5.7%	29,841	1,701

Source: CRS, compiled from OMB, *Reports to the Congress on the Joint Committee Reductions,* various fiscal years. Available for FY2018-FY2021 at https://www.whitehouse.gov/omb/legislative/sequestration-reports-orders and for FY2013-FY2017 at https://obamawhitehouse.archives.gov/omb/legislative_reports/sequestration.

a. Sequestration rates listed here are for nonexempt, nondefense accounts as determined by OMB.

b. Totals are computed by CRS for accounts and programs that are in the jurisdiction of Agriculture appropriations.

Sequestration is triggered as a budget enforcement mechanism when federal spending exceeds statutory budget goals.[34] Sequestration is currently authorized by the Budget Control Act (BCA, P.L. 112-25). For

---

[33] CRS Report R43411, *The Budget Control Act of 2011: Legislative Changes to the Law and Their Budgetary Effects*.

[34] CRS Report R42972, *Sequestration as a Budget Enforcement Process: Frequently Asked Questions*.

discretionary accounts, sequestration continues through FY2021. For mandatory accounts, Congress amended the BCA to continue sequestration through FY2029.

OMB computes sequestration rates annually. Table C.1 shows the rates of sequestration that have been announced and the total amounts of budget authority that have been cancelled from accounts in Agriculture appropriations.

Discretionary Spending

For discretionary spending, sequestration is authorized through FY2021 if discretionary defense and nondefense spending exceed caps that are specified in statute (2 U.S.C. §901(c)).

In FY2013, the timing of the appropriations acts and the first year of sequestration resulted in triggering sequestration on discretionary spending.

In FY2014-FY2020, BBAs in 2013, 2015, 2018, and 2019 (P.L. 113-67, P.L. 114-74, P.L. 115- 123, and P.L. 116-37, respectively) have avoided sequestration on discretionary spending. These BBAs raised the discretionary budget caps that were placed in statute by the BCA and allowed Congress to enact larger appropriations than would have been allowed. The enacted appropriations in FY2014-FY2020 met the spending limitations of the revised budget caps, and therefore no sequestration on discretionary accounts was necessary.

For FY2021, the BBA of 2019 (P.L. 116-37) similarly provides a higher discretionary cap that may avoid sequestration (see "Budget Caps, Subcommittee Allocations").[35]

---

[35] CRS Insight IN11148, *The Bipartisan Budget Act of 2019: Changes to the BCA and Debt Limit.*

Mandatory Spending

Sequestration Occurs and Continues

For mandatory spending, sequestration is presently authorized and scheduled to continue through FY2029, having been amended and extended by budget acts that were subsequent to the BCA (2 U.S.C. §901a(6)). That is, sequestration of mandatory spending has not been avoided by the BBAs and continues to apply annually to certain accounts.

The original FY2021 sunset on the sequestration of *mandatory* accounts has been extended five times as an offset to pay for raising the caps on *discretionary* spending to avoid sequestration in the near term (or as a general budgetary offset for other authorization acts):

1. Congress extended the duration of mandatory sequestration by two years (until FY2023) as an offset in BBA 2013.[36]
2. Congress extended it by another year (until FY2024) to maintain retirement benefits for certain military personnel (P.L. 113-82).
3. Congress extended sequestration on nonexempt mandatory accounts another year (until FY2025) as an offset in BBA 2015.[37]
4. Congress extended sequestration on nonexempt mandatory accounts for two years (until FY2027) as an offset in BBA 2018 (P.L. 115-123, §30101(c)).[38]
5. Congress extended sequestration on nonexempt mandatory accounts by another two years (until FY2029) as an offset in BBA 2019 (P.L. 116-37, §402).[39]

---

[36] CBO, *Bipartisan Budget Act of 2013*, December 11, 2013, https://www.cbo.gov/publication/44964.

[37] CBO, *Bipartisan Budget Act of 2015*, October 28, 2015, https://www.cbo.gov/publication/50938. In addition to extending sequestration and related to mandatory spending in agriculture, crop insurance was used as an additional budgetary offset in BBA 2015. The effect was temporary, however, and the crop insurance reduction was restored. For more background, see the section on crop insurance and the Standard Reinsurance Agreement in CRS Report R44240, *Agriculture and Related Agencies: FY2016 Appropriations*.

[38] CBO, *Bipartisan Budget Act of 2018*, February 8, 2018, https://www.cbo.gov/publication/53556.

[39] CBO, *Bipartisan Budget Act of 2019*, July 23, 2019, https://www.cbo.gov/publication/55478.

Exemptions from Sequestration

Some USDA mandatory programs are statutorily exempt from sequestration. Those expressly exempt by statute are the nutrition programs (SNAP, the child nutrition programs, and the Commodity Supplemental Food Program)[40] and the Conservation Reserve Program.[41] Some prior legal obligations in the Federal Crop Insurance Corporation[42] and the farm commodity programs may be exempt[43] as determined by OMB.[44]

Generally speaking, the experience since FY2013 is that OMB has ruled that most of crop insurance is exempt from sequestration, while the farm commodity programs, disaster assistance, and most conservation programs have been subject to it.[45]

Implementation of Sequestration

Nonexempt mandatory spending in FY2021 is to be reduced by a 5.7% sequestration rate (Table C.1) and thus would be paid at 94.3% of what would otherwise have been provided. This is projected to result in a reduction of about $1.7 billion from mandatory agriculture accounts in FY2021, including over $1.3 billion from the CCC that has $23 billion of budget authority subject to sequestration.

For example, for the farm commodity programs that support farm income such as the Agricultural Risk Coverage and Price Loss Coverage programs,[46] payments to farmers are computed by a regular formula authorized in the farm bill, and the final actual payment to the farmer is reduced by the sequestration rate. For programs that operate on a fixed budget authority, such as the Environmental Quality Incentives Program

---

[40] 2 U.S.C. §905(h).

[41] 2 U.S.C. §905(g)(1)(A).

[42] 2 U.S.C. §905(g)(2).

[43] 2 U.S.C. §906(j).

[44] Some administrative expenses may be subject to sequestration, and therefore programs that are otherwise exempt may have a relatively small sequestration-eligible amount compared to their total budget authority. This has been the case for the Federal Crop Insurance Corporation, SNAP, child nutrition programs, and WIC.

[45] CRS Report R42050, *Budget "Sequestration" and Selected Program Exemptions and Special Rules*.

[46] CRS In Focus IF11161, *2018 Farm Bill Primer: ARC and PLC Support Programs*.

and the Market Assistance Program, the sequestration rate is applied to the available budget authority for the fiscal year.

APPENDIX D. ACTION ON AGRICULTURE APPROPRIATIONS, FY1996-FY2021

Table D.1. Congressional Action on Agriculture Appropriations since FY1996

Fiscal Year	House Action			Senate Action			Final Appropriation			CRS Report
	Subcmte.	Cmte.	Floor	Subcmte.	Cmte.	Floor	Enacted		Public Law	
1996	6/14/ 1995	6/27/ 1995	7/21/ 1995	9/13/ 1995	9/14/ 1995	9/20/ 1995	10/21/ 1995	E	P.L. 104-37	95-624
1997	5/30/ 1996	6/6/ 1996	6/12/ 1996	7/10/ 1996	7/11 1996	7/24/ 1996	8/6/ 1996	E	P.L. 104-180	IB96015
1998	6/25/ 1997	7/14/ 1997	7/24/ 1997	7/15/ 1997	7/17/ 1997	7/24/ 1997	11/18/ 1997	E	P.L. 105-86	97-201
1999	6/10/ 1998	6/16/ 1998	6/24/ 1998	6/9/ 1998	6/11/ 1998	7/16/ 1998	10/21/ 1998	O	P.L. 105-277	98-201
2000	5/13/ 1999	5/24/ 1999	6/8/ 1999	6/15/ 1999	6/17/ 1999	8/4/ 1999	10/22/ 1999	E	P.L. 106-78	RL30201
2001	5/4/ 2000	5/16/ 2000	7/11/ 2000	5/4/ 2000	5/10/ 2000	7/20/ 2000	10/28/ 2000	E	P.L. 106-387	RL30501
2002	6/6/ 2001	6/27/ 2001	7/11/ 2001	Polled out	7/18/ 2001	10/25/ 2001	11/28/ 2001	E	P.L. 107-76	RL31001
2003	6/26/ 2002	7/26/ 2002	—	7/23/ 2002	7/25/ 2002	—	2/20/ 2003	O	P.L. 108-7	RL31301
2004	6/17/ 2003	7/9/ 2003	7/14/ 2003	7/17/ 2003	11/6/ 2003	11/6/2003	1/23/ 2004	O	P.L. 108-199	RL31801
2005	6/14/ 2004	7/7/ 2004	7/13/ 2004	9/8/ 2004	9/14/ 2004	—	12/8/ 2004	O	P.L. 108-447	RL32301
2006	5/16/ 2005	6/2/ 2005	6/8/ 2005	6/21/ 2005	6/27/ 2005	9/22/ 2005	11/10/ 2005	E	P.L. 109-97	RL32904
2007	5/3/ 2006	5/9/ 2006	5/23/ 2006	6/20/ 2006	6/22/ 2006	—	2/15/ 2007	Y	P.L. 110-5	RL33412
2008	7/12/ 2007	7/19/ 2007	8/2/ 2007	7/17/ 2007	7/19/ 2007	—	12/26/ 2007	O	P.L. 110-161	RL34132
2009	6/19/ 2008	—	—	Polled out	7/17/ 2008	—	3/11/ 2009	O	P.L. 111-8	R40000

Table D.1. (Continued)

Fiscal Year	House Action			Senate Action			Final Appropriation			CRS Report
	Subcmte.	Cmte.	Floor	Subcmte.	Cmte.	Floor	Enacted		Public Law	
2010	6/11/2009	6/18/2009	7/9/2009	Polled out	7/7/2009	8/4/2009	10/21/2009	E	P.L. 111-80	R40721
2011	6/30/2010	—	—	Polled out	7/15/2010	—	4/15/2011	Y	P.L. 112-10	R41475
2012	5/24/2011	5/31/2011	6/16/2011	Polled out	9/7/2011	11/1/2011	11/18/2011	O	P.L. 112-55	R41964
2013	6/6/2012	6/19/2012	—	Polled out	4/26/2012	—	3/26/2013	O	P.L. 113-6	R43110
2014	6/5/2013	6/13/2013	—	6/18/2013	6/20/2013	—	1/17/2014	O	P.L. 113-76	R43110
2015	5/20/2014	5/29/2014	—	5/20/2014	5/22/2014	—	12/16/2014	O	P.L. 113-235	R43669
2016	6/18/2015	7/8/2015	—	7/14/2015	7/16/2015	—	12/18/2015	O	P.L. 114-113	R44240
2017	4/13/2016	4/19/2016	—	5/17/2016	5/19/2016	—	5/5/2017	O	P.L. 115-31	R44588
2018	6/28/2017	7/12/2017	9/14/2017	7/18/2017	7/20/2017	—	3/23/2018	O	P.L. 115-141	R45128
2019	5/9/2018	5/16/2018	—	5/22/2018	5/24/2018	8/1/2018	2/15/2019	O	P.L. 116-6	R45230
2020	5/23/2019	6/4/2019	6/25/2019	9/17/2019	9/19/2019	10/31/2019	12/20/2019	O	P.L. 116-94	R45974
2021	—	—	—	—	—	—	—	—	—	—

Source: CRS.

Notes: E = enacted as standalone appropriation (eight times over 25 years); O = omnibus appropriation (15 times); Y = year-long CR (two times). "Polled out" refers to a procedure that permits a Senate subcommittee to transmit a bill to its full committee without a formal markup session. See CRS Report RS22952, *Proxy Voting and Polling in Senate Committee*.

In: COVID 19: Agricultural and Food Issues ISBN: 978-1-53618-561-4
Editor: Kristian Goldschmidt

Chapter 2

MAJOR AGRICULTURAL TRADE ISSUES IN 2020*

Anita Regmi, Tadlock Cowan, Genevieve K. Croft, Joel L. Greene, Renée Johnson and Randy Schnepf

SUMMARY

Sales of U.S. agricultural products to foreign markets absorb about one-fifth of U.S. agricultural production, thus contributing significantly to the health of the farm economy. Farm product exports, which totaled $136 billion in FY2019 (see chart), make up about 8% of total U.S. exports and contribute positively to the U.S. balance of trade. The economic benefits of agricultural exports also extend across rural communities, while overseas farm sales help to buoy a wide array of industries linked to agriculture, including transportation, processing, and farm input suppliers.

---

* This is an edited, reformatted and augmented version of a Congressional Research Service publication R46242, prepared for Members and Committees of Congress dated February 27, 2020.

A major area of interest for the 116th Congress during its first session was the loss of export demand for agricultural products in the wake of tariff increases imposed by the Trump Administration on U.S. imports of steel and aluminum from certain countries and other imported products from China. Some of the affected countries levied retaliatory tariffs on U.S. agricultural products, contributing to a 53% decline in value of U.S. agricultural exports to China in 2018 and a broader decline in exports across countries imposing retaliatory tariffs in 2019. To help mitigate the economic impact from export losses, the U.S. Department of Agriculture (USDA) authorized two short-term assistance ("trade aid") programs to producers of affected agricultural commodities, valued at up to $12 billion in 2019 and $16 billion in 2019.

Other major agricultural trade developments in 2019 included efforts to ratify the U.S.-Mexico-Canada Agreement (USMCA), trade negotiations with China, Japan, and the European Union, and continued review of U.S. participation in the World Trade Organization (WTO). The USMCA was ratified by Mexico and the U.S. Congress, and awaits ratification by Canada before it can enter into force. The United States and Japan signed an agreement increasing market access for many U.S. agricultural exports to Japan. This agreement, which does not require congressional approval, excludes provisions pertaining to non-tariff measures that could become future trade barriers for U.S. agricultural exporters. A second-stage negotiation toward a more comprehensive pact could commence in 2020.

In January 2020, President Trump signed a "Phase One" executive agreement (that also does not require congressional approval) with the Chinese government on trade and investment issues, including agriculture. Under the agreement, China is not required to repeal any tariffs, but it has reduced certain retaliatory tariffs and is granting tariff exclusions for various agricultural products in order to reach a target level of U.S. imports—$32 billion (relative to a 2017 base of $24 billion) over a two-year period. The coronavirus outbreak since January 2020 may affect China's ability to meet these commitments.

In addition to further negotiations with Japan and China, the Administration has stated its intent to pursue trade agreements with the

European Union, India, Kenya, the United Kingdom, and possibly other countries. The Trump Administration has also indicated that reforming the WTO is a priority for 2020. The WTO Ministerial Conference in June 2020 presents an opportunity to address pressing concerns over agricultural reform efforts.

Among other agricultural trade issues that may arise in the 116th Congress are proposed changes to U.S. trade remedy laws to address imports of seasonal produce affecting growers in the Southeast, the establishment of a common international framework for approval, trade, and marketing of the products of agricultural biotechnology, and foreign restrictions on U.S. exports of meat that are inconsistent with international trade protocols. Additionally, U.S. beef and pork face trade barriers in several markets because of U.S. producers' use of growth promotants and the feed additive ractopamine.

INTRODUCTION

This chapter identifies selected current major trade issues for U.S. agriculture that may be of interest in the second session of the 116th Congress. It provides background on individual trade issues and attempts to bring perspective on the significance of each for U.S. agricultural trade. Each trade issue summary concludes with an assessment of its status.

The report begins by examining a series of overarching issues. These include U.S. agricultural trade and its importance to the sector; a brief description of the trade policy being pursued by the Trump Administration in 2020 and its ramifications for U.S. agricultural exports; an update on the Administration's 2019 trade policy actions; a discussion of the ongoing and proposed new trade negotiations planned for 2020; and an update on World Trade Organization (WTO) agricultural issues related to the United States—including the Administration's 2020 plans to engage in reforming the institution. The report then reviews a number of ongoing trade policy concerns to U.S. agriculture, including non-tariff measures, and trade barriers and disputes involving specialty crops, livestock, and dairy issues.

The format for these trade issues is similar, consisting of background and perspective on the issue at hand and an assessment of their current status.

OVERVIEW OF U.S. AGRICULTURAL TRADE[1]

U.S. agricultural exports have long been a bright spot in the U.S. balance of trade, with exports exceeding imports in every year since 1960.[2] In recent years, the value of farm exports has remained below the record level of $152 billion reached in FY2014.

The U.S. Department of Agriculture (USDA) reports U.S. agricultural exports in FY2019 of $136 billion (see Figure 1).[3] The FY2019 export total represents an $8 billion decline from FY2018.[4] The decline in the value of farm exports since FY2014 initially reflected lower market prices for bulk commodities, such as soybeans and corn. Agricultural prices and U.S. exports of certain commodities, such as soybeans, were further affected by retaliatory tariffs imposed on U.S. agricultural imports by China and some other countries since 2018 in response to the Trump Administration's imposition of tariffs on certain imports from China and on U.S. imports of steel and aluminum from selected countries.[5]

In FY2019, U.S. agricultural imports were $131 billion, up $3 billion from FY2018, resulting in an agricultural trade surplus of $5 billion. This is below the surplus of $16 billion in FY2018 and below the record high in nominal dollars of $43 billion in FY2014.

---

[1] Prepared by Anita Regmi, Analyst in Agricultural Policy, CRS.

[2] U.S. Department of Agriculture (USDA), Foreign Agricultural Service (FAS), Global Agricultural Trade System (GATS), February 2020.

[3] CRS calculation based on Census Bureau Trade Statistics, accessed via Global Trade Atlas, February 2020.

[4] USDA, Economic Research Service (ERS), U.S. Agricultural Trade Data Update, updated January 8, 2020, https://www.ers.usda.gov/data-products/foreign-agricultural-trade-of-the-united-states-fatus/us-agricultural-trade-data- update/.

[5] For more information on this issue, see CRS Report R45903, *Retaliatory Tariffs and U.S. Agriculture*.

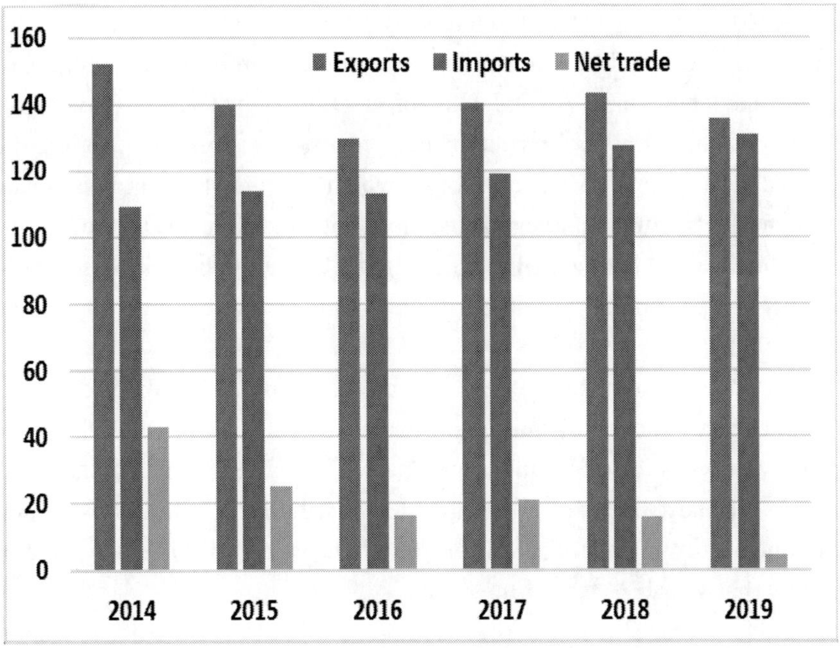

Source: USDA, Economic Research service (ERS), U.S. Agricultural Trade Data Update, updated January 8, 2020, https://www.ers.usda.gov/data-products/foreign-agricultural-trade- of-the-united-states-fatus/us-agricultural-trade-data-update/.

Notes: Data are not adjusted for inflation and pertain to fiscal years. 'Net trade' denotes the trade surplus, which is the difference between U.S. exports and U.S. imports.

Figure 1. U.S. Agricultural Trade, 2014-19. Billion U.S. Dollars

Agricultural exports are important both to farmers and to the U.S. economy. During the calendar years 2017 and 2018, the value of U.S. agricultural exports accounted for 8% and 9% of total U.S. exports, respectively.[6] USDA's Economic Research Service (ERS) estimates that in 2017 U.S. agricultural exports generated about 1,161,000 full-time civilian jobs, including 795,000 jobs outside the farm sector.[7] Exports account for

---

[6] U.S. Census, https://www.census.gov/foreign-trade/Press-Release/current_press_release/exh15.pdf. USDA generally expresses agricultural trade forecasts on a fiscal year basis, but is expressed here on a calendar year basis to allow for a comparison with Census Bureau data of all U.S. merchandise trade.

[7] ERS, Effects of Trade on the U.S. Economy, 2017 Data Overview.

around 20% of total farm production by value[8] and are a major outlet for many farm commodities, absorbing over three-fourths of U.S. output of cotton and about half of total U.S. production of wheat and soybeans.[9] Although feed crops and wheat account for most exports by volume, the high value product (HVPs) category—which includes live animals, meat, dairy products, fruits and vegetables, nuts, fats, hides, manufactured feeds, sugar products, processed fruits and vegetables, and other processed food products— accounted for 68% of the value of agricultural exports in FY2019.[10]

All states export agricultural commodities, but a minority of states account for a majority of farm export sales. In calendar year 2018, the 10 leading agricultural exporting states based on value— California, Iowa, Illinois, Minnesota, Texas, Nebraska, Kansas, Indiana, North Dakota, and Missouri—accounted for 58% of the total value of U.S. agricultural exports that year.[11]

In December 2018, Congress reauthorized major agricultural export promotion programs through FY2023 with the 2018 farm bill (P.L. 115-334).[12] Title III of the farm bill includes provisions covering export credit guarantee programs, export market development programs, and international science and technical exchange programs designed to develop agricultural export markets in emerging economies. Among other provisions, the 2018 farm bill permits funding to operate two U.S. agricultural export promotion programs in Cuba—the Market Access Program and the Foreign Market Development Cooperator Program.[13]

---

[8] ERS, "U.S. Agricultural Trade, Export Share of Production," https://www.ers.usda.gov/topics/international-markets- us-trade/us-agricultural-trade/data/.

[9] CRS calculations based on FAS, Production Supply and Demand (PSD) Online, https://apps.fas.usda.gov/psdonline/app/index.html#/app/home, accessed January 27, 2020.

[10] CRS calculations based on GATS, https://www.ers.usda.gov/data-products/foreign-agricultural-trade-of-the-united- states-fatus/fiscal-year/, accessed January 27, 2020.

[11] ERS, Data Products, https://www.ers.usda.gov/data-products/state-export-data/, accessed January 2020.

[12] For more information on this issue, see CRS Report R45525, *The 2018 Farm Bill (P.L. 115-334): Summary and Side-by-Side Comparison*.

[13] The Agricultural Act of 2018, P.L. 115-334, Title III, Sec. 203.

TRUMP ADMINISTRATION TRADE PRIORITIES FOR 2020[14]

In establishing policy for U.S. participation in international trade, the Trump Administration has emphasized reducing U.S. bilateral trade deficits;[15] focusing on renegotiating existing trade agreements that it viewed as being "unfair;" initiating new bilateral agreements; and responding to the trade practices of U.S. trading partners that it viewed as unfair, in violation of international trading commitments, or threatening to U.S. industry.[16] Under various provisions of law, the Administration imposed punitive tariffs on U.S. imports of steel and aluminum from certain countries and on U.S. imports of selected products from China.[17] These countries in turn, responded with retaliatory tariffs on U.S. exports, particularly agricultural products.[18]

During the second session of the 116th Congress, the Trump Administration's agenda may focus on the following priorities:

Trade Agreement Implementation and Monitoring

U.S.-Mexico-Canada Agreement (USMCA)

Legislation implementing a new trade agreement among the United States, Mexico, and Canada was enacted on January 29, 2020. The agreement awaits ratification by Canada, and certification by the United States that all parties have completed the necessary steps for entry into force. The U.S.-Mexico-Canada agreement replaces the North American Free Trade Agreement (NAFTA), which took effect in 1994.

---

[14] Prepared by Anita Regmi, Analyst in Agricultural Policy, CRS.

[15] A bilateral trade deficit represents an imbalance whereby the value of U.S. imports from a particular trading partner exceed the value of U.S. exports to that same country during a particular time period, usually a year. A bilateral trade surplus occurs when U.S. exports exceed imports from a particular country.

[16] For more on this issue, see CRS Report R45474, *International Trade and Finance: Overview and Issues for the 116th Congress*.

[17] For more information on this issue, see CRS Insight IN10971, *Escalating U.S. Tariffs: Affected Trade*; CRS In Focus IF11346, *Section 301 of the Trade Act of 1974*; and CRS In Focus IF11284, *U.S.-China Trade and Economic Relations: Overview*.

[18] For more on this issue, see CRS Report R45903, *Retaliatory Tariffs and U.S. Agriculture*.

"Stage One" U.S.-Japan Trade Agreement (USJTA)

On October 7, 2019, the Trump Administration signed the "Stage One" trade agreement with Japan, which included significant market access improvements in Japan for U.S. agricultural exports. The agreement took effect on January 1, 2020. Because it dealt only with tariffs and other market access issues, pursuant to P.L. 114-26, the agreement did not require congressional approval.[19] The Administration has indicated that it hopes to negotiate a second trade agreement with Japan that addresses a broader range of issues. Such an agreement might require congressional approval.

U.S.-China Phase One Agreement

On January 15, 2020, President Trump signed a "Phase One" executive agreement with China on trade and investment issues, including agriculture.[20] This agreement, which entered into force on February 14, 2020, did not require congressional approval as it consisted largely of commitments by China. The Administration has stated its intent to negotiate a second phase of the agreement with China.[21] Depending on the scope of such a negotiation, the Administration could be required under law to consult with Congress in advance and to submit an eventual agreement for congressional approval.

Ongoing and Proposed Negotiations

The Office of the U.S. Trade Representative (USTR) has indicated that the United States may also pursue new trade agreements with the European Union (EU), India, Kenya, the United Kingdom (UK), and a number of other countries. The Administration has stated that the U.S.- Kenya and the U.S.-UK negotiations will be "comprehensive," dealing with other trade-

---

[19] For more on this, see CRS Report R46140, *"Stage One" U.S.-Japan Trade Agreements*.
[20] For more on this issue, see CRS In Focus IF11412, *U.S.-China Phase I Deal: Agriculture*.
[21] I. Isco, "Mnuchin: Administration Has Phase-Two China Pact Chapters 'Dealt With,'" *Inside U.S. Trade*, February 12, 2020.

related issues in addition to market access.[22] In those cases, the Administration might be required to consult with Congress in advance of negotiations and to submit any agreements for congressional approval.

Multilateral Trading System Reforms

USTR has indicated interest in WTO institutional reform.[23] The upcoming WTO Ministerial Conference in June 2020 in Kazakhstan presents the United States and WTO members with an opportunity to address reform efforts, which are expected to include consideration of the WTO's treatment of agricultural trade.[24] Some Members of Congress have indicated WTO reform to be a priority for 2020.[25]

Agricultural Trade Disputes and Negotiations[26]

Since early 2018, Canada, China, the EU, India, Mexico, and Turkey targeted U.S. food and agricultural products with retaliatory tariffs in response to tariffs imposed by the United States on imports of steel and aluminum and certain imports from China. To facilitate ratification of USMCA, the United States removed tariffs on steel and aluminum imports from Canada and Mexico and these countries removed their retaliatory tariffs on U. S. agricultural imports in May 2019. The retaliatory tariffs made imports of U.S. agricultural products relatively more expensive compared to similar products from competitor nations.

---

[22] Office of the United States Trade Representative (USTR) Fiscal Year 2021 Budget, February 2020.

[23] USTR, *2019 Trade Policy Agenda and 2018 Annual Report*, March 2019; and USTR, *USTR Issues Report on the WTO Appellate Body*, February 11, 2020.

[24] WTO, "DDG Wolff: It is Time to Update the WTO Rulebook for Agriculture," January 18, 2020.

[25] See H.Res. 746; and Icso, I., "Blumenauer Cites USMCA Enforcement, WTO Reform Among Trade Panel's 2020 Priorities," *Inside World Trade,* January 31, 2020.

[26] Prepared by Anita Regmi, Analyst in Agricultural Policy, CRS.

Initially, the announcements of retaliatory tariffs led to an increase in U.S. agricultural exports as importing countries built stocks in anticipation of the tariffs. U.S. agricultural exports increased slightly in 2018. In 2019, however, U.S. agricultural exports declined about 2%, due to lower global demand for affected U.S. agricultural products and downward pressure on prices of some commodities.[27]

In the short run, retaliatory tariffs contributed to price declines for certain U.S. agricultural commodities and to a reduction in exports, particularly for soybeans. Declining prices and export sales, combined with rising input and farm machinery costs, contributed to a 16% decrease in U.S. net farm income in 2018, which prompted USDA to provide trade aid payments to the farm sector in 2018 and 2019.

Negotiations with China

Imports from China have been subject to U.S. tariff increases on steel and aluminum under Section 232 of the Trade Expansion Act, which allows the President to impose tariffs on imports that "threaten to impair the national security." Additionally, U.S. imports of certain other Chinese products are subject to tariff increases under Section 301 of the Trade Act of 1974, which allows tariffs in response to trade practices that are determined to be unfair and injurious to a U.S. industry. China first retaliated in April 2018, by raising tariffs on certain U.S. imports, including agricultural products such as pork, fruit, and tree nuts.[28] These retaliatory tariffs are in addition to existing Most Favored Nation (MFN) tariffs that China levies on imports from all countries including the United

---

[27] U.S. Census Bureau Trade Data, accessed via GATS, February 6, 2020, https:// apps.fas.usda.gov/gats/ ExpressQuery1.aspx.

[28] FAS, "China Responds to U.S. 301 Announcement with Revised Product List," GAIN Report Number: CH 18034, June 21, 2018.

States.[29] By September 2019, China had levied retaliatory tariffs on almost all U.S. agricultural products, ranging from 5% to 60%.[30]

After the imposition of retaliatory tariffs on U.S. products, U.S. agricultural exports to China experienced a 53% decline from $19.5 billion in 2017 to $9.2 billion in 2018. The Chinese market is important for several U.S. agricultural products. For example, in 2016 and 2017, the United States supplied over one-third of China's total soybean imports, almost all of China's distillers' grain imports (primarily used as animal feed), and most of China's sorghum imports.[31] With the retaliatory tariffs in effect, U.S. soybean exports to China in 2018 declined in value to $3 billion (8 billion metric tons [MT]) from $12 billion (32 billion MT) in 2017. Similarly, the value of U.S. exports of sorghum and distillers dry grain declined about 40% and 30% respectively from 2017 to 2018. Most other U.S. agricultural exports to China also declined in 2018.[32]

Negotiations to resolve the U.S.-China dispute began in the fall of 2019 and resulted in a "Phase One" executive agreement (that does not require congressional approval) on trade and investment issues, including agriculture, signed in January 2020.[33] Under the agreement, China is to import $32 billion worth of additional U.S. agricultural products over a two-year period. This implies an average annual increase of two-thirds from a 2017 base of $24 billion.[34] Products mentioned in the agreement include oilseeds, meat, cereals, cotton, and seafood. China has not committed to tariff exemptions or import levels for any specific products, but it may grant tariff exclusions on U.S. imports on a case-by-case basis.

---

[29] MFN tariffs must be levied in a non-discriminatory manner, but lower levels of tariffs can be applied to imports from countries with which a nation has a preferential trade agreement.

[30] For more on this, see CRS Report R45929, *China's Retaliatory Tariffs on U.S. Agriculture: In Brief.*

[31] Chinese customs data, accessed via Global Trade Atlas, August 2019.

[32] For more on this issue, see CRS Report R45903, *Retaliatory Tariffs and U.S. Agriculture*.

[33] For more on this issue, see CRS In Focus IF11412, *U.S.-China Phase I Deal: Agriculture*.

[34] Chinese commitments of $24 billion includes products defined by USDA as agricultural products, plus agriculture- related products such as distilled spirits and fish and seafood products.

On February 18, 2020, China released a list indicating that it may be willing to grant one-year tariff exemptions on most agricultural products.[35]

China agreed to improve its administration of tariff-rate quotas (TRQs) on wheat, corn, and rice to comply with a WTO ruling in favor of the United States in a dispute case regarding China's TRQ administration.[36] Changes in China's TRQ administration would be expected to improve market access for these U.S. grains.

Other Provisions of the Phase One agreement

Domestic Support

China agreed to improve the transparency of its domestic agricultural support measures.

Sanitary and Phytosanitary Measures

China agreed to implement science- and risk-based food safety regulations. China also agreed to finalize phytosanitary protocols for U.S. avocadoes, blueberries, potatoes, barley, alfalfa pellets and cubes, almond meal pellets and cubes, hay, and California nectarines, and to implement a transparent, predictable, efficient, science- and risk- based regulatory process for the evaluation and authorization of products of agricultural biotechnology. In exchange, the United States agreed to complete its regulatory notice process for imports of Chinese fragrant pears, citrus, and jujube, and to complete a phytosanitary protocol for bonsai.

Livestock and Fish

China agreed to improve access for U.S. beef products, including eliminating age restrictions on cattle slaughtered for export, eliminating traceability requirements, and establishing maximum residue levels for three hormones that are approved for use in livestock in the United States.

---

[35] State Council Customs Tariff Commission—Exclusions Criteria Circular, February 18, 2020, http://www.gov.cn/zhengce/zhengceku/2020-02/18/content_5480381.htm.

[36] Tariff-rate quotas provide for a comparatively low tariff rate on a specified quota of imports and a higher tariff rate on imports of the relevant commodity above the quota.

It agreed to engage in technical discussions to import U.S. live cattle for breeding. China agreed to broaden the list of pork products that are eligible for importation, and to conduct a risk assessment for the veterinary drug ractopamine, which is allowed in U.S. beef and pork production. With respect to poultry, after having lifted a five-year ban on imports of U.S. poultry in November 2019, China agreed to adopt import regulations consistent with the World Organization for Animal Health Terrestrial Animal Health Code; this would potentially limit future import bans imposed due to avian influenza to poultry from the affected U.S. region rather than the entire country. China also agreed to approve for importation 26 aquatic species from the United States, and to streamline its procedures for registering U.S. seafood facilities and products.

Technical Barriers to Trade

China agreed to implement the USDA Public Health Information System, an electronic system to provide export health certificates to an importing country in advance of shipment arrival. It also made commitments to provide regulatory certainty and market stability regarding U.S. dairy and infant formula products, rice, distillers' dried grains with solubles, feed additives, and pet foods. It agreed not to undermine market access for U.S. exports that use trademarks and generic terms by recognizing geographical indications (GI) in international agreements. GIs are place names used to identify products that come from certain regions or locations.

Status and Outlook

The U.S.-China Phase One agreement is expected to improve opportunities for certain U.S. exporters; however, it may not create notable new market demand. Instead, it may produce a rearrangement of trading patterns between China and its various import suppliers, in which case the market price effects may be limited. Additionally, the coronavirus outbreak is expected to slow China's economic growth in the near-term, and may reduce Chinese overall import demand for agricultural products. It has also been disrupting global supply chains going in and out of China.[37] Therefore, U.S. agricultural exports to China could fall short of the target of $32 billion additional exports to the 2017 base over a two-year period. The agreement provides China some flexibility to meet its purchase commitment. Both the United States and China "acknowledge that purchases will be made at market prices based on commercial considerations and that market conditions, particularly in the case of agricultural goods, may dictate the timing of purchases within any given year" (Chapter 6, Article 6.2.1 of the Phase One agreement).

Under the agreement, China is not required to repeal any tariffs, but it has reduced certain retaliatory tariffs and will grant one-year tariff exclusions for various agricultural products in order to reach a target level of U.S. imports.[38] Effective February 14, 2020, China halved the additional 5% and 10% retaliatory tariffs that it had imposed on U.S. products in August 2019.[39] Nevertheless, tariffs imposed in April and July 2018, ranging from 2.5% to 55%, remain in place.[40] USDA and USTR have stated that China has also taken a number of other actions to begin implementing its agriculture related commitments.[41] Both China and the United States have indicated they expect to engage in further negotiations on trade during 2020.

---

[37] For example, see I. Almeida, "U.S. Farmers Need to Wait a While for China's Buying Spree," *Bloomberg*, February 3, 2020; A. Behsudi, "USTR: No Chinese Request to Ease Purchase Demands," *Politico*, February 3, 2020; and CRS In Focus IF11434, *The Coronavirus: U.S.-China Economic Considerations*.

[38] State Council Customs Tariff Commission—Exclusions Criteria Circular, February 18, 2020, http://www.gov.cn/ zhengce/zhengceku/2020-02/18/content_5480381.htm; and FAS, "China Announces a New Round of Tariff Exclusions," GAIN Report: CH2020-0017, February 26, 2020.

[39] FAS, "China Announces Reductions in Certain Additional Tariffs," GAIN Report Number: CH2020-0016, February 14, 2020.

[40] For more on this, see CRS Report R45929, *China's Retaliatory Tariffs on U.S. Agriculture: In Brief*.

[41] USTR, "USDA and USTR Announce Progress on Implementation of U.S.-China Phase One Agreement," February 25, 2020.

Negotiations with Canada and Mexico[42]

Soon after taking office in January 2017, the Trump Administration announced its desire to renegotiate the North America Free Trade Agreement (NAFTA) among the three countries. Nonetheless, the United States imposed tariffs on steel and aluminum imports from Canada and Mexico in 2017. The United States also threatened tariffs on imported passenger vehicles, an action that would have a significant impact on both Canada and Mexico. In June 2018, Mexico retaliated against the steel and aluminum tariffs with a 15% tariff on U.S. sausage imports; a 20% tariff on other pork products, certain cheeses, apples, potatoes, and cranberries; and a 25% tariff on whey, blue-veined cheese, and whiskies.[43] The following month, Canada imposed a retaliatory tariff of 10% on certain U.S. products, including dairy, poultry and beef products; coffee, chocolate, sugar and confectionery; prepared food products; condiments; bottled water; and whiskies.[44]

A new trade agreement, referred to as the United States-Mexico-Canada Agreement (USMCA), was announced in 2018. The U.S. implementing legislation was enacted on January 29, 2020. Mexico has ratified the USMCA and the Canadian Parliament has begun deliberations on the agreement.[45] After ratification by all three countries, and certification by the United States that all parties have taken actions required under the agreement, the agreement would enter into force. The agricultural provisions of USMCA are summarized below.[46]

- All food and agricultural products that had zero tariffs under NAFTA is to remain at zero under USMCA. This includes all

---

[42] Prepared by Anita Regmi, Analyst in Agricultural Policy, CRS.
[43] FAS, "The Phasing In Of Mexican Retaliatory Tariffs," GAIN Report Number: MX8028, July 11, 2018.
[44] FAS, "Canada Announces Final List of Ag Products in Response to U.S. Tariffs," GAIN Report Number: CA18046, June 29, 2018.
[45] Inside U.S. Trade, "Canada Begins USMCA Ratification Process; Trump to Ink Implementing Bill," January 27, 2020.
[46] For more on this issue, see CRS Report R45661, *Agricultural Provisions of the U.S.-Mexico-Canada Agreement*.

agricultural imports from Mexico and almost all from Canada—excepting certain dairy and poultry products.

- Canada is to increase market access for U.S. dairy products via TRQs. U.S. dairy imports within a TRQ is to enter Canada duty-free, while imports beyond the quota level face higher over-quota tariff rates of over 200% in many cases.
- Canada is to replace poultry TRQs under NAFTA with new TRQs. These are expected to lead to greater imports of U.S. eggs, turkey meat, and eggs, but reduce the quantity of U.S. chicken meat that can be imported into Canada duty free. Imports of U.S. poultry products above the set quotas is to face tariffs exceeding 200%.
- The United States, agreed to provide additional access to Canadian dairy products, sugar, peanuts and peanut products.
- Canada is to provide treatment and price to U.S. wheat equivalent to those of Canadian wheat if the U.S. wheat variety is registered as being similar to a Canadian variety. Currently, U.S. wheat exports to Canada are graded as feed wheat, and as such command a lower price. Four Members of Congress have requested USTR to work closely with Canada, through the Consultative Committee on Agriculture, to expedite the process for the registration of U.S. wheat varieties in Canada.[47]
- The United States, Canada, and Mexico are required to treat the distribution of each other's spirits, wine, beer, and other alcoholic beverages as they do for products of national origin. The agreement establishes listing requirements for a product to be sold, along with specific limits on cost markups.
- Regarding sanitary and phytosanitary measures (SPS), USMCA requires greater transparency in rules and regulatory alignment among the three countries. It also would establish a new mechanism for technical consultations to resolve SPS issues.
- USMCA includes procedural safeguards for recognition of new geographical indications. USMCA would protect the GIs for food

[47] Letter to USTR from Senators Kevin Cramer, John Hoeven, Tina Smith and Steve Daines, July 8, 2019.

products that Canada and Mexico have already agreed to in trade negotiations with the EU, and would lay out transparency and notification requirements for recognition of any proposed new GIs. In a side letter accompanying the agreement, Mexico confirmed a list of 33 terms for cheese that would remain available as common names for U.S. cheese producers to use in exporting cheeses to Mexico. The list includes some terms that are protected as GIs by the EU. USMCA provisions also would protect certain U.S., Canadian, and Mexican spirits as distinctive products.

- USMCA signatories agreed to protect the confidentiality of proprietary formula information in the same manner for domestic and imported products.
- USMCA includes provisions for a Working Group for Cooperation on Agricultural Biotechnology to facilitate information exchange on policy and trade-related matters associated with agricultural biotechnology, an issue that was not covered under NAFTA.

Status

The United States removed the tariffs it had imposed on steel and aluminum imports from Canada and Mexico on May 17, 2019, and, in turn, these countries removed their retaliatory tariffs on U.S. imports.[48] USMCA requires ratification by Canada to enter into force.

"Stage One" U.S. Japan Trade Agreement (USJTA)[49]

On October 7, 2019, the United States and Japan signed the U.S.-Japan Trade Agreement (USJTA), which provides for limited tariff reductions and quota expansions to improve U.S. access to Japan's market, including for agricultural products. The agreement, which entered into force January 1, 2020, also provides for reciprocal U.S. tariff reductions, largely on

---

[48] USTR, "United States Announces Deal with Canada and Mexico to Lift Retaliatory Tariffs," press release, May 17, 2019.

[49] Prepared by Anita Regmi, Analyst in Agricultural Policy, CRS.

industrial goods. Japan previously negotiated agricultural market access provisions with the United States in the context of the Trans-Pacific Partnership (TPP), a 2016 agreement among 12 Pacific-facing nations[50] that the United States did not ratify. Those provisions were folded into the agreement that the remaining TPP countries agreed upon—TPP-11—that went into force for Japan on December 30, 2018.[51] As Japan began to improve market access for TPP-11 countries, various U.S. agricultural exports to Japan became less competitive compared to products from TPP-11 countries.

Under the USJTA, Japan provides the same level of market access to U.S. products included in the USJTA as it provides to exports from TPP-11 member countries. Japan agreed to eliminate or reduce tariffs for certain U.S. agricultural exports and to provide preferential quotas for other U.S. agricultural products. Some products included in TPP-11 such as rice and certain dairy products are not included in the USJTA. Key agricultural provisions of USJTA are provided below.

- Japan is to reduce tariffs on meat products such as beef and pork or gradually eliminate them.
- Upon entry into force, tariffs were eliminated for certain products, including almonds, walnuts, blueberries, cranberries, corn, sorghum, and broccoli.[52]
- Japan is to phase out tariffs in stages for products such as cheeses, processed pork, poultry, beef offal, ethanol, wine, frozen potatoes, oranges, fresh cherries, egg products, and tomato paste.
- Japan agreed to provide country-specific quotas (CSQ) to all products that the United States had negotiated CSQs for under TPP, excepting for rice. Products covered by CSQs include wheat,

[50] The countries include Australia, Brunei, Canada, Chile, Japan, Malaysia, Mexico, New Zealand, Peru, Singapore, Vietnam and the United States.

[51] Comprehensive and Progressive Agreement for Trans-Pacific Partnership text and resources, February 21, 2018, https://www.mfat.govt.nz/en/trade/free-trade-agreements/free-trade-agreements-in-force/cptpp/comprehensive-and- progressive-agreement-for-trans-pacific-partnership-text-and-resources/.

[52] Note that Japan's current tariff on soybeans, another important export commodity for the United States, is zero.

wheat products, malt, whey, processed cheese, glucose, fructose, corn starch, potato starch, and inulin.

- Japan agreed to reduce the mark-ups on U.S. products that Japanese state trading enterprises import under quotas and sell in the domestic market with an additional price mark-up that makes them more expensive that the domestic product.
- Under Japan's WTO market access schedule, it reserves the right to temporarily increase tariffs on imports of sensitive agricultural products when they exceed a set threshold, or when the price of the imported product is below a set threshold. Under USJTA, Japan agreed to restrict the use of these additional tariffs (known as safeguards) on U.S. beef, pork, whey, oranges and race horses.
- Under TPP, the United States had negotiated market access under TRQs that were open to all TPP members, for barley and barley products other than malt; butter; skim and other milk powder; cocoa products; evaporated and condensed milk; edible fats and oils; vegetable preparations; coffee, tea and other preparations; chocolate, candies and confectionary; and sugar. No corresponding U.S. access to these TPP-wide TRQs is included in USJTA.
- The United States agreed to reduce tariffs on imports of certain perennial plants and cut flowers, persimmons, green tea, chewing gum, certain confectionary products, and soy sauce. The United States also agreed to provide Japan the opportunity to export more beef by folding a country-specific quota for Japan of 200 MT into a larger TRQ designated for "other countries."

Status

The Administration took a staged approach to U.S. negotiations with Japan in order to facilitate expedited market access improvements for U.S. agricultural products in Japan. The first stage agreement (USJTA) is much more limited than a traditional U.S. free trade agreement, allowing the USJTA (P.L. 114-26) to take effect without approval by Congress.[53] In

---

[53] For more on this issue, see CRS Report R46140, *"Stage One" U.S.-Japan Trade Agreements*.

consequence, the text does not address non-tariff issues such as sanitary and phytosanitary measures, agricultural biotechnology, technical barriers to trade, or geographical indications. These issues are expected to be covered in a further negotiation, which may commence in 2020.

In February 2019, after the USJTA entered into force, Japan reached a trade agreement with the EU under which Japan agreed to recognize more than 200 EU GIs. If USTR were to determine that any of these European GIs poses a barrier to U.S. agricultural exports to Japan, the lack of legal text regarding geographical indications and the absence of a formal dispute settlement mechanism could limit U.S. ability to challenge such a barrier under the USJTA. Both the United States and Japan are members of the WTO, so the United States could challenge potential new trade barriers as inconsistent with Japan's WTO commitments.

U.S.-EU Agricultural Trade[54]

The Trump Administration's decision to impose tariffs on steel and aluminum affected imports from the EU. In June 2018, the EU responded to the steel and aluminum tariffs by imposing a 25% tariff on imports of U.S. corn, rice, sweetcorn, kidney beans, certain breakfast cereals, peanut butter, orange juice, cranberry juice, whiskies, cigars, and other tobacco products, and a 10% tariff on certain essential oils.[55] The EU also could be affected if the United States were to impose tariffs on passenger vehicles, and could respond with further punitive tariffs against U.S. exports.

On October 18, 2019, the United States imposed additional tariffs on $7.5 billion worth of U.S. imports from the EU. The action, authorized by WTO dispute settlement procedures, came after USTR determined that the EU and certain EU member states had not complied with a WTO Dispute

---

[54] Prepared by Renée Johnson, Specialist in Agricultural Policy, CRS, with contributions from Andres B. Schwarzenberg, Analyst in International Trade and Finance, CRS.

[55] FAS, "EU Imposes Additional Tariffs on U.S. Products," GAIN Report Number: E18045, June 21, 2018.

Settlement Body ruling recommending the withdrawal of subsidies on the manufacture of large civil aircraft.[56]

USTR has indicated that additional tariffs initially will be limited to 10% of the product value on large civil aircraft and 25% on agricultural and other products from the EU. In total, 561 agricultural tariff lines are affected,[57] including cheeses, biscuits, pork products, fish products, fruit products, olives, whiskies, liquors, and wine. The UK, which left the EU in January 2020, is included among the affected countries, and 56 tariff lines of UK products are subject to additional 25% tariffs.

Limited Expected Role of Agricultural Issues in Upcoming Trade Talks

Against this background, in October 2018, USTR officially notified the Congress of the Trump Administration's plans to enter into formal trade negotiations with the EU.[58] This action followed a July 2018 U.S.-EU Joint Statement by President Trump and then-European Commission President Jean-Claude Juncker announcing that they would work to reduce tariffs and other trade barriers, address unfair trading practices, and increase U.S. exports of soybeans and certain other products. Previously, in 2016, U.S.-EU negotiations to create a Transatlantic Trade and Investment Partnership (T-TIP) under the Obama Administration stalled after 15 rounds. Among the areas of contention were certain regulatory and administrative differences between the United States and the EU on issues of food safety, public health, and product naming schemes for some types of food and agricultural products.

The United States and the EU are the world's largest trade and investment partners.[59] While food and agricultural trade between the

---

[56] For more on this, see CRS In Focus IF11364, *Boeing-Airbus Subsidy Dispute: Recent Developments*.

[57] CRS compilation by Andres Schwarzenberg, Analyst in International Trade and Finance, based on information from USTR, February 2020.

[58] Letter from Robert Lighthizer to then Speaker of the House of Representatives, Paul Ryan, October 16, 2018.

[59] For more information, see CRS In Focus IF11209, *Proposed U.S.-EU Trade Agreement Negotiations*.

United States and the EU27[60] accounts for less than 1% of the value of overall trade in total goods and services, the EU27 remains a leading export market for U.S. agricultural exports. It accounted for about 8% of the value of all U.S. exports and ranked as the fifth largest market for U.S. food and farm exports in 2019—after Canada, Mexico, China, and Japan. In 2019, U.S. exports of agricultural and related product exports to the EU27 totaled $12.4 billion, while U.S. imports of agricultural and related product imports from the EU27 totaled $29.7 billion, resulting in a U.S. trade deficit of approximately $17.3 billion.[61] This is the reverse of U.S. trade surpluses with the EU27 during the 1990s. Leading U.S. agricultural exports to the EU27 were corn and soybeans, tree nuts, distilled spirits, fish products, wine and beer, planting seeds, tobacco products, and processed foods. Leading U.S. agricultural imports from the EU27 were wine, distilled spirits, beer, drinking waters, olive oil, cheese, baked goods, processed foods, and cocoa products.

In January 2019, USTR announced its negotiating objectives for the agricultural portion of a U.S.-EU trade agreement following a public comment period and a hearing involving several leading U.S. agricultural trade associations.[62] The objectives include greater market access, changes to EU administration of tariff-rate quotas, and changes to a variety of EU regulations. Among regulatory issues, key U.S. objectives include harmonizing regulatory processes and standards to facilitate trade, including sanitary and phytosanitary standards, and establishing specific commitments for trade regarding agricultural biotechnologies. The U.S. objectives also include addressing geographical indications by protecting generic terms for common use.[63] U.S. agricultural interests generally support including agriculture as part of the U.S. negotiating objectives for

---

[60] This excludes the UK, which officially exited EU as of January 2020, but will remain a member of the EU customs union until December 31, 2020.

[61] USDA trade statistics for "Agricultural and Related Products," which includes agricultural products (including bulk and intermediate products and also consumer-oriented products) and agricultural-related products (including fish and shellfish products, distilled spirits, forest products, and ethanol and biodiesel blends.

[62] USTR, "United States-European Union Negotiations, Summary of Specific Negotiating Objectives," January 2019.

[63] For more on this, see section on "Geographical Indications (GIs)."

a U.S.-EU trade agreement. The EU negotiating mandate, however, states that a key EU goal is "a trade agreement limited to the elimination of tariffs for industrial goods only, excluding agricultural products."[64] Several Members of Congress have stated their opposition to the EU's decision to exclude agricultural policies in their negotiating mandate.[65]

The U.S.-EU trade negotiations come amid heightened U.S.-EU trade frictions. In response to U.S. Section 232 tariffs on steel and aluminum imports, the EU had retaliated in June 2018 by imposing a tariff increase of 25% on imports of certain U.S. food and beverage products.[66] The value of U.S. agricultural exports to the EU28 (included the UK) targeted by these additional tariffs is approximately $1.2 billion in 2018, or about 9% of total U.S. agricultural exports to the EU28. In October 2019, U.S.-EU trade tensions escalated further when the United States imposed additional tariffs on $7.5 billion worth of certain U.S. imports from the EU, including food products. This action—authorized by the WTO—followed a USTR investigation initiated in April 2019 under Section 301 of the Trade Act of 1974.[67]

Aside from ongoing trade tension, some of the same issues that stalled U.S.-EU agricultural talks in the T-TIP negotiations could prove to be equally intractable today. For food and agricultural products, a series of non-tariff issues stem in part from commercial and cultural practices often enshrined in EU laws and regulations that vary from those of the United States—namely differences involving SPS and technical barriers to trade, broadly covering laws and regulations measures intended to protect public health—as well as differences involving GIs.[68]

---

[64] Council of the European Union, "Trade with the United States: Council Authorizes Negotiations on Elimination of Tariffs for Industrial Goods and on Conformity Assessment," press release, April 15, 2019.

[65] See, for example, letter to USTR Robert E. Lighthizer from 114 House Members, March 14, 2019, and U.S. Senate Finance Committee press release, April 15, 2019.

[66] For a full list of product codes subject to higher duties, see Commission Implementing Regulation (EU) 2018/886 available at https://eur-lex.europa.eu/legal-content/EN/TXT/HTML/?uri=CELEX:32018R0886&from=EN. This includes the UK.

[67] For a list of product codes, see 84 *Federal Register* 32248, July 5, 2019.

[68] For more on this, see section "Sanitary and Phytosanitary (SPS) and Other Non-Tariff Barriers;" and "Geographical Indications (GIs)".

Status

The outlook for the new U.S.-EU trade talks remains uncertain, given ongoing trade tensions. Whether or not the talks will include food and agriculture is also uncertain, as there continues to be disagreement between the two trading partners about the scope of the negotiations, particularly the EU's intent to exclude agriculture from the talks. Perhaps the overarching goal for the U.S. side is addressing the U.S. trade deficit in agricultural products with the EU.[69]

Public statements by U.S. and EU officials in early 2020 signaled that the U.S.-EU trade talks might include SPS and regulatory barriers to agricultural trade. It is not clear, however, that both sides agree which specific types of non-tariff trade barriers might actually be part of the talks. Some press reports indicate that USDA officials have said that selected SPS barriers as well as GIs would need to be addressed.[70] Specific SPS issues important to the U.S. side include the EU's prohibitions on the use of hormones in meat production (see "U.S.-EU Beef Hormone Dispute") and pathogen reduction treatments for poultry (see section "U.S.-EU Dispute Over Pathogen Reduction Treatments (PRTs)"), and EU restrictions on the use of biotechnology (see section "Agricultural Biotechnology"). Other press reports, however, indicate that some EU officials have downplayed the extent that certain non-tariff barriers—such as biotechnology product permits, approval of certain pathogen rinses for poultry, regulations on pesticides or food standards— would be part of the

---

[69] USTR, "United States-European Union Negotiations, Summary of Specific Negotiating Objectives," January 2019.

[70] See R. McCrimmon, "Perdue Lays Out Ag Objectives in U.S. Trade Talks," *Politico*, January 29, 2020; S. Chase, "Perdue Eyes SPS, GI barriers as Key Issues in Potential US-EU Deal," *Agri-Pulse*, January 29, 2020; and S. Michalopoulos, "US Agriculture Chief Urges EU to Listen to Science, Not Fear-Mongering NGOs," *Euractiv*, January 27, 2020; and *World Trade Online*, "Hogan Hopes SPS Solutions Can Break EU-U.S. Ag Impasse," January 17, 2020.

talks.[71] The United States continues to push for additional concessions from the EU.[72] More formal discussions are expected in spring 2020.[73]

Limited Expected Role of Agricultural Issues in Upcoming Trade Talks

Against this background, in October 2018, USTR officially notified the Congress of the Trump Administration's plans to enter into formal trade negotiations with the EU.[74] This action followed a July 2018 U.S.-EU Joint Statement by President Trump and then-European Commission President Jean-Claude Juncker announcing that they would work to reduce tariffs and other trade barriers, address unfair trading practices, and increase U.S. exports of soybeans and certain other products. Previously, in 2016, U.S.-EU negotiations to create a Transatlantic Trade and Investment Partnership under the Obama Administration stalled after 15 rounds. Among the areas of contention were certain regulatory and administrative differences between the United States and the EU on issues of food safety, public health, and product naming schemes for some types of food and agricultural products.

The United States and the EU are the world's largest trade and investment partners.[75] While food and agricultural trade between the United States and the EU27 accounts for less than 1% of the value of overall trade in total goods and services, the EU27 remains a leading export market for U.S. agricultural exports. It accounted for about 8% of the value of all U.S. exports and ranked as the fifth largest market for U.S. food and farm exports in 2019—after Canada, Mexico, China, and Japan.

---

[71] See, for example, A. Shalal and D. Lawder, "As Trump Takes Aim at EU Trade, European Officials Brace for Fight," *Reuters Business News*, February 11, 2020; and *World Trade Online*, "Hogan Doubles Down on EU Regulations as U.S. Officials Demand Ag Concessions," February 20, 2020.

[72] *World Trade Online*, "U.S., EU Negotiators Accelerating Talks, Eyeing Monthly High-level Meetings," February 13, 2020.

[73] European Parliament press release, "Trade MEPs in Washington, DC, to Discuss EU-US Trade Relations," February 21, 2020; and *World Trade Online*, "U.S., EU Negotiators Accelerating Talks, Eyeing Monthly High-Level Meetings," February 13, 2020.

[74] Letter from Robert Lighthizer to then Speaker of the House of Representatives, Paul Ryan, October 16, 2018.

[75] For more information, see CRS In Focus IF11209, *Proposed U.S.-EU Trade Agreement Negotiations*.

In 2019, U.S. exports of agricultural and related product exports to the EU27 totaled $12.4 billion, while U.S. imports of agricultural and related product imports from the EU27 totaled $29.7 billion, resulting in a U.S. trade deficit of approximately $17.3 billion.[76] This is the reverse of U.S. trade surpluses with the EU27 during the 1990s. Leading U.S. agricultural exports to the EU27 were corn and soybeans, tree nuts, distilled spirits, fish products, wine and beer, planting seeds, tobacco products, and processed foods. Leading U.S. agricultural imports from the EU27 were wine, distilled spirits, beer, drinking waters, olive oil, cheese, baked goods, processed foods, and cocoa products.

In January 2019, USTR announced its negotiating objectives for the agricultural portion of a U.S.-EU trade agreement following a public comment period and a hearing involving several leading U.S. agricultural trade associations.[77] The objectives include greater market access, changes to EU administration of tariff-rate quotas, and changes to a variety of EU regulations. Among regulatory issues, key U.S. objectives include harmonizing regulatory processes and standards to facilitate trade, including sanitary and phytosanitary standards, and establishing specific commitments for trade regarding agricultural biotechnologies. The U.S. objectives also include addressing geographical indications by protecting generic terms for common use.[78] U.S. agricultural interests generally support including agriculture as part of the U.S. negotiating objectives for a U.S.-EU trade agreement. The EU negotiating mandate, however, states that a key EU goal is "a trade agreement limited to the elimination of tariffs for industrial goods only, excluding agricultural products."[79] Several

---

[76] USDA trade statistics for "Agricultural and Related Products," which includes agricultural products (including bulk and intermediate products and also consumer-oriented products) and agricultural-related products (including fish and shellfish products, distilled spirits, forest products, and ethanol and biodiesel blends.

[77] USTR, "United States-European Union Negotiations, Summary of Specific Negotiating Objectives," January 2019.

[78] For more on this, see section on "Geographical Indications (GIs)."

[79] Council of the European Union, "Trade with the United States: Council Authorizes Negotiations on Elimination of Tariffs for Industrial Goods and on Conformity Assessment," press release, April 15, 2019.

Members of Congress have stated their opposition to the EU's decision to exclude agricultural policies in their negotiating mandate.[80]

The U.S.-EU trade negotiations come amid heightened U.S.-EU trade frictions. In response to U.S. Section 232 tariffs on steel and aluminum imports, the EU had retaliated in June 2018 by imposing a tariff increase of 25% on imports of certain U.S. food and beverage products.[81] The value of U.S. agricultural exports to the EU28 (included the UK) targeted by these additional tariffs is approximately $1.2 billion in 2018, or about 9% of total U.S. agricultural exports to the EU28. In October 2019, U.S.-EU trade tensions escalated further when the United States imposed additional tariffs on $7.5 billion worth of certain U.S. imports from the EU, including food products. This action—authorized by the WTO—followed a USTR investigation initiated in April 2019 under Section 301 of the Trade Act of 1974.[82]

Aside from ongoing trade tension, some of the same issues that stalled U.S.-EU agricultural talks in the T-TIP negotiations could prove to be equally intractable today. For food and agricultural products, a series of non-tariff issues stem in part from commercial and cultural practices often enshrined in EU laws and regulations that vary from those of the United States—namely differences involving SPS and technical barriers to trade, broadly covering laws and regulations measures intended to protect public health—as well as differences involving GIs.[83]

Status

The outlook for the new U.S.-EU trade talks remains uncertain, given ongoing trade tensions. Whether or not the talks will include food and agriculture is also uncertain, as there continues to be disagreement between the two trading partners about the scope of the negotiations, particularly

[80] See, for example, letter to USTR Robert E. Lighthizer from 114 House Members, March 14, 2019, and U.S. Senate Finance Committee press release, April 15, 2019.

[81] For a full list of product codes subject to higher duties, see Commission Implementing Regulation (EU) 2018/886 available at https://eur-lex.europa.eu/legal-content/EN/TXT/HTML/?uri=CELEX:32018R0886&from=EN. This includes the UK.

[82] For a list of product codes, see 84 *Federal Register* 32248, July 5, 2019.

[83] For more background, see section "Sanitary and Phytosanitary (SPS) and Other Non-Tariff Barriers" and section "Geographical Indications (GIs)."

the EU's intent to exclude agriculture from the talks. Perhaps the overarching goal for the U.S. side is addressing the U.S. trade deficit in agricultural products with the EU.[84]

Public statements by U.S. and EU officials in early 2020 signaled that the U.S.-EU trade talks might include SPS and regulatory barriers to agricultural trade. It is not clear, however, that both sides agree which specific types of non-tariff trade barriers might actually be part of the talks. Some press reports indicate that USDA officials have said that selected SPS barriers as well as GIs would need to be addressed.[85] Specific SPS issues important to the U.S. side include the EU's prohibitions on the use of hormones in meat production (see "U.S.-EU Beef Hormone Dispute") and pathogen reduction treatments for poultry (see section "U.S.-EU Dispute Over Pathogen Reduction Treatments (PRTs)"), and EU restrictions on the use of biotechnology (see section "Agricultural Biotechnology"). Other press reports, however, indicate that some EU officials have downplayed the extent that certain non-tariff barriers—such as biotechnology product permits, approval of certain pathogen rinses for poultry, regulations on pesticides or food standards— would be part of the talks.[86] The United States continues to push for additional concessions from the EU.[87] More formal discussions are expected in spring 2020.[88]

---

[84] USTR, "United States-European Union Negotiations, Summary of Specific Negotiating Objectives," January 2019.

[85] See R. McCrimmon, "Perdue Lays Out Ag Objectives in U.S. Trade Talks," *Politico*, January 29, 2020; S. Chase, "Perdue Eyes SPS, GI barriers as Key Issues in Potential US-EU Deal," *Agri-Pulse*, January 29, 2020; and S. Michalopoulos, "US Agriculture Chief Urges EU to Listen to Science, Not Fear-Mongering NGOs," *Euractiv*, January 27, 2020; and *World Trade Online*, "Hogan Hopes SPS Solutions Can Break EU-U.S. Ag Impasse," January 17, 2020.

[86] See, for example, A. Shalal and D. Lawder, "As Trump Takes Aim at EU Trade, European Officials Brace for Fight," *Reuters Business News*, February 11, 2020; and *World Trade Online*, "Hogan Doubles Down on EU Regulations as U.S. Officials Demand Ag Concessions," February 20, 2020.

[87] *World Trade Online*, "U.S., EU Negotiators Accelerating Talks, Eyeing Monthly High-Level Meetings," February 13, 2020.

[88] European Parliament, "Trade MEPs in Washington, DC, to Discuss EU-US Trade Relations," press release, February 21, 2020; and *World Trade Online*, "U.S., EU Negotiators Accelerating Talks, Eyeing Monthly High-Level Meetings," February 13, 2020.

Trade Aid in Response to Trade Retaliation[89]

During 2018 and 2019, the Secretary of Agriculture used his authority under the Commodity Credit Corporation Charter Act[90] to initiate two ad hoc trade assistance programs in response to foreign trade retaliation targeting U.S. agricultural products. The trade aid packages were part of the Administration's effort to provide short-term assistance to farmers for the temporary loss of important international markets.[91] On July 24, 2018, USDA announced the first "trade aid" package, which targeted production of selected agricultural commodities in 2018 and was valued at up to $12 billion. On May 23, 2019, USDA announced a second package, which targeted production of an expanded list of commodities and was valued at up to an additional $16 billion. Thus, the two years of combined trade assistance were valued at up to $28 billion.

Both trade aid packages included (1) a Market Facilitation Program (MFP) of direct payments to producers of commodities most affected by the trade retaliation, (2) a Food Purchase and Distribution Program (FPDP) designed to partially offset lost export sales of affected commodities, and (3) an Agricultural Trade Promotion (ATP) program to expand foreign markets. The largest part of the aid is two years of MFP payments initially valued at a combined $24.5 billion (up to $10 billion in 2018 and $14.5 billion in 2019).

Status

As of February 10, 2020, USDA estimates that it has spent $8.6 billion under the 2018 MFP and $14.2 billion under the 2019 MFP.[92] Payments of this magnitude could attract international attention about whether they are consistent with WTO rules and U.S. commitments on domestic support,[93]

---

89 Prepared by Randy Schnepf, Specialist in Agricultural Policy.

90 CRS Report R44606, *The Commodity Credit Corporation: In Brief*, by Megan Stubbs.

91 See CRS Report R45903, *Retaliatory Tariffs and U.S. Agriculture*, by Anita Regmi.

92 USDA, MFP payment data, accessed February 24, 2020, https://www.farmers.gov/manage/mfp.

93 For a discussion, see section "2018 Farm Bill, Trade Aid, and WTO Compliance," and also see CRS Report R45310, *Farm Policy: USDA's 2018 Trade Aid Package*.

as some WTO member countries are questioning whether this additional aid violates U.S. spending limits under the WTO.[94] The trade aid packages raise other potential questions as well. For instance, if the U.S.-China Phase One trade agreement does not produce the commodity purchases promised by China, or if commodity prices remain relatively low, should another trade aid package, or some alternative compensatory measure, be provided in 2020, and possibly beyond? If MFP payments are provided in the future, should USDA revise its payment formulation to provide a broader distribution of payments across the U.S. agricultural sector?

FUTURE TRADE NEGOTIATIONS[95]

India

India is the world's second most populous country after China. Since 2000, its economy has been the fastest growing in the world. Given the rapid growth in population and income among a large segment of the population, demand for higher-value food products such as fruits, nuts, dairy products, and other livestock products, is expected to increase among Indian consumers. While India is among the world's largest producers and consumers of a range of crop and livestock commodities, USDA projects India will continue to be an important importer of dairy products, vegetable oils, pulses, tree nuts, and fruit, and that it will continue to be a major exporter of rice, cotton and buffalo meat.[96]

U.S. agricultural exports to India have increased since 2015, reaching $1.6 billion in 2017 (Figure 2). In 2018, U.S. exports declined to $1.5 billion, coinciding with India's imposition of retaliatory tariffs on imports of U.S. almonds, walnuts, apples, chickpeas, and lentils, but U.S. exports rebounded to $1.8 billion in 2019 due to increased sales of cotton and tree

[94] For more on this issue, see CRS Report R45940, *U.S. Farm Support: Compliance with WTO Commitments*.

[95] Prepared by Anita Regmi, Analyst in Agricultural Policy, CRS.

[96] Landes and Hjort, "Food Policy and Productivity Key to India Outlook," *Amber Waves*, ERS, July 2015.

nuts (largely pecans, pistachios, and dried coconut). Tree nuts (mainly almonds), cotton, and fresh fruit are key U.S. exports to India. However, other U.S. high-value products are registering rapid growth. For example, U.S. dairy exports to India grew by almost 300% from $16 million in 2015 to $60 million in 2019.

In 2019, the United States imported agricultural products valued at $2.6 billion from India.[97] Spices, rice, essential oils, tea, processed fruit and vegetables, and other vegetable oils are the leading U.S. imports from India.

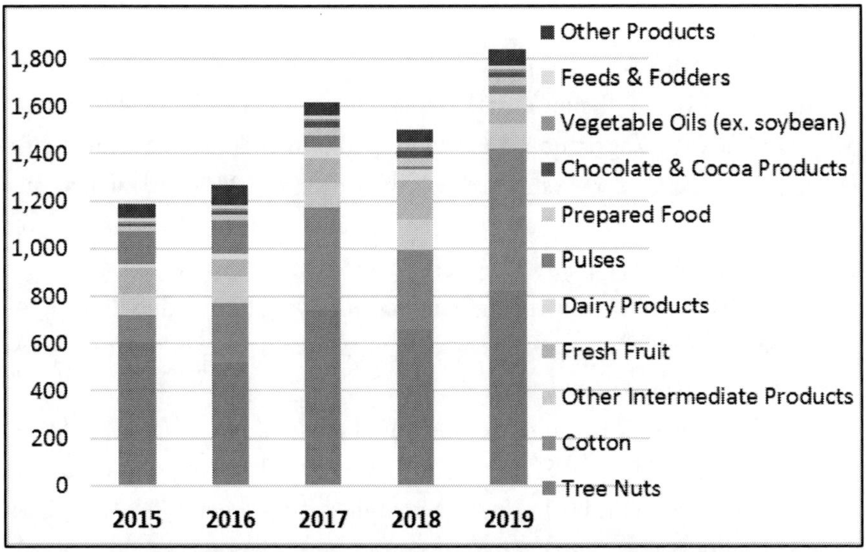

Source: U.S. Census Bureau trade data, accessed via U.S. Department of Agriculture (USDA), Foreign Agricultural Service (FAS), BICO-HS-10 grouping, February 2019, https://apps.fas.usda.gov/gats/ default.aspx.

Notes: Based on USDA's definition of *agriculture*.

Figure 2. U.S. Exports to India. Millions of U.S. Dollars

[97] U.S. Census Bureau Trade Data, accessed from GATS, February 2020, https:// apps.fas.usda.gov/gats/ ExpressQuery1.aspx.

U.S.-India trade negotiations follow a period of trade tensions. In March 2018, the United States levied additional tariffs on steel and aluminum imports from India. India responded by identifying certain U.S. food products for retaliatory tariffs[98] but did not levy them until June 16, 2019, after the United States terminated preferential treatment for India under the Generalized System of Preferences (GSP).[99] India's retaliatory tariffs range from 10% to 25% on imports of U.S. chickpeas, shelled almonds, walnuts, apples, and lentils.[100] Both countries' tariffs are likely to become an issue if the United States and India undertake a major trade negotiation, as USTR has proposed.

Trade Policy Issues

India's tariffs and non-tariff barriers have prevented greater market penetration of U.S. agricultural products. India maintains very high tariffs on many products, for example 60% on flowers, 100% on raisins, and 150% on alcoholic beverages.[101] Since 2017, a system of annual import quotas on pulses has restricted U.S. exports of pulses to India.[102] U.S. exports of wheat and barley to India are currently restricted due to its zero-tolerance standard for certain pests and weeds, and restrictions also exist on imports of livestock genetic material.

Similarly, processed products, including ethanol, are subject to various restrictions that prevent U.S. exports to India. India bans imports of tallow, fat, and oils of animal origin. India's complex requirements for U.S. dairy products have been a barrier for expanding U.S. exports. In 2015, India

---

98 India, Immediate Notification Under Article 12.5 of the Agreement on Safeguards to the Council for Trade in Goods of Proposed Suspension of Concessions and other Obligations Referred to in Paragraph 2 of Article 8 of the Agreement on Safeguards, WTO, May 18, 2018.

99 The GSP provides duty-free tariff treatment for certain products from designated developing countries.

100 Data from Ministry of Commerce and Industry of India, February 2020.

101 USTR, *2019 National Trade Estimate Report on foreign Trade Barriers*, 2019.

102 Senators Cramer (North Dakota) and Daines (Montana) requested in a February 29, 2020 letter to President Trump that the Administration seek a favorable pulse crop provision in negotiations with India, February 19, 2020, ttps://senatorkevincramer.app.box.com/s/1lc5yt7ja6w9ttr9oeph34x8e3ik5u7c.

revised its health certificate requirement for pork imports. Since then, the United States has been seeking approval to export pork to India.

USTR asserts that India's customs regulations are not transparent or predictable.[103] India's approval process for genetically engineered products are slow and not transparent.[104]

India maintains a large and complex program for public food stockholding, both to distribute food to poor consumers and to stabilize market prices, essentially subsidizing domestic production. India provides a broad range of support to its agricultural sector. In May 2018, the United States argued at the WTO that India was under-reporting its price supports for rice and wheat.[105] In November 2018, the United States questioned India's price support for cotton,[106] while Australia has questioned India's price support for sugarcane.[107]

Status

In 2019, in response to various U.S. concerns over India's trade barriers, the United States revoked India's eligibility for preferential tariff treatment under the U.S. GSP.[108] Total value of U.S. imports of agricultural products from India were down 1% in 2019 from $2.7 billion in 2018 to $2.6 billion in 2019. USTR has stated that it hopes to reach an agreement in 2020 that will, among other things, provide greater access to the Indian market for U.S. agricultural products, potentially in exchange for U.S. restoration of India's eligibility under GSP.[109]

---

[103] USTR, *2019 National Trade Estimate Report on foreign Trade Barriers*, 2019.

[104] FAS, "India: Agricultural Biotechnology Annual," IN2019-0109, February 04,2020.

[105] For more on this, see CRS Report R45728, *Major Agricultural Trade Issues in the 116th Congress*.

[106] WTO, "Certain Measures of India Providing Market Price Support to Cotton," G/AG/W/188, November 9, 2018.

[107] WTO, "India's Measures to Provide Market Price Support to Sugarcane," G/AG/W/189, November 16, 2018.

[108] For more, see CRS Report RL33663, *Generalized System of Preferences (GSP): Overview and Issues for Congress*.

[109] A. Behsudi, "U.S., India in Final Stages of Limited Trade Deal," *Politico*, January 28, 2020.

Kenya

On February 6, 2020, the Trump Administration announced that the United States intends to negotiate a comprehensive trade agreement with Kenya using the authority under P.L. 114-26.[110] The Administration asserts that such a trade agreement will complement Africa's regional integration efforts, including as part of the African Continental Free Trade Area (AfCFTA), to which the United States has pledged support.[111]

Kenya hosts three international agricultural research centers that focus on innovations, including agricultural biotechnology, to sustainably improve global food security. These institutions are the International Livestock Research Institute, the World Agroforestry Center, and the International Centre of Insect Physiology and Ecology.

Kenya is an emerging middle-income country, home to more than 47 million people with an estimated population growth rate of 2.5% in 2017.[112] USDA projects Kenya's real GDP per capita to grow at an annual rate of about 4% though 2031.[113] With anticipated growth in population and per capita income, Kenya has the potential to increase its imports of food and other agricultural products. Kenya's top five agricultural imports are wheat, palm oil, sugar, corn and rice. Its top exports from the United States are wheat, vegetable oils excluding soybean oil, pulses, coarse grains, and other products that include many prepared food products (Figure 3).

Trade Policy Issues

Kenya is a beneficiary of the African Growth and Opportunity Act, most recently extended in P.L. 114-27, under which it has duty-free access to the U.S. market for 6,400 products including agricultural products. In

[110] USTR, "President Trump Announces Intent to Negotiate Trade Agreement with Kenya", press release, February 6, 2020.

[111] USTR, "U.S.-Kenya Trade and Investment Relationship," February 2020, https://ustr.gov/about-us/policy-offices/press-office/fact-sheets/2020/february/fact-sheet-us-kenya-trade-and-investment-relationship.

[112] USTR, "U.S.-Kenya Trade and Investment Relationship."

[113] ERS, International Macroeconomic Data Set, Real Per Capita GDP Projections in 2010 U.S. Dollars, January 3, 2020.

2019, the United States imported agricultural products valued at $126 million from Kenya,[114] with major products being macadamia and cashew nuts, coffee, tea, roses, and non- edible vegetable and nut oils.[115]

Kenya's MFN tariffs—rates that apply to imports from the United States—are relatively high. For example, simple average MFN tariffs for animal products are 23.1%, dairy products are 51.7%, fruit and vegetables are 22%, cereals and preparations are 22.2%, sugar is 40%, and fish products are 24.8%.[116] Other concerns raised by USDA include a Kenyan ban on imports of genetically engineered (GE) agricultural products (although it has approved field trials for GE cotton[117] and drought and insect resistant corn), bans on imports of U.S. whole peas and lentils,[118] and had a ban on wheat from the U.S. Pacific Northwest over concerns regarding a certain fungus. In February 2020, Kenya adopted a phytosanitary protocol that allows wheat growers in Washington State, Oregon, and Idaho access to Kenya's wheat market, potentially allowing increased U.S. wheat exports to Kenya.[119]

Status

USTR has said it plans to officially notify Congress of its intent to start negotiations following consultations with Congress as required by the Bipartisan Congressional Trade Priorities and Accountability Act of 2015 (P.L. 114-26). Subsequently, USTR is to publish notices in the *Federal Register* requesting public comment on the direction, focus, and content of the trade negotiations with Kenya. USTR is to publish objectives for the negotiations at least 30 days before trade negotiations begin. Some

---

[114] U.S. Census Bureau Trade Data, accessed from GATS, February 2020, https://apps.fas.usda.gov/gats/ ExpressQuery1.aspx.

[115] FAS, "Strengthening the U.S.-Kenya Trade Relationship to Grow U.S. Agricultural Exports to East Asia," *International Agricultural Trade Report*, June 2019.

[116] WTO, Kenya and the WTO, accessed February 4, 2020, https://www.wto.org/english/thewto_e/countries_e/kenya_e.htm. The average tariffs are not trade weighted.

[117] In 2020, Kenya is planning its first production of Bt cotton for commercialization, for more on this see FAS, "Kenya: Agricultural Biotechnology Annual," GAIN Report KE2019-0008, February 14, 2020.

[118] USTR, *2019 National Trade Estimate Report on foreign Trade Barriers*, 2019.

[119] FAS, "USDA Expands Market for U.S. Wheat: Adds Idaho, Oregon, and Washington to List of States That Can Export Wheat to Kenya," press release, February 25, 2020.

Members of Congress have expressed their support for a free trade agreement with Kenya.[120]

United Kingdom (UK)

In January 2020, the UK left the EU. It remains a member of the EU customs union, so U.S.-UK trade continues to be governed by agreements between the United States and the EU in addition to WTO rules. However, the UK has announced its intention to withdraw from the EU customs union on December 31, 2020. Thereafter, U.S.-UK trade will occur under WTO rules unless a separate agreement is reached between the United States and the UK. The UK entered the WTO as a member of the EU, and does not have its own schedule of commitments under the WTO.[121] U.S.-UK trade would thus continue to be governed by the EU WTO schedule, with some confusion regarding what share of quota and subsidy commitments made by the EU will henceforth apply to the UK. Therefore, some Members of Congress have indicated that a comprehensive U.S.-UK trade agreement should be a priority for the United States.[122]

The UK has accounted for about 1.3% of total U.S. agricultural exports from 2015 to2019. Major U.S. exports are wine and beer, tree nuts, prepared food, soybeans, live animals and other products (Figure 4). The United States does not export notable quantities of meat products to the UK, and the Trump Administration and some Members of Congress and U.S. agricultural industry would like to expand exports of these products in the post-Brexit environment.[123]

---

[120] Office of Senator Mike Enzi, "Inhofe, Enzi, Coons, Rounds, Kaine, Boozman Applaud First Steps to US-Kenya Free Trade Agreement," press release, February 7, 2020.

[121] This raises issues for the UK and the EU regarding how to apportion subsidies and quotas between EU-27 and the UK and would require the UK to establish its own WTO schedule. For more on this see, CRS Report R45944, *Brexit: Status and Outlook*.

[122] Twelve U.S. Senators' Letter to USTR, February 14, 2020, https://www.portman.senate.gov/sites/default/files/2020-02/Signed%20UK%20Trade%20Letter.pdf.

[123] See for example, A. Behsudi, "Trump's U.K. Trade Deal Could Depend on Whether the Brits Can Stomach 'Chlorine Chicken'," *Politico*, January 6, 2020.

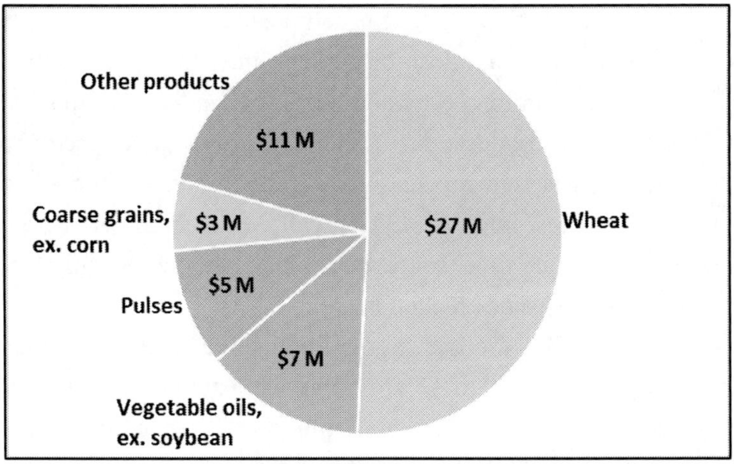

Source: U.S. Census Bureau Trade Data, BICO-10 grouping, accessed from FAS, USDA, February 7, 2020, https://apps.fas.usda.gov/gats/ ExpressQuery1.aspx.
Notes: USDA definition of *agriculture* is used; 'ex.'=excluding.

Figure 3. U.S. Agricultural Exports to Kenya. $53 Million in 2019

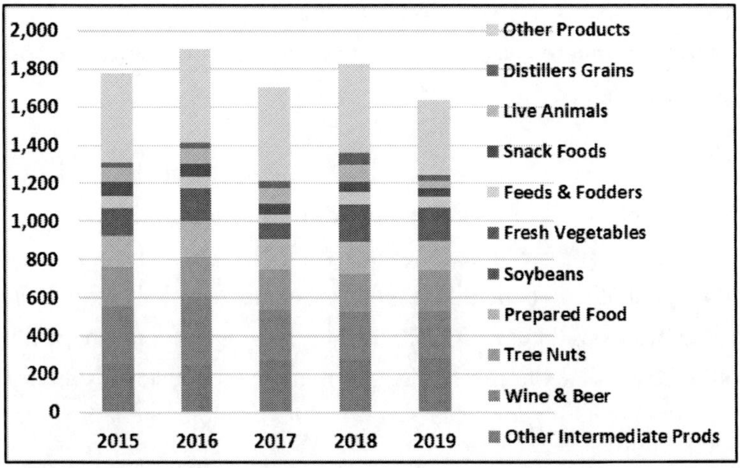

Source: U.S. Census Bureau Trade Data, BICO-10 grouping, accessed from FAS, USDA, February 7, 2020, https://apps.fas.usda.gov/gats/ExpressQuery1.aspx.
Notes: USDA definition of *agriculture* is used.

Figure 4. U.S. Agricultural Exports to the UK. Millions of U.S. Dollars

As a member of the EU, the UK posed the same set of trade barriers to U.S. agricultural exports as those discussed under "U.S.-EU Agricultural Trade". In particular, hormone treated beef, chlorine-washed poultry, and bio-engineered food products have faced restrictions in accessing EU markets. The UK has sent mixed signals regarding these issues and has hinted that it may allow imports of genetically engineered U.S. agricultural products.[124] At the same time, some reports indicate the UK will not allow imports of chlorine-washed chicken meat.[125]

Among other goals for U.S. agricultural trade, USTR has identified reducing or eliminating tariffs, providing adjustment periods for U.S. import-sensitive products before initiating tariff reduction, eliminating non-tariff barriers that discriminate against U.S. agricultural goods, improving UK's TRQ administration, promoting regulatory compatibility, and establishing commitments for trade in agricultural biotechnology products.[126] USTR has also articulated specific goals regarding sanitary and phytosanitary provisions, customs and trade facilitation, rules of origin, and technical barriers to trade. Some Members of Congress have requested that improved market access for U.S. rice be an objective of U.S. negotiators.[127]

Status

On October 16, 2018, the Trump Administration notified Congress of proposed trade agreement negotiations with the UK. The UK could not formally negotiate or conclude a new agreement until it exited the EU, which occurred on January 31, 2020, and any agreement could not take effect until the UK exits the EU single market and customs union. Given

---

[124] R. Mason "Boris Johnson Hints at Allowing GM Food Imports from U.S.," *The Guardian*, February 3, 2020.

[125] E. Courea, "Pompeo: Agriculture a Sticking Point in U.K.-U.S. Trade Talks," *Politico*, January 30, 2020.

[126] USTR, "United States-United Kingdom Negotiations: Summary of Specific Negotiating Objectives," February 2019.

[127] U.S. Senate, Letter to Ambassador Lighthizer, Signed by Senators Wicker (MI), Hyde-Smith (MI), Boozman (AK), Blunt (MO), Cornyn (TX), Cotton (AK), Cruz (TX), and Hawley (MO), February 10, 2020, https://www.hydesmith.senate.gov/sites/default/files/2020-02/021020%20USTR%20Ambassador%20Lighthizer%20UK%20Rice%20Letter.pdf.

the proposed scope of the negotiations, any resulting agreement would likely be subject to ratification by Congress.

WTO AND U.S. AGRICULTURE[128]

The World Trade Organization is an international organization that administers the rules and agreements negotiated among its 164 members to eliminate trade barriers and govern trade.[129] It also serves as an important forum for resolving trade disputes through its committee structures and its Dispute Settlement Body, which approves reports issued by panels of legal experts and a separate Appellate Body. The United States was a major force behind the establishment of the WTO in 1995.

Under the WTO's Agreement on Agriculture (AoA), agreed in 1995, national agricultural policies—including domestic farm support, agricultural export subsidies, and restrictive import controls—were placed under a multilaterally agreed-upon set of disciplines for the first time.[130] WTO members agreed to reform their domestic agricultural support policies, increase access to imports, and reduce export subsidies. The disciplines on these three "pillars" of agricultural policy involved freezing (or "binding") protective measures and subsidies at base period levels, then instituting annual reductions from the bound levels. Article 15 of the AoA granted developing and least-developed countries special rights or extra leniency—termed "special and differential treatment"—in the implementation of their policy commitments. Specifically, they had longer periods over which to reduce subsidies and to improve market access. They were also allowed to retain certain subsidies that were prohibited for other countries.

---

[128] Prepared by Anita Regmi, Analyst in Agricultural Policy, CRS.

[129] CRS Report R45417, *World Trade Organization: Overview and Future Direction*.

[130] WTO, Agreement on Agriculture, Legal Text 1995, https://www.wto.org/english/docs_e/legal_e/14-ag_01_e.htm.

During the AoA's early years, Article 13, known as the Peace Clause or "due restraint" clause, provided additional impetus for reform. The Peace Clause provided temporary protection for market-distorting domestic support and export subsidy measures from challenges under other WTO provisions, as long as these measures complied with certain requirements.[131] However, such subsidies would be open to challenge after the Peace Clause expired around January 2004.[132]

The AoA was envisioned as a first step in the process of global market liberalization in the agricultural sector. The impending expiration of the Peace Clause coupled with Article 20's directive to continue the reform process led WTO members to launch the Doha Round of negotiations in 2001. But, the Doha Round failed to reach consensus on formulas to reduce tariffs and agricultural subsidies, due in part to disagreements among developing countries that wished to retain their special and differential treatment under the AoA and wealthier countries that wanted to limit such preferences. The Doha Round has been at an impasse since 2009.[133]

The WTO's effectiveness as a negotiating body for broad-based trade liberalization and its role in resolving trade disputes therefore have come under intensified scrutiny in recent years. The WTO has struggled to address newer issues, such as digital trade and regulations affecting services. In addition, the Appellate Body is effectively non-functional due to the United States' decision to block the nomination of members, which prevents it from having a quorum needed to resolve disputes.

---

[131] Exemption was allowed provided that cumulative outlays on such measures did not grant support to a specific commodity in excess of that decided during the 1992 marketing year.

[132] WTO, AoA, Article 1.f, 1995. There has never been a definitive statement as to when the Peace Clause expired, with the only WTO panel to address it finding that, at the earliest, it expired on January 1, 2004, but could have expired at later points in 2004.

[133] For more on this issue, see CRS Report RS22927, *WTO Doha Round: Implications for U.S. Agriculture*.

Status

USTR has stated that WTO institutional reform is a priority in 2020.[134] Some Mof Congress have voiced their agreement.[135] The WTO's chair for agricultural negotiations may circulate a negotiating framework for the June 2020 meeting of WTO trade ministers in Kazakhstan that includes rules designed to increase sustainable agricultural production.[136] The meeting may also consider a proposal by a group representing 19 countries, known as the Cairns Group, to "cap and reduce by at least half the current sum of global agricultural trade- and production-distorting domestic support entitlements by 2030."[137]

2018 Farm Bill, Trade Aid, and WTO Compliance[138]

Under the AoA, the United States has committed to limit its domestic support program spending deemed most trade-distorting (referred to as "amber box" outlays) to $19.1 billion per year. The AoA spells out the rules for countries to determine whether their policies are potentially trade-distorting, how to calculate the costs of any distortion using a specially defined indicator, the "Aggregate Measure of Support" (AMS), and how to report those costs to the WTO in a public and transparent manner.[139] While the AMS is subject to a spending limit, the AoA provides four potential exemptions from the AMS spending limit.

---

[134] For more on this issue see, CRS Report R45474, *International Trade and Finance: Overview and Issues for the 116th Congress*; USTR, *2019 Trade Policy Agenda and 2018 Annual Report*, March 2019; and USTR, *USTR Issues Report on the WTO Appellate Body*, February 11, 2020.

[135] See H.Res. 746; and Icso, "Blumenauer Cites USMCA Enforcement, WTO Reform Among Trade Panel's 2020 Priorities," *Inside World Trade,* January 31, 2020.

[136] WTO, "DDG Wolff: It is Time to Update the WTO Rulebook for Agriculture," January 18, 2020.

[137] The Cairns Group, "41st Cairns Group Ministerial Meeting Statement and Framework for Negotiations," January 23, 2020, https://cairnsgroup.org/Pages/Statement-of-the-41st-Cairns-Group-Ministerial-Meeting.aspx.

[138] Prepared by Randy Schnepf, Specialist in Agricultural Policy, CRS.

[139] For more information, see CRS Report R45940, *U.S. Farm Support: Compliance with WTO Commitments*, by Randy Schnepf.

First, if a program's outlays are considered to be minimally trade distorting or non-trade distorting (in accordance with specific criteria listed in Annex 2 of the AoA), then they may qualify as "green box" programs and not be included in the AMS. Second, if program spending is trade-distorting but has offsetting features that limit the production associated with support payments, then they may qualify as "blue box" programs and not be included in the AMS. Third, if AMS outlays for a specific commodity are sufficiently small relative to the output value of that commodity (product- specific de minimis), they may be exempted. Finally, if aggregate AMS outlays are small relative to the value of total agricultural production (non-product-specific de minimis)—then they may be exempted. Any AMS left over after applying these four exemptions constitutes the amber box.

Since the WTO's establishment, the United States has generally met its WTO amber box spending commitment. However, in some years U.S. compliance has hinged on judicious use of de minimis exemptions, which permit it to exclude certain spending from being considered under its amber box limit (see Figure 5). To date, no WTO member has challenged these exemptions.

Since 2010, U.S. outlays on potentially market-distorting farm programs have been trending upward (Figure 5). From 2011 through 2016, AMS outlays (amber box plus de minimis exemptions) averaged $14.6 billion per year.[140] However, several policy developments since 2016 have created uncertainty about whether the United States will remain in compliance with the rules and spending limits for domestic support programs that it has agreed to in the WTO. These developments are, first, farm program changes under both the 2018 farm bill (P.L. 115-334), which expanded payment eligibility and eliminated certain programs from payment limits,[141] and, second, USDA trade aid programs implemented in 2018 and 2019 under other statutory authorities in response to foreign trade

---

[140] Compiled by CRS from U.S. official notifications to the WTO.

[141] See CRS Report R45730, *Farm Commodity Provisions in the 2018 Farm Bill (P.L. 115-334)*.

retaliation targeting U.S. agricultural products (see "Trade Aid in Response to Trade Retaliation").[142]

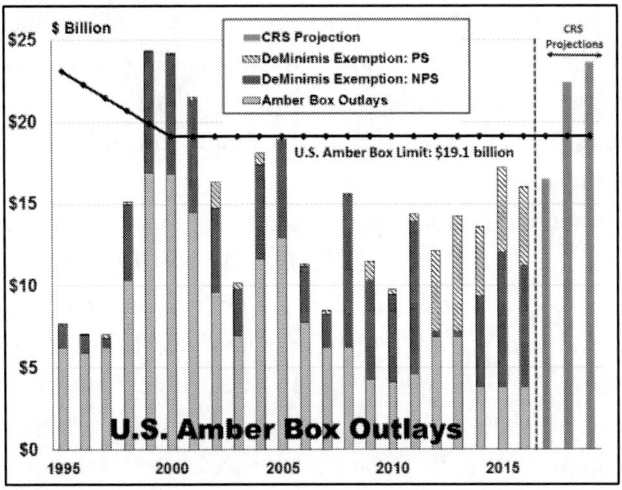

Source: PS = product specific; NPS = non-product- specific. Compiled by CRS from U.S. notification to the WTO for 1995-2016. Years 2017-2019 are compiled by CRS from USDA payment and farm income data and Congressional Budget Office May 2019 baseline projections for farm program outlays.

Notes: The United States has yet to notify domestic support outlays beyond 2016. Thus, it is unknown how USDA will categorize new spending programs such as the Market Facilitation Program (MFP). As a result, CRS does not distinguish between amber box and de minimis spending for the projected years 2017, 2018, and 2019; for a projected breakout, see CRS Report R45940, *U.S. Farm Support: Compliance with WTO Commitments*.

Figure 5. U.S. Amber Box Compliance, Historical and Projected.

U.S. AMS spending is estimated to have been higher in 2017 through 2019, based on CRS compilation of USDA program data. Outlays in 2017 are estimated to have been $16.5 billion; however, the classification of $10.1 billion in program spending as de minimis exemptions would limit amber box outlays to $6.3 billion. The addition of the Administration's two

[142] For details regarding trade-aid payments, see CRS Report R45310, *Farm Policy: USDA's 2018 Trade Aid Package*; and CRS Report R45865, *Farm Policy: USDA's 2019 Trade Aid Package*.

MFP "trade aid" payments, valued at $8.6 billion in 2018 and approximately $10.7 billion in 2019, are estimated to push total AMS outlays above the U.S. amber box spending limit—to $22.4 billion in 2018 and $23.6 billion in 2019. Whether the United States will violate its spending commitment or not would be expected to depend on the extent that de minimis exemptions apply for those two years.[143]

The United States has yet to notify spending to the WTO under any of the trade assistance programs, so the exact WTO spending classification is currently unknown. However, past practice can serve as a guide for the likely notification. The FPDP and ATP programs for 2018 and 2019 are expected to have been implemented in a similar manner during both years. USDA outlays under food purchase and distribution programs have historically been notified to the WTO as green box compliant and thus not subject to any spending limit. Trade promotion programs, such as ATP, are not notified under domestic support, because they do not involve direct payments to producers. Thus, the FPDP and ATP programs are not expected to affect the United States' ability to meet its WTO commitments.

Payments under the two MFP programs were structured differently during 2018 and 2019. As a result, they are likely to be notified under different WTO classifications. The specific manner of determining how payments are made to individual producers is likely to determine their WTO status. Potential AMS classifications are:[144]

- USDA's MFP payments for 2018 were based on each farm's harvested production of eligible crops during 2018 times a fixed per-unit payment rate. Payments to dairy were based on historical production, while hog payments used mid-year inventory data.

---

[143] These projections hinge on several as-yet-unknown factors, including market prices, output values, and program outlays under traditional countercyclical ARC and PLC programs. If the final prices are higher than currently projected, then program payments under ARC and PLC could be smaller than those used in this analysis. This could decrease both aggregate non-product-specific outlays and the possibility of exceeding the amber box spending limit.

[144] These potential notifications are CRS projections based on analysis of the design of the 2018 and 2019 MFP programs and how they correspond with previous U.S. notifications. USDA may use a different line of reasoning and notify 2018 and 2019 MFP payments under different WTO classifications.

Under this specification, 2018 MFP payments are likely to be notified as coupled, product-specific AMS and would count against the U.S. annual spending limit of $19.1 billion (unless they are exempted under the product-specific de minimis exemption).

- USDA's MFP payments for 2019 were coupled to a producer having planted at least one eligible commodity within the county, but they are independent of which commodity or commodities were planted. Under this specification, the 2019 MFP payments would appear to be coupled to planted acres—a producer has to plant an eligible crop to get a payment—but are non-product-specific, thus possibly notifiable as non-product-specific AMS.

Status

Most recent studies suggest that, for U.S. program spending to exceed the $19.1 billion cumulative spending limit, even with the addition of large MFP payments and higher traditional program support levels, a combination of events would have to occur that would broadly depress commodity prices. Perhaps more relevant to U.S. agricultural trade is the concern that, because the United States plays such a prominent role in most international markets for agricultural products, any distortion resulting from U.S. policy could be both visible and potentially vulnerable to challenge under WTO rules.[145]

U.S. Challenges to Farm Support Spending of WTO Members[146]

Since the inception of the WTO in 1995, the United States has initiated 46 WTO dispute cases related to agriculture. Of these cases, 34 were fully or partially decided in favor of the United States by the WTO panel hearing the case.[147]

---

[145] See CRS Report RS22522, *Potential Challenges to U.S. Farm Subsidies in the WTO: A Brief Overview*.

[146] Prepared by Anita Regmi, Analyst in Agricultural Policy, CRS.

[147] Extracted from WTO, Disputes by Member, case total reported as of April 23, 2019. https://www.wto.org/english/tratop_e/dispu_e/dispu_by_country_e.htm.

U.S. Challenges of China's Agricultural Domestic Support

In September 2016, USTR filed a dispute settlement case (DS511) at the WTO over China's domestic agricultural support policies, alleging they were inconsistent with WTO rules and commitments.[148] USTR contended that the level of support that China provided for rice, wheat, and corn had exceeded—by nearly $100 million from 2012 through 2015—the level to which China had committed to when it joined the WTO. USTR also asserted that China's price support for domestic production had been above the world market prices since 2012, thereby creating an incentive for Chinese farmers to increase production of the subsidized crops, which in turn displaced imports from the United States and elsewhere.[149] In December 2016, USTR requested that the WTO establish a dispute settlement panel to examine China's domestic support levels for these crops.

On February 28, 2019, the WTO dispute settlement panel found that China had exceeded its domestic support limits for wheat and rice in each year between 2012 and 2015 and therefore was not in compliance with its WTO commitment. The panel made recommendations that China change its calculations of reference prices and domestic support in order to comply with its WTO commitments. The panel did not make a ruling on corn because China had already made changes to its support for corn that were found to be less trade distorting than the method used prior to 2015.

Status

Under the U.S.-China Phase One trade agreement, China stated that it will respect its WTO obligations and publish in its official journal its laws, regulations and other measures pertaining to its domestic support programs and policies.

---

[148] See WTO, Dispute Settlement, DS511: China-Domestic Support for Agricultural Producers; https://www.wto.org/english/tratop_e/dispu_e/cases_e/ds511_e.htm.

[149] See USTR, "United States Challenges Excessive Chinese Support for Rice, Wheat, and Corn," press release, September 13, 2016.

U.S. Challenges to China's Agricultural Market Access Policy

On December 15, 2016, USTR filed another WTO dispute settlement case (DS517) against China, alleging that China administered its TRQs for wheat, rice, and corn in such a way that the duty-free quotas were never filled, even when imported grains were priced lower than domestic grains.[150]

USTR stated that China's TRQ administration appeared to restrict imports and failed to provide sufficient information to permit the processing of quota applications and importation.

On September 22, 2017, a WTO dispute settlement panel was established on *China – Tariff Rate Quotas for Certain Agricultural Products*. On April 18, 2019, the panel ruled in favor of the United States, stating that "China's administration of its TRQs for wheat, rice and corn were inconsistent with its obligations under the WTO to administer TRQs on a transparent, predictable and fair basis." The panel recommended that China make changes to its TRQ administration to conform to its WTO obligations.[151]

Status

In the U.S.-China Phase One trade agreement, China stated that it will ensure that its TRQ measures conform with the WTO panel ruling.

U.S. Challenges to India's Domestic Agricultural Support

In May 2018, the United States asserted at the WTO that India had not accurately notified the WTO of its spending on its market price support for rice and wheat for the marketing years 2010/11 through 2013/14.[152] The United States alleged that India's market price support for wheat and rice exceeded its allowable levels of trade distorting domestic support under the WTO.

---

150 See USTR, "United States Challenges Chinese Grain Tariff Rate Quotas for Rice, Wheat, and Corn," press release, December 15, 2016.

151 WTO, China—Tariff Rate Quotas for Certain Agricultural Products, Report of the Panel, April 18, 2019.

152 WTO, Certain Measures of India Providing Market Price Support to Rice and Wheat, G/AG/W/174, May 9, 2018.

In November 2018, the United States also challenged India's domestic support for cotton at the WTO, stating that it exceeded its allowable level under its WTO commitments.[153] At about the same time, Australia, Brazil, and Guatemala challenged India's level of domestic support for sugar, charging that India had violated its WTO commitment levels.[154]

In February 2019, the United States further challenged India at the WTO, stating that it had substantially underreported its market price support for chickpeas, pigeon peas, black matpe (a type of black lentil), mung beans, and lentils. According to USTR, when calculated using the AoA methodology, India's market price support for each of these pulses has exceeded the allowable levels of trade-distorting domestic support under India's WTO commitments.[155]

The United States' challenge to India's domestic support for rice and wheat was raised at the May 2018 WTO Committee on Agriculture meeting. USTR raised the issue concerning India's cotton price support during the November 2018 committee meeting, and the challenge against India's domestic support for pulses was raised at the February 2019 meeting.

Status

USTR may continue challenging India's domestic support for agriculture at upcoming WTO Committee on Agriculture (COA) meetings and, if necessary, could pursue these concerns through WTO's dispute settlement mechanism. India's domestic support for agriculture could be an issue during U.S.-India trade negotiations or during the discussions related to WTO reform on agriculture.

---

[153] WTO, Certain Measure of India Providing Market Price Support to Cotton, G/AG/W/188, November 9, 2018.

[154] WTO, India's Measures to Provide Market Price Support to Sugarcane, G/AG/W/189, November 16, 2018; WTO, DS580: India—Measures Concerning Sugar and Sugarcane; WTO, DS579: India—Measures Concerning Sugar and Sugarcane; WTO, DS581: India—Measures Concerning Sugar and Sugarcane.

[155] USTR, "United States Issues WTO Counter Notification Concerning India's Market Price Support for Various Pulses," February 15, 2019.

Foreign Challenges to U.S. Farm Support[156]

The U.S. shift toward greater use of domestic trade laws and less reliance on the WTO to address concerns about other countries' trade policies could also produce unintended consequences as trading partners consider responding to a pattern of increasing U.S. farm support outlays over the past decade. For example, in lieu of using the WTO's dispute settlement process to have an independent panel resolve disputes, countries may choose to use trade remedy investigations performed by their national authorities to impose anti-dumping (AD) duties on products found to be sold below cost and countervailing duties (CVD) on imports found to be unfairly subsidized or otherwise traded unfairly.

Under the Article 13 of the 1995 WTO Agreement on Agriculture (AoA), a provision known as the Peace Clause kept members from taking action against domestic subsidies of WTO members who complied with their AoA commitments. Article 13's protection expired in January 2004, making countries with subsidies to their agricultural sectors vulnerable to AD or CVD actions by their trading partners. Since then, a number of challenges to U.S. imports have involved repeated or multiple investigations into the same products (examples include Mexican investigations into apples and the Peruvian investigation into corn).[157] Large trade aid payments to the U.S. farm sector in 2018 and 2019 have raised new questions from some WTO members, who may perceive these payments as providing an unfair advantage for the U.S. agricultural sector.

When a country initiates an AD or a CVD investigation of U.S. agricultural exports, the U.S. government and the affected industries may participate in the investigation by providing evidence, such as showing that

---

[156] Prepared by Anita Regmi, Analyst in Agricultural Policy, CRS. This section includes contributions from Nina Hart, Legislative Attorney, CRS.

[157] NAFTA, *Binational Panel Report, In the Matter of the Review of the Final Determination of the Anti-dumping Duty Investigation on Imports of Certain Red Delicious Apples and Golden Delicious Apples from the United States of America*, paras. 54-55, MEX-USA-2006-1904-2, October 15, 2009; and Technical Secretariat of the Supervision Commission of Dumping and Subsidies, *Informe No. 026-2018/CDB-INDECOPI*, July 12, 2018, para. 7 n.7 and para. 107, https://www.indecopi.gob.pe/documents/1902049/4099489/INFORME+026-2018.PDF.pdf/3df6d1c7-49d3-d75c-0a16-29e7aaa435bd.

any subsidies were permissible under WTO rules or that the imposition of duties is not justified. U.S. exporters may also challenge an AD or CVD ruling under free trade agreements, such as NAFTA or USMCA in the future. A third option is for the United States to bring a claim via the WTO dispute settlement process, alleging that the trading partner has violated the WTO Anti-Dumping Agreement or the Agreement on Subsidies and Countervailing Measures. However, the WTO Appellate Body, which hears appeals of cases from WTO dispute settlement panels, currently lacks a sufficient number of judges to issue rulings, because the United States has blocked the appointment of judges to replace those whose terms have expired. This means that the Appellate Body is unable to adjudicate disputes.

Peru currently imposes countervailing duties on U.S. ethanol imports. In May 2019, Colombia imposed preliminary duties on U.S. ethanol for a four-month period during a countervailing duty investigation.[158] In 2018, Peru initiated a similar investigation into U.S. corn, and China launched an investigation into U.S. sorghum, although neither case has resulted in countervailing duties to date.

Status

Over the years, trading partners have expanded the scope of U.S. programs that they considered to be "actionable"—that is, potentially subject to punitive duties.[159] In some cases, programs other than those that the United States reports to the WTO under its amber box commitments have been the subject of foreign government investigations. These have included direct payments to farmers, subsidies for biodiesel and ethanol, export credit guarantees, farm ownership and operating loans, and Market Access and Foreign Market Development Programs operated by the Foreign Agricultural Service.[160] In 2019, a European Parliament report suggested that perhaps the U.S. Environmental Quality Incentives Program

---

[158] FAS, "Colombia: Biofuels Annual," Attaché Report, July 23, 2019.

[159] Technical Secretariat of the Supervision Commission of Dumping and Subsidies, *Informe No. 002-202/CDB- INDECOPI*, Jan. 8, 2020; and FAS, "MOFCOM Drops AD and CVD Investigations of Imports of U.S. Sorghum," GAIN Report No. CH:18029, May 18, 2018.

[160] The latter two were included in the China CVD investigation into U.S. sorghum.

could be considered an unfair subsidy to the U.S. farm sector.[161] Given the WTO's limited ability to resolve disputes though legal procedures at present, the United States may have difficulty challenging duties levied on U.S. agricultural products by a country with which the United States does not have a trade agreement that includes dispute resolution provisions.

NON-TARIFF TRADE BARRIERS

Sanitary and Phytosanitary (SPS) and Other Non-Tariff Barriers[162]

SPS measures are laws, regulations, standards, and procedures that governments employ as "necessary to protect human, animal or plant life or health" from the risks associated with the spread of pests, diseases, or disease-carrying and causing organisms, or from additives, toxins, or contaminants in food, beverages, or feedstuffs. Examples include product standards, requirements that products be produced in disease-free areas, quarantine and inspection procedures, sampling and testing requirements, residue limits for pesticides and drugs in foods, and limits on food additives. Technical barriers to trade (TBTs) cover both food and non-food traded products. TBTs in agriculture include SPS measures, but also include other types of measures related to health and quality standards, testing, registration, and certification requirements, as well as packaging and labeling regulations. Both SPS and TBT measures regarding food safety and related public health protection are addressed in various multilateral trade agreements and are regularly notified to and debated within both the SPS Agreement and TBT Agreement within the WTO.[163] Under the agreements, countries are encouraged to observe established and

---

[161] European Parliament, "US Duties on Imports of Spanish Ripe Olives," March 2019, http://www.europarl.europa.eu/RegData/etudes/ATAG/2019/635558/EPRS_ATA(2019)635558_EN.pdf.

[162] Prepared by Renée Johnson, Specialist in Agricultural Policy, CRS.

[163] See CRS Report R43450, *Sanitary and Phytosanitary (SPS) and Related Non-Tariff Barriers to Agricultural Trade*.

recognized international standards, and avoid any improper use of SPS and TBT measures that might create barriers to trade that are not supported by science.

Examples of prominent U.S. trade concerns involving SPS and TBT issues include restrictions in some global markets on the use of agricultural biotechnology (see section "Agricultural Biotechnology"), EU prohibitions on the use of hormones in meat production (see "U.S.-EU Beef Hormone Dispute"), and the use of pathogen reduction treatments for poultry (see section "U.S.- EU Dispute Over Pathogen Reduction Treatments (PRTs)").

Bilateral and regional free trade agreements (FTAs) between the United States and other countries address SPS and TBT matters. Provisions in most U.S. FTAs have generally reaffirmed rights and obligations of both parties under the WTO SPS and TBT agreements. Some FTAs have resulted in the establishment of a standing bilateral committee to enhance understanding of each other's measures and to consult regularly on related matters. Some FTAs have included side letters or agreements for the parties to continue to cooperate on scientific and technical issues, which in some cases may be related to certain specific market access concerns. However, to date, most FTAs have not addressed specific non-tariff trade concerns directly.

In the early 2010s, as part of the lead up to negotiations with the EU and with Asia-Pacific countries, there were active efforts to "go beyond" the rules, rights, and obligations in the WTO SPS and TBT Agreements, as well as beyond commitments in existing U.S. FTAs. These efforts were often referred to as "WTO-Plus" rules, or alternatively, as "SPS-Plus" and "TBT-Plus" rules, and they were intended to address concerns that trade negotiations might not adequately address SPS concerns and cover "all significant barriers in a single comprehensive agreement."[164] Related issues involved the need to more effectively address enhanced regulatory cooperation and coherence between trading partners in an FTA.[165] Many in

---

[164] See, for example, letter to former U.S. Trade Representative Ron Kirk from several U.S. agriculture and food groups, March 4, 2013.

[165] See CRS Report R44564, *Agriculture and the Transatlantic Trade and Investment Partnership (T-TIP) Negotiations*. Regulatory cooperation generally refers to enhanced

Congress also continued to call for "effective rules and enforceable rules to strengthen the role of science" to resolve international trade differences in FTA negotiations.[166]

Status

Statements by USDA and EU officials in early 2020 signaled that issues involving SPS barriers and regulatory cooperation could become part of the U.S.-EU Trade Agreement negotiations.[167] Other statements by USDA officials further indicated that certain long-standing SPS disputes—including the EU's continued ban on the use of hormones and certain pathogen reduction treatments in meat production—might also be part of the negotiations.[168] These and other non-tariff barriers continue to be actively debated as part of the official U.S. trade agenda.[169] Among U.S. concerns involving the application of such measures in some countries is the perception that their use may not be based on accepted science or on international standards, and that they instead constitute disguised protectionist barriers to U.S. exports.

In recent developments, both USMCA and the U.S.-China Phase One trade agreement incorporated policy changes regarding SPS and TBT measures that go beyond the rules, rights, and obligations in the WTO. Those changes also go beyond commitments in existing U.S. trade agreements. Specifically, according to the U.S. International Trade Commission, USMCA "goes further [than previous agreements] in requiring transparency and encouraging harmonization or equivalence of

---

partnerships and interactions among regulators in each country, while regulatory coherence refers to the practices, transparency, and stakeholder engagement in the regulatory process.

[166] See, for example, letter to former U.S. Trade Representative Michael Froman from Members of the House Agriculture Committee and House Ways and Means Committee, August 7, 2013.

[167] S. Chase, "Perdue Eyes SPS, GI Barriers as Key Issues in Potential US-EU Deal," *Agri-Pulse*, January 29, 2020; and *World Trade Online*, "Hogan Hopes SPS Solutions Can Break EU-U.S. Ag Impasse," January 17, 2020.

[168] *World Trade Online*, "Perdue: EU Should Accept U.S. Chicken, Beef as Part of U.S.-EU Reset," January 28, 2020; and K. Good, "Sec. Perdue Highlights EU Ag Trade Issues," *Farm Policy News*, February 3, 2020.

[169] USTR, "2019 Trade Policy Agenda," March 2019, pp. 44-45; and USTR, *2019 National Trade Estimate Report on Foreign Trade Barriers* (which provides specific examples of SPS and TBT measures in selected countries.).

SPS measures" and incorporates all of the proposed enhanced TPP disciplines "in the areas of equivalence, science and risk analysis, transparency, and cooperative technical consultations."[170] Some industry representatives claim USMCA "goes beyond TPP in establishing deadlines for 'import checks,' by requiring importing parties to inform exporters or importers within five days of shipments being denied entry."[171] The final U.S.-China Phase One trade agreement also requires both parties to "engage each other cooperatively" on agriculture-related technical and SPS measures, including "risk communication."[172] It further requires that China implement a phytosanitary protocol to allow the importation of U.S. agricultural crops, and establish various protocols and certificate requirements. Both of these U.S. FTAs are notable in that they specifically address agricultural biotechnology in the agreement.[173]

Ongoing Trade Issues Involving SPS Measures[174]

Outside of the FTA negotiation process, various U.S. federal agencies regularly address trade concerns involving SPS and TBT measures as part of their day-to-day oversight and regulatory responsibilities. For example, the United States maintains ongoing interagency processes and mechanisms to identify, review, analyze, and address foreign government standards-related measures that may function as barriers to trade. These activities are coordinated through the USTR-led Trade Policy Staff Committee, which comprises representatives from several federal agencies, including USDA, the Department of Commerce (DOC), and the State Department. USTR also chairs an interagency group (i.e., both USDA and

---

[170] USITC, "U.S.-Mexico-Canada Trade Agreement: Likely Impact on the U.S. Economy and on Specific Industry Sectors," Publication# 4889, Investigation Number: TPA 105-003, April 2019, pp. 132-133.

[171] USITC, "U.S.-Mexico-Canada Trade Agreement," p. 133.

[172] Economic and Trade Agreement Between the Government of the United States of America and the Government of the People's Republic of China, Chapter 3 (Trade in Food and Agricultural Products).

[173] For more on this, see section on "Agricultural Biotechnology."

[174] Prepared by Renée Johnson, Specialist in Agricultural Policy, CRS.

non-USDA agencies with SPS and TBT responsibilities) that meets weekly to review SPS and TBT measures involving globally traded goods that are notified to the WTO, as required under the SPS and TBT agreements. These agency officials also work with their international counterparts on an ongoing basis on various trade concerns involving SPS and TBT measures.[175] USTR tracks issues related to SPS and TBT measures as part of a series of ongoing annual reports.[176] In addition, USDA's Animal and Plant Health Inspection Service (APHIS) administers various regulatory and control programs pertaining to animal and plant health and quarantine, humane treatment of animals, and the control and eradication of pests and diseases. APHIS also oversees SPS certification requirements for imported and exported agricultural goods.[177] This work is ongoing.[178]

Status

While specific SPS and TBT issues regarding individual agricultural commodities generally fall outside most formal FTA negotiations, statements by USDA officials in early 2020 have signaled that certain issues that arise from normal day-to-day operations within the Executive Branch could become part of the U.S.-EU trade agreement negotiations. Press reports indicate that such issues could include EU concerns involving phytosanitary certificates for U.S. imports of apples and pears from some EU countries as well as post-arrival requirements for U.S. imports of sheep

---

[175] Both the SPS and TBT agreements encourage the international harmonization of food standards, recognizing three international standard-setting organizations including the Codex Alimentarius Commission, the World Organisation for Animal Health, and the International Plant Protection Convention. The SPS Agreement recognizes each of these organizations as the primary "relevant" (or reference) organizations for developing international standards, guidelines, and recommendations on animal health, food safety, and plant health.

[176] USTR reports include Trade Policy Agenda Annual Report of the President of the United States on the Trade Agreements Program; the annual *National Trade Estimate Report on Foreign Trade Barriers* (or NTE report) and periodic SPS and TBT reports (last published covering 2014).

[177] For more background, see CRS Report R45267, *Animal and Plant Export Health Certificates in U.S. Agricultural Trade* and CRS Report R45457, *Animal and Plant Health Import Permits in U.S. Agricultural Trade*.

[178] See USDA's website at https://www.federalregister.gov/agencies/animal-and-plant-health-inspection-service.

and goat semen from the EU.[179] U.S. concerns include the EU's restrictions on the use of agricultural chemicals and biotechnology, animal cloning, pesticide maximum residues limits, and import requirements for live cattle and animal byproducts.[180]

Agricultural Biotechnology[181]

Agricultural biotechnology refers primarily to the commercial development of plants and animals through recombinant DNA techniques to provide certain desired characteristics, primarily herbicide tolerance and pest resistance. More recently, the term has come to encompass a range of new technologies that manipulate genetic material through targeted *in vivo* or *in vitro* techniques, popularly referred to as genomic "editing" (e.g., CRISPR-Cas9) rather than just recombinant DNA techniques. U.S. soybean, corn, cotton, and sugar beet producers have rapidly adopted genetically engineered (GE) varieties of these crops since commercialization began in the mid-1990s. Globally, the United States leads in cultivating GE crops, accounting for nearly 40% of total acres growing GE crops worldwide.[182]

Elsewhere in the world, the adoption and cultivation of GE crops by both producers and consumers are mixed.[183] Argentina and Brazil, for example, are major cultivators and exporters of GE corn and soybeans. India is a major cultivator of GE cotton. EU policy is more complicated. Through labeling requirements, strict traceability rules for imported food and commodities, and comparatively strong democratic pressures from the

---

[179] *World Trade Online*, "Perdue: US Could Address Apples, Goats, Other Irritants in Deal with EU," January 28, 2020.

[180] For example, based on issues highlighted in USTR's "2019 Trade Policy Agenda," March 2019, and USTR's *2019 National Trade Estimate Report on Foreign Trade Barriers.*

[181] Prepared by Tadlock Cowan, Analyst in Natural Resources and Rural Development, and Genevieve Croft, Analyst in Agriculture, CRS.

[182] International Service for the Acquisition of Agri-Biotech Applications, ISAAA Brief 54-2018. http://www.isaaa.org/resources/publications/briefs/54/executivesummary/default.asp.

[183] For a review of restrictions on GE organisms in other countries, see Law Library of Congress, Global Legal Research Center, *Restrictions on Genetically Modified Organisms*. March 2014.

public at local levels, the EU has made cultivation and sale of GE foods and crops very difficult. Moreover, while the European Commission (EC) has approved varieties of GE commodities for import and marketing, individual member states may maintain bans. This opposition in the EU has also been a factor in opposition to GE crops in less developed countries. Many African countries have largely followed the EU in restricting or banning the commercial cultivation of GE crops, confining cultivation mostly to field trials and greenhouse containment.

In March 2018, the U.S. Secretary of Agriculture stated that the United States will not regulate plants created through genomic editing as long as they are developed without using a plant pest as the donor or vector, and are not plant pests themselves.[184] In contrast, the European Court of Justice ruled in July 2018 that organisms obtained by mutagenesis are genetically modified organisms (GMOs) and are, in principle, within the scope of the GMO Directive, which governs the deliberate release of GMOs into the environment. The European Court considers the risks posed by new mutagenic techniques such as gene editing (CRISPR-Cas9), to be similar to crops created from transgenesis, where GE crops have genetic material from other, unrelated organisms introduced into the host plant.

China's reluctance to approve GE crops or GE imports remains a source of frustration for U.S. agricultural interests. Nonetheless, U.S.-developed GE varieties appear to be grown in China despite Chinese laws banning their cultivation. In September 2016, China agreed to improve its agricultural biotechnology approval process and, in January 2019, it announced approval of five new GE traits in imported crops for processing, the first new approvals since June 2017. At the same time, the ministry amended regulations on safety assessment, import approval, and labeling of agricultural GMOs without notifying the changes to the WTO, nor soliciting comments from stakeholders. In the U.S.-China Phase One trade agreement, China agreed to establish a predictable and risk-based regulatory regime with respect to its safety evaluation of agricultural

---

[184] USDA Press Release No. 0070.18. March 28, 2018.

biotechnology.[185] With respect to GE products for animal feed or further processing, China also agreed to reduce the time between submission of applications for authorization and a final decision to approve or disapprove.

For the first time in an FTA, the USMCA specifically includes provisions to improve transparency and coordination in approving and bringing to market products of agricultural biotechnology. USMCA provisions will cover crops produced with all biotechnology methods, including recombinant DNA and gene editing.[186]

Trade negotiations concerning agricultural biotechnology also involve labeling issues and other provisions that address the unintended presence of unapproved GE products in food and commodity imports. In 2016, Congress enacted P.L. 114-216, comprehensive legislation to govern the mandatory labeling of *bioengineered* foods, a term defined in the act and similar to the terms *GE foods* and *GMOs*. USDA's Agricultural Marketing Service established the National Bioengineered Food Disclosure Standard to regulate the mandatory disclosure of bioengineered foods and food ingredients to consumers. Food manufacturers, retailers, and importers are responsible for making disclosures. Importers are responsible for ensuring that all imported bioengineered foods comply with the new regulation.[187]

Implementation of the labeling standard began on January 1, 2020, and compliance is voluntary until January 1, 2022, when it becomes mandatory. The labeling standard does not require refined products derived from bioengineered crops (e.g., refined soy oil, high-fructose corn syrup) to be labeled if the modified genetic material is not detectable in the food

---

[185] Economic and Trade Agreement Between the Government of the United States of America and the Government of the People's Republic of China, Chapter 3. https://ustr.gov/countries-regions/china-mongolia-taiwan/peoples-republic-china/phase-one-trade-agreement/text.

[186] USMCA, Chapter 3, Section B. https://ustr.gov/trade-agreements/free-trade-agreements/united-states-mexico- canada-agreement/agreement-between.

[187] For more information, see CRS Report R46183, *The National Bioengineered Food Disclosure Standard: Overview and Select Considerations*.

product. The Agricultural Marketing Service stated that it does not expect the new regulation to disrupt foreign trade.[188]

Status

A key objective of U.S. trade negotiations has been to establish a common framework for GE approvals and adoption. This includes labeling practices consistent with the U.S. guidelines and harmonized regulatory procedures concerning GE presence in products that are consistent with the Codex Alimentarius Commission Annex on Food Safety Assessment in Situations of Low-Level Presence of Recombinant-DNA Plant Material in Food. This general policy was reiterated through publication of the June 2019 *Executive Order on Modernizing the Regulatory Framework for Agricultural Products*.[189] For the first time in an FTA, the USMCA specifically includes provisions to improve transparency in approving and bringing to market products of agricultural biotechnology. The Phase One trade agreement with China has resulted in China's agreement to establish a predictable and risk-based regulatory regime regarding its safety evaluation of agricultural biotechnology.

Geographical Indications (GIs)[190]

GIs are geographical names that act to protect the quality and reputation of a distinctive product originating in a certain region. The term GI is most often applied to wines, spirits, and agricultural products. Some food producers benefit from the use of GIs because their products gain recognition for their distinctiveness, thereby differentiating them in the marketplace. In this manner, GIs can be commercially valuable. GIs may

---

[188] AMS, "BE Frequently Asked Questions—General," at https://www.ams.usda.gov/rules-regulations/be/faq/general. See also 7 U.S.C. §1639c(a), a provision in the act that states, "This subchapter shall be applied in a manner consistent with United States obligations under international agreements."

[189] E.O. 13874. June 2019. See https://www.whitehouse.gov/presidential-actions/executive-order-modernizing- regulatory-framework-agricultural-biotechnology-products/.

[190] Prepared by Renée Johnson, Specialist in Agricultural Policy, CRS.

also be eligible for relief from acts of infringement or unfair competition. While the use of GIs may protect consumers from deceptive or misleading labels, they also have the potential to impair trade when the use of names that are considered common or generic in one market are protected in another. Examples of registered or established GIs include Parmigiano Reggiano cheese and Prosciutto di Parma ham from the Parma region of Italy, Roquefort cheese from France, Champagne from the region of the same name in France, Irish whiskey, Darjeeling tea, Florida oranges, Idaho potatoes, Vidalia onions, Washington State apples, and Napa Valley wines.[191]

GIs are protected by the WTO Agreement on Trade-Related Aspects of Intellectual Property Rights (TRIPS), which obligates WTO members to recognize and protect GIs as intellectual property. The United States is a signatory of TRIPS and is subject to its rights and obligations. Accordingly, under TRIPS, the United States and EU have committed to providing a minimum standard of protection for GIs (i.e., protecting GI products to avoid misleading the public and prevent unfair competition) and an "enhanced level of protection" to wines and spirits that carry a GI, subject to certain exceptions. However, the United States considers some EU GIs to be generic or semi-generic terms. For example, in the United States, *feta* is considered the generic name for a type of cheese; however, it is protected as a GI in Europe. As such, cheese produced in the United States may not be exported for sale as feta cheese in the EU, since only feta produced in countries or regions currently holding GI registrations may be sold there commercially.

Laws and regulations governing GIs differ markedly between the United States and EU, which further complicates this issue. More than 3,300 product names registered and protected in the EU for foods, wine, and spirits originating in both EU member states and other countries. In addition, registered products often fall under GI protections in certain third-country markets, and some EU GIs have been trademarked in some

---

[191] For more information, see CRS Report R44556, *Geographical Indications (GIs) in U.S. Food and Agricultural Trade*; and CRS Report R43658, *The U.S. Wine Industry and Selected Trade Issues with the European Union*.

non-EU countries pursuant to those countries' trade agreements with the EU.[192] For example, Canada has agreed to recognize a list of 143 EU GIs in Canada,[193] and Japan has agreed to recognize more than 200 EU GIs in Japan.[194] These GI protections could limit U.S. sales of certain products to these countries.

The EU is in the process of negotiating trade agreements with several other U.S. trading partners, including Mexico, Australia, New Zealand, and the Mercosur states (Argentina, Brazil, Paraguay, and Uruguay). Each of these efforts include a selected list of GIs that would become protected under the proposed trade agreement.[195] In December 2019, the EU also entered into an agreement with China regarding GIs that would protect a reported 100 EU GIs in China.[196]

Some Members of Congress, particularly those with dairy constituencies, have claimed that EU protections for GIs are being misused to create market and trade barriers.[197] Much of this debate is focused on expanding restrictions on the use of certain terms used by cheesemakers, such as "parmesan," "asiago," and "feta," which are generally regarded as generic names in the United States.[198] Some U.S. industry groups, however, are trying to institute GI protections to promote distinctive American agricultural products. For example, the American Origin Products Association, which represents certain U.S. potato, maple syrup, ginseng, coffee, and chili pepper producers and certain U.S. winemakers, seeks to

---

192 A listing of EU FTAs is at https://ec.europa.eu/trade/policy/countries-and-regions/negotiations-and-agreements/.

193 USDA, "U.S.-Canada Agricultural Trade Implications of Canada-EU CETA," *GAIN* Report CA17004, March 3, 2017.

194 EU Business in Japan, "EPA & Geographical Indications," EU-Japan Center for Industrial Cooperation, accessed September 2019, https://www.eubusinessinjapan.eu/issues/economic-partnership-agreement/epa-geographical- indications. See also EU-Japan Economic Partnership Agreement, Annex 14-B, List of Geographical Indications.

195 Based on agreement text for Mexico, Singapore, and Vietnam, and also other released FTA language for Australia.

196 European Commission, "EU-China Geographical Indications Agreement—Factsheet," December 13, 2019.

197 See, for example, comments during a House Committee on Ways and Means trade policy hearing, January 27, 2015, and testimony during a Senate Finance Committee trade policy hearing, January 27, 2015.

198 Informa Agribusiness Consulting, "Assessing the Potential Impact of Geographical Indications for Common Cheeses on the U.S. Dairy Sector," February 2019.

work with federal authorities to "create of a list of qualified U.S. distinctive product names, which correspond to the GI definition."[199]

Status

Statements by USDA officials in early 2020 have signaled that concerns about GIs could resurface as part of the U.S.-EU trade talks.[200] In addition, both USMCA and the U.S.-China Phase One trade agreement address GIs in ways that could further complicate future U.S.-EU discussions. Specifically, USMCA includes language regarding the transparency of GI applications, approvals, and cancellations, along with guidelines for determining whether a term is customary in common use.[201] USMCA also includes a side letter between the United States and Mexico regarding more than 30 cheese terms.[202] These provisions may prove to be incompatible with GI provisions that are likely to be part of a trade agreement between the EU and Mexico, as well as existing provisions in the EU-Canada Comprehensive Economic and Trade Agreement. The U.S.-China Phase One trade agreement requires China to "not undermine market access for U.S. exports to China of goods," and provides the United States with "necessary opportunities to raise disagreement" regarding GIs, among other provisions.[203] These provisions may also prove to be incompatible with provisions agreed to in the 2019 EU-China agreement which protect certain EU GIs in China.[204]

---

[199] American Origin Products Association, "AOPA Policy Agenda," http://www.aop-us.org/aopa-policy-agenda.html.

[200] S. Chase, "Perdue Eyes SPS, GI Barriers as Key Issues in Potential US-EU Deal," *Agri-Pulse*, January 29, 2020.

[201] USITC, "U.S.-Mexico-Canada Trade Agreement: Likely Impact on the U.S. Economy and on Specific Industry Sectors," Publication# 4889, Investigation Number: TPA 105-003, April 2019, pp. 118.

[202] USMCA side letter from U.S. Trade Representative Robert E. Lighthizer to Mexico's Secretary of the Economy, Ildefonso Guajardo Villarreal, November 30, 2018. Another side letter is an agreement to protect certain distilled spirits.

[203] Economic and Trade Agreement Between the Government of the United States of America and the Government of the People's Republic of China, Chapter 1 (Intellectual Property), Section F (Geographical indications), pp. 1-9, 1-10.

[204] European Commission, "EU-China Geographical Indications Agreement—Factsheet," December 13, 2019.

U.S.-EU Beef Hormone Dispute[205]

The United States and the EU have engaged in a long-standing trade dispute over the EU's ban on hormone-treated meat. The EU adopted restrictions on livestock production in the early 1980s, limiting the use of natural hormones to therapeutic purposes, banning the use of synthetic hormones, and prohibiting imports of animals and meat from animals that have been administered the hormones. In response, the United States, which maintains that beef produced using hormones is safe for consumers, suspended trade concessions with the EU in 1999 by imposing retaliatory tariffs of 100% *ad valorem*[206] on selected EU food products. Despite an ongoing series of WTO dispute settlement proceedings and decisions, the United States and the EU continue to disagree on a range of legal and procedural issues, as well as the scientific evidence and consensus affirming the safety of hormone-treated beef.[207]

In January 2009, USTR announced its intent to make changes to the list of EU products subject to increased tariffs under the dispute, including changes to the EU countries and products affected, with additional tariffs on some products. The EU claimed that this action constituted an "escalation" of the dispute. In May 2009, following a series of negotiations, the United States and the EU signed a memorandum implementing an agreement specifying actions intended to resolve this dispute over the next several years, and the United States suspended its retaliatory tariffs for imported EU products under the dispute.

As part of the 2009 memorandum, the EU agreed to expand market access to U.S. exports of beef raised without hormones as part of its High-Quality Beef (HQB) TRQ. The EU's HQB quota is set at 45,000 MT annually and assessed a tariff of 20%.[208] However, as the HQB quota is

---

205 Prepared by Renée Johnson, Specialist in Agricultural Policy, CRS.

206 *Ad valorem* tariffs refer to tariff rates charged as a percentage of the price or value of the traded product.

207 For more information on this issue, see CRS Report R40449, *The U.S.-EU Beef Hormone Dispute*. See also WTO, "DS26: EC—Measures Concerning Meat and Meat Products (Hormones)," https://www.wto.org/english/tratop_e/ dispu_e/cases_e/ds26_e.htm.

208 As part of the EU-Canada Comprehensive Economic and Trade Agreement, Canada is granted an additional access of 3,200 MT of hormone-free beef, which is duty-free (i.e., 0% tariff).

open to other beef-exporting nations, this has effectively limited the ability for U.S. beef producers to fully benefit under the quota. According to USTR and the U.S. beef industry, most of the HQB quota was being filled by countries other than the United States, and the EU has been unwilling to consider an allocation that would reserve a significant part of the HQB quota for the United States.[209]

In December 2016, USTR proposed reinstating retaliatory tariffs on EU products under the U.S.- EU beef hormone dispute, given the U.S. contention that the U.S.-specific allocation of the EU's HQB import quota for hormone-free beef had not expanded pursuant to the 2009 memorandum. In February 2017, USTR convened a hearing to review this possible retaliatory action.[210] In late 2018, the EU agreed to review its existing HQB quota and renegotiate its quota with the United States with the expectation that a revised HQB agreement would be implemented in early 2019.[211] The United States ultimately did impose retaliatory tariffs in connection with the dispute.

Status

The U.S. and the EU reached an agreement in principle regarding U.S.-specific allocation of the EU's HQB import quota for hormone-free beef in June 2019. The agreement provides that the United States would be allocated 35,000 MT of the 45,000 HQB quota (about 78%), phased- in over a seven year period.[212] Starting January 1, 2020, the phased-in quota allocations are as follows: 18,500 MT (2020), 23,000 MT (2021), 25,400 MT (2022), 27,800 (2023), 30,200 MT (2024), 32,600 MT (2025), 35,000 (2026 and subsequent years).[213] During this time, the remaining amount of the quota each year would be available to other exporting countries.

---

[209] See, for example, National Cattlemen's Beef Association, "NCBA Applauds USTR for Defending U.S. Beef from European [sic]," December 22, 2016.

[210] 81 *Federal Register* 95724, December 28, 2016 (Docket# USTR-2016-0025-0001).

[211] USDA, "EU-28: 2018-19 HQB Q2 Fill Maintains Record Pace as U.S. and EU Renegotiate," *GAIN* Report E18064, November 12, 2018; and EC, "European Commission Recommends Settling Longstanding WTO Dispute," September 3, 2018.

[212] EC, "The European Union and the United States Reach an Agreement on Imports of Hormone-Free Beef," press release, June 14, 2019.

[213] USDA, "New High Quality Beef TRQ Allocation Starting January 1 of 2020," GAIN Report E42019-0049, December 23, 2019.

Current substantial users of EU's HQB quota—Australia, Argentina, and Uruguay—all had to agree to the reallocation in order for the agreement to be compliant with WTO rules.[214]

The EU continues to impose bans and restrictions on meat produced using hormones, beta agonists, and other growth promotants, and it allows only imports of beef produced without hormones subject to the EU's HQB quota. The EU's restrictions involving meat production continues to be actively debated as part of the official U.S. trade agenda, as these types of practices are common in U.S. meat production.[215] Statements by USDA officials in early 2020 have signaled that this issue could resurface as part of the U.S.-EU trade agreement negotiations.[216]

U.S.-EU Dispute Over Pathogen Reduction Treatments (PRTs)[217]

In January 2009, the United States escalated a long-running dispute with the EU over its refusal to accept imports of U.S. poultry that are subject to certain pathogen reduction treatments (PRTs). PRTs are antimicrobial rinses that have been approved for use by the USDA in poultry production to reduce the amount of microbes on meat. Meat and poultry products processed with PRTs are judged safe by the United States and also by European food safety authorities. However, the EU prohibits the use of PRTs and the importation of poultry treated with these substances. The EU generally opposes such chemical interventions and asserts that its own poultry producers follow much stricter production and processing rules that are more effective in reducing microbiological contamination than simply washing poultry products. In general, EU

---

214 H. von der Burchard, "Australia, Argentina, and Uruguay Accept EU-U.S. Beef Deal," *Politico*, June 14, 2019.

215 USTR, "2019 Trade Policy Agenda," March 2019, pp. 48 and 87. USTR's *2019 National Trade Estimate Report on Foreign Trade Barriers*, pp. 184-185.

216 *World Trade Online*, "Perdue: EU Should Accept U.S. Chicken, Beef as Part of U.S.-EU Reset," January 28, 2020.

217 Prepared by Renée Johnson, Specialist in Agricultural Policy, CRS.

consumer groups argue that the use of such treatments compensates for poor hygiene in the supply chain.[218] The United States requested WTO consultations with the EU on the matter, a prerequisite first step toward the establishment of a formal WTO dispute settlement panel. A WTO panel was subsequently established in November 2009, but this case has not moved forward.[219]

In 2013, USDA submitted an application for the approval of peroxyacetic acid as a PRT for poultry. Although the EU initially put forward a proposal to authorize the PRT, it withdrew its proposal in December 2015, citing the European Food Safety Authority's (EFSA) opinion of insufficient evidence of peroxyacetic acid's efficacy against campylobacter.[220]

EFSA cleared lactic acid for reducing pathogens on beef carcasses, cuts, and trimmings in 2011.[221] In 2013, the EU lifted its ban on the use of lactic acid in beef PRTs on beef carcasses, half-carcasses, and beef quarters in the slaughterhouse.[222] In 2017, the National Pork Producers Council submitted an application to EFSA to approve organic lactic and acetic acid for use on pork carcasses and cuts. EFSA's panel report, issued in October 2018, concluded that use of the treatments does not pose a safety concern provided that the substances comply with EU specifications for food additives and that their use is efficacious compared to untreated meat.[223]

---

[218] See, for example, European Consumer Organization, "Peroxyacetic Acid Rinses on Poultry Meat: The Consumer Perspective," BEUC Position Paper BEUC-X-2014-052, July 7, 2019.

[219] For more information on this issue, see CRS Report R40199, *U.S.-EU Poultry Dispute on the Use of Pathogen Reduction Treatments (PRTs)*. See also WTO, "DS389: EC—Certain Measures Affecting Poultry Meat and Poultry Meat Products from the United States," https://www.wto.org/english/tratop_e/dispu_e/cases_e/ds389_e.htm.

[220] USTR, "2018 National Trade Estimate Report," p. 168.

[221] EFSA, "Scientific Opinion on the Evaluation of the Safety and Efficacy of Lactic Acid for the Removal of Microbial Surface Contamination of Beef Carcasses, Cuts and Trimmings," *EFSA Journal* 2011:9 (7):2317, July 26, 2011.

[222] Commission Regulation (EU) No 101/2013 of 4 February 2013 concerning the use of lactic acid to reduce microbiological surface contamination on bovine carcases (sic). See also EU Food Law, "EU to Allow Lactic Acid Opening Door to US Beef Imports," November 30, 2012.

[223] EFSA, "Evaluation of the Safety and Efficacy of the Organic Acids Lactic and Acetic Acids to Reduce Microbiological Surface Contamination on Pork Carcasses and Pork Cuts," *EFSA Journal*, October 25, 2018.

However, EFSA raised questions about whether lactic and acetic acid were more efficacious than water treatment for certain applications.

Status

The United States continues to maintain that PRTs are a "critical tool during meat processing that helps further the safety of products being placed on the market" and continues to seek EU approval of certain PRTs for beef, pork, and poultry.[224] To date, the United States and the EU have not been able to agree on a number of issues related to veterinary equivalency, and the EU continues to prohibit any substance other than water to remove contamination from animal products unless the EU approves the substance. Statements by USDA officials in early 2020 have signaled that this issue could resurface as part of the U.S.-EU trade agreement negotiations.[225]

Trade Restrictions on Ractopamine Use[226]

Ractopamine, an animal drug that increases animal weight gain and meat yield, is approved by the U.S. Food and Drug Administration (FDA) for use in U.S. cattle, hog, and turkey production. It is also approved for use in countries such as Canada, Japan, Mexico, and South Korea, but many other countries ban the use of ractopamine in meat production. In 2012, the Codex Alimentarius—the international food standards organization that sets guidelines to protect public health and ensure fair practices in the food trade—set maximum residue levels for ractopamine in beef and pork. However, several of the largest markets for U.S. meat

---

[224] USTR, *2019 National Trade Estimate Report*, p. 188. See also testimony by William Roenigk, National Chicken Council, at a Senate Finance Committee hearing on T-TIP, October 30, 2013; and C. Perkins, "U.S. Poultry Industry Raises Concerns About TTIP," *Global Meat*, June 4, 2013.

[225] *World Trade Online*, "Perdue: EU Should Accept U.S. Chicken, Beef as Part of U.S.-EU Reset," January 28, 2020.

[226] Prepared by Joel Greene, Analyst in Agricultural Policy, CRS.

exports have restricted imports of meat produced with ractopamine, despite U.S. adherence to the residue standards established by Codex.

USTR, in its "2019 National Trade Estimate Report on Foreign Trade Barriers," states that the EU, China, Taiwan, and Thailand continue to restrict U.S. meat exports produced with ractopamine.[227] According to USDA's Food Safety and Inspection Service, U.S. meat exports—particularly pork—may be shipped to markets with ractopamine restrictions if the exported product is raised without ractopamine and is certified through USDA's Never Fed Beta Agonists Program.[228] U.S. exports to markets that have ractopamine restrictions are subject to increased certification and testing costs, potentially affecting competitiveness and dampening market opportunities.

Status

USDA and USTR continue to encourage trading partners to accept international standards on the use of ractopamine. Under the U.S.-China Phase One trade agreement, China agreed to consult with U.S. experts and conduct a risk assessment of ractopamine that is consistent with Codex standards. The assessment is to be based on conditions and use in the United States. The countries are to set up a working group to discuss steps to follow based on a risk assessment of ractopamine.[229] The United States exported 250% more pork to China in 2019 than 2018 largely because of China's African Swine Fever outbreak. An agreement on a ractopamine maximum residue limit (MRL) should facilitate more U.S. pork shipments to China going forward.

---

[227] USTR, "2019 National Trade Estimate Report on Foreign Trade Barriers," March 29, 2019, https://ustr.gov/about- us/policy-offices/press-office/reports-and-publications/2019/2019-national-trade-estimate.

[228] USDA, "Never Fed Beta Agonists Program," https://www.ams.usda.gov/services/imports-exports/beta-agonists.

[229] U.S.-China trade agreement, Chapter 3, Annex 7.5, January 15, 2020.

SELECTED TRADE ISSUES INVOLVING SPECIALTY CROPS

The United States has gone from being a net exporter of fresh and processed fruits and vegetables in the early 1970s to being a net importer of fruits and vegetables today. Although U.S. fruit and vegetable exports totaled $9.2 billion in 2018, U.S. imports of fruits and vegetables were $24.8 billion, resulting in a gap between imports and exports of $15.6 billion (excludes nuts). Several factors have contributed to this trade imbalance including a relatively open import regime and lower average tariffs in the United States, increased competition from low-cost or government- subsidized producing countries, and non-tariff trade barriers to U.S. exports in some countries. Additionally, other market factors, such as exchange rate fluctuations and structural changes in the U.S. food industry, as well as increased U.S. overseas investment and diversification in market sourcing by U.S. companies, have contributed to the trade imbalance. Increased domestic and year-round demand for fruits and vegetables as well as opportunities for counter-seasonal supplies through imports have also contributed to this trade situation.[230] Despite U.S. efforts to address some of these issues as part of recent FTA discussion, a number of these issues are unresolved. Other U.S. concerns include import competition regarding seasonal produce from Mexico, long-standing suspensions agreements between the U.S. and Mexico involving fresh tomatoes, and regulatory requirements regarding retail wine sales in Canada.

Import Competition of Seasonal Produce from Mexico[231]

Mexico remains the largest foreign supplier of U.S. imports of vegetables and fruits (excluding bananas). Production of some Mexican fruits and vegetables—tomatoes, peppers, cucumbers, berries, and

---

[230] For more information, see CRS Report RL34468, *The U.S. Trade Situation for Fruit and Vegetable Products*.

[231] Prepared by Renée Johnson, Specialist in Agricultural Policy, CRS.

melons—has increased in recent years in part due to Mexico's investment in large- scale greenhouse production facilities and other types of technological innovations. Reportedly, protected (greenhouse/shade) production in Mexico has risen to nearly 101,000 acres in 2016, up from about 19,500 acres in 2000.[232] According to researchers, Mexican growers benefit from a combination of relatively lower labor costs and subsidies invested in the specialty crop sector under various government programs, including Mexico's Agriculture Promotion Program and its AgriFood Productivity and Competiveness Program.[233] These programs are generally focused on increasing the infrastructure capacity of Mexico's agricultural sector. The Florida Fruit and Vegetable Association (FFVA) claims that Mexico's produce industry benefits from subsidies paid by the Mexican government and that it prices its products below fair market value, and therefore should be subject to both AD duties and CVD on U.S. imports of some fruits and vegetables.[234] Trade concerns by U.S. growers have primarily centered on imported tomatoes, peppers, and berries.

One of the Trump Administration's initial agriculture-related objectives in the renegotiation of NAFTA included a proposal to establish new rules for seasonal and perishable products, such as fruits and vegetables.[235] The proposal would have established a separate domestic industry provision for perishable and seasonal products in AD and CVD proceedings, making it easier for a group of regional producers to initiate an injury case and to prove injury, thereby resulting in CVD or AD duties on the imported products responsible for the injury. This could protect certain U.S. seasonal produce growers in some regions by making it easier to initiate trade remedy cases.[236] The U.S. International Trade Commission

---

[232] F. Wu et al., "Government Support in Mexican Agriculture," *CHOICES* magazine, 3rd Quarter (2018).

[233] E. Canales et al., "Mexico's Agricultural Sector: Production Potential and Implications for Trade," *CHOICES* magazine, 3rd Quarter 2019.

[234] FFVA, "Renegotiating NAFTA: Opportunities for Agriculture," statement at a House Agriculture Committee hearing, July 26, 2017; and comments from FFVA to USTR Robert E. Lighthizer, Docket No. 2017-0006, June 12, 2017.

[235] USTR, "Summary of Objectives for the NAFTA Renegotiation," November 2017.

[236] See CRS Report R45038, *Efforts to Address Seasonal Agricultural Import Competition in the NAFTA Renegotiation*. Information on the renegotiation is in CRS In Focus IF10682, *NAFTA Renegotiation: Issues for U.S. Agriculture*.

(USITC) has previously reviewed trade remedy cases involving perishable agricultural products—namely, Fall-harvested Round White Potatoes from Canada and Spring Table Grapes from Chile—that proved difficult to settle. As noted by USTR, current trade laws "are really not set up for seasonal product," making it difficult to prove injury over a period of time.[237]

Support for seasonal produce protections through changes to U.S. trade laws is mixed. Some Members of Congress supported including seasonal protections as part of NAFTA's renegotiation.[238] Others opposed including such protections, contending that seasonal production complements rather than competes with U.S. growing seasons.[239] Others worried it could open the door to an "uncontrolled proliferation of regional, seasonal, perishable remedies against U.S. exports."[240] Most U.S. food and agricultural sectors, including some fruit and vegetable producer groups, opposed including seasonal protections as part of the renegotiation.[241] Some worried that efforts to push for seasonal protections would derail the renegotiation. Others claimed that such efforts would favor a few "politically-connected, wealthy agribusiness firms from Florida" at the expense of others in the U.S. produce industry[242] and at the expense of both consumers and growers in other fruit and vegetable producing states, such as California.[243] The Agricultural Technical Advisory Committee for Trade in Fruits and Vegetables (F&V ATAC) supported not including seasonal provisions in

---

237 Comments by U.S. Trade Representative Robert E. Lighthizer to the House Ways and Means Committee trade hearing, June 19, 2019.

238 Letters from the Florida and Georgia congressional delegations to USTR, dated April 4, 2019, August 31, 2017, and September 1, 2017. See also press release from Senator Marco Rubio, "Rubio, Colleagues Raise Concerns to Lighthizer Regarding Lack of Effective Trade Enforcement for Seasonal Produce in USMCA," April 4, 2019.

239 See, for example, statements from Members of Congress at a House Agriculture Committee hearing, "Renegotiating NAFTA: Opportunities for Agriculture," July 26, 2017.

240 Letter from several Members of Congress to U.S. Trade Representative Robert E. Lighthizer, August 17, 2017.

241 Letters from U.S. agricultural, including produce industry, groups, to USTR, USDA, DOC, and National Economic Council, August 30, 2017, and August 31, 2017.

242 Fresh Produce Association of the Americas, "To Favor a Few Agribusiness, U.S. NAFTA Objective Would Hurt All Consumers," August 9, 2017.

243 G. C. Hufbauer and E. Jung, "NAFTA Mischief in Fruits and Vegetables," Peterson Institute for International Economics, July 26, 2017.

the NAFTA renegotiation.[244] In January 2018, F&V ATAC passed a resolution supporting the withdrawal of the seasonal and perishable trade remedy proposal from the U.S. negotiating objectives.[245]

Changes to USMCA released in October 2018 did not alter U.S. trade remedy laws to address seasonal produce trade. USTR claimed it tried to include such provisions but was unable to do so.[246] In response, the Agricultural Trade Improvement Act of 2018 (S. 3510; H.R. 7015) was introduced in the House and the Senate. These bills were reintroduced in the 116th Congress but renamed as Defending Domestic Produce Production Act of 2019 (S. 16; H.R. 101).

Status

USMCA does not include changes to U.S. trade remedy laws to address seasonal produce trade. Although lawmakers from Florida and Georgia continued to push USTR for seasonal produce provisions in USMCA, others in Congress continued to oppose such changes.[247] In January 2020, USTR announced that it planned to investigate trade practices by Mexico's produce industry, hold field hearings in Florida and Georgia, and engage the help of U.S. International Trade Commission (USITC) and DOC to monitor imports, among other actions.[248] One Member of Congress claimed USTR's plan would "sidestep the issue and install policies" that could result in future trade conflicts;[249] another encouraged USTR to "consider data from a variety of sources" when

---

[244] F&V ATAC, letter to U.S. Trade Representative Robert E. Lighthizer reflecting consensus advisory opinion, September 27, 2018.

[245] See, for example, letter from Senators to U.S. Trade Representative Robert E. Lighthizer April 6, 2018, and letter from some Members of the U.S. Senate to U.S. Trade Representative Robert E. Lighthizer, August 27, 2018.

[246] Comments by U.S. Trade Representative Robert E. Lighthizer to the House Ways and Means Committee trade hearing, June 19, 2019.

[247] See, for example, a bipartisan letter to U.S. Trade Representative Robert E. Lighthizer from several House and Senate members representing Arizona, Texas, and California, June 14, 2019.

[248] See letters from U.S. Trade Representative Robert E. Lighthizer to Members of Congress from Florida and Georgia, January 9, 2020.

[249] Letter to U.S. Trade Representative Robert E. Lighthizer from Senator Martha McSally, January 17, 2020.

examining the issue.[250] Some in Congress have raised concerns about the possible negative impacts of imported fruits and vegetables on U.S. growers more broadly.[251] Legislation introduced in the 116th Congress (S. 564) would establish a task force to identify countervailable subsidies and dumping practices to counter perceived unfair trade practices involving imports within the U.S. produce market.[252]

U.S.-Mexico Tomato Suspension Agreements[253]

The U.S.-Mexico Tomato Suspension Agreement is an agreement between DOC and signatory producers/exporters[254] of fresh tomatoes grown in Mexico that suspends the U.S. AD investigation into whether Mexican fresh tomatoes were sold into the U.S. market at less than fair value.[255] Fresh tomatoes imported from Mexico have been governed by suspension agreements since 1996.[256] The first suspension agreement became effective in November 1996. The Mexican signatory growers and the United States entered into new agreements in 2002, 2008, and 2013. Under the 2013 agreement, the signatories agreed to suspend the AD investigation and monitor compliance with the agreement. The basis for the suspension agreement was a commitment by each signatory

---

250 Letter to U.S. Trade Representative Robert E. Lighthizer from Senator Kyrsten Sinema, January 16, 2020.

251 See, for example, letter to U.S. Trade Representative Robert E. Lighthizer from Representative Bill Huizenga, January 17, 2020.

252 See also Representative Gary Peters, "Peters Testifies Before International Trade Commission in Support of Michigan Cherry Growers, Tariffs Against Turkish Cherry Exporters," press release, December 3, 2019.

253 Prepared by Renée Johnson, Specialist in Agricultural Policy, CRS.

254 Reportedly, more than 600 Mexican growers and exporters signed the most recent agreement.

255 61 *Federal Register* 18377, April 25, 1996. In general, under a suspension agreement, the exporters and producers or the foreign government agree to modify their behavior in a manner that eliminates dumping or subsidization and injury. 19 C.F.R. Section 351.208 specifies procedures for suspending a trade remedy investigation.

256 The text of the agreement is at http://ia.ita.doc.gov/tomato/2013-agreement/2013-agreement.html. See also USDA, "Section 8e Regulations and the Tomato Suspension Agreement—FAQs," https://www.ams.usda.gov/sites/default/ files/media/ Tomato_Suspension_FAQs%5B1%5D_0.pdf. For more background, see USDA and DOC, "Tomato Suspension Agreement Explained," March 4, 2013.

producer/exporter to sell tomatoes at or above the stated reference price in order to eliminate the injurious effects of exports of fresh tomatoes to the United States. The agreement set different floor prices for Mexican fresh tomatoes during the summer and winter and specifies prices for open field/adapted-environment and controlled-environment production. These price floors covered all types of fresh or chilled tomatoes from Mexico. The agreement did not cover tomatoes that are for processing.

In early 2018, DOC initiated consultations with the Mexican tomato growers and exporters to negotiate possible revisions to the 2013 agreement. DOC also initiated its five-year sunset review of the suspended AD investigation and published the preliminary and final results of its analysis in late 2018. DOC's analysis indicated that dumping of fresh tomatoes was likely to occur/recur and calculated weighted-average dumping margins of up to 188%.[257] In November 2018, the Florida Tomato Exchange requested that the United States withdraw from the suspension agreement, eliminate the reference prices, and resume the related initial 1996 AD investigation.[258] They claim the pricing agreements failed to ensure that Mexico did not undercut U.S. growers, costing the Florida tomato industry $3.4 billion to $6.8 billion per year in lost sales.[259] Several Members of Congress expressed support for withdrawing from the agreement.[260] Among the groups that opposed withdrawal were the Fresh Produce Association of the Americas and other groups representing Mexican growers and exporters as well as businesses, various associations,

---

[257] 83 *Federal Register* 43642, August 27, 2018 (preliminary); 83 *Federal Register* 66680, December 27, 2018 (final). 19 C.F.R. Section 351.218 specifies procedures for conducting a sunset review. The weighted average dumping margin refers to "the percentage determined by dividing the aggregate dumping margins determined for a specific exporter or producer by the aggregate export prices and constructed export prices of such exporter or producer" (19 USCS §1677 35 (B)).

[258] T. Thompson, "Tomato Suspension Agreement Under Florida Assault," *Produce News*, November 26, 2018.

[259] E. Ferguson, "A Mexican Tomato Beef Could Lead to a Bigger Trade Battle," *Roll Call*, May 7, 2019.

[260] Letter from several Members of Congress to Wilbur Ross, Secretary of Commerce, February 1, 2019. See also Office of Senator Marco Rubio, "Rubio, Yoho Applaud U.S. Commerce Department Decision to Withdraw from Tomato Suspension Agreement with Mexico," press release, February 7, 2019.

and local and county governments.[261] These groups claim the U.S. lost sales because Mexico offers more variety of tomatoes that appeal to consumers and commercial users.

DOC initially announced its intention to withdraw from the agreement in February 2019 following its periodic review of the agreement, which concluded that Mexican fresh tomatoes have been sold into the U.S. market at less than fair value.[262] In May 2019, the United States terminated the 2013 agreement and announced it would resume collecting tariffs on chilled and fresh tomatoes from Mexico, and later set a preliminary dumping margin of 25.28%.[263] Mexican tomato grower filed a suit at the Court of international Trade requesting an injunction against the reimposed tariffs.[264] The Mexican government claimed that the new duties would cost its tomato industry more than $350 million annually.[265] USITC resumed its AD investigation of Mexican tomatoes, and concluded that U.S. growers are "threatened with material injury" from imports.[266]

Status

Between May and September 2019, the United States and Mexican tomato growers considered various proposals regarding a possible revised agreement. On September 19, 2019, DOC signed a new suspension agreement with Mexico's growers and exporters of fresh tomatoes. DOC and USITC suspended their respective AD investigations.[267] The new suspension agreement sets increased minimum prices for specialty and organic tomatoes at certain times of the year, and establishes new

---

[261] See, for example, letter from the Border Trade Alliance to Wilbur Ross, Secretary of Commerce, March 6, 2019. See also 84 *Federal Register* 7872, March 5, 2019.

[262] 84 *Federal Register* 7872, March 5, 2019. See also DOC, "U.S. Department of Commerce Announces Intent to Withdraw from Suspension Agreement on Fresh Tomatoes from Mexico," press release, February 6, 2019.

[263] DOC, "U.S. Department of Commerce Announces the Termination of the Suspension Agreement on Fresh Tomatoes from Mexico," press release, May 7, 2019.

[264] Confederación de Asociaciones Agrícolas del Estado de Sinaloa, A.C. v. United States, 2019 CIT 69, June 6, 2019.

[265] Mexico's Secretary of the Economy, "Comunicado de Prensa Sobre el Acuerdo de Suspensión del Tomate Mexicano en Estados Unidos," press release, Comicado No. 045, May 7, 2109.

[266] 84 *Federal Register* 67958, December 12, 2019.

[267] 84 *Federal Register* 49987, September 24, 2019, and 84 *Federal Register* 54639, September 24, 2019, respectively.

inspections requirements of tomato shipments crossing the border to prevent low- quality tomatoes from entering the United States where they might undercut domestic prices.[268]

More recently, there have been growing concerns that a virus (brown rugose) found in tomatoes imported from Mexico could be harmful to U.S.-grown tomatoes and peppers. Increased inspections have reportedly caused border delays of product shipments, and have led to complaints from Mexican officials that such detentions are "unjustified."[269] During the last two months of 2019, the United States reportedly returned 43 tomato shipments inspected at the U.S.- Mexico border.

Regulatory Requirements Regarding Retail Wine Sales in Canada[270]

In Canada, the authority to import and distribute alcohol rests with the provincial governments. Starting in 2015, British Columbia (BC) initiated a series of policies and regulations that provide BC wine exclusive access to retail channels and grocery store shelves, while imported wine may be sold in grocery stores only through a "store within a store"[271]—that is, a space that is physically separated from the main retail outlet with separate cash registers. In 2016, Quebec—the largest wine-importing province in Canada—enacted policies that would streamline provincial approval for Quebec wines. Most wine in Quebec is distributed through retail outlets owned by its provincial liquor authority, the Société des alcools du

[268] DOC, "U.S. Department of Commerce Finalizes Suspension Agreement on Fresh Tomatoes from Mexico," September 19, 2019; and *Produce Blue Book*, "Importers Prepare for Mandatory Tomato Inspections," January 17, 2020.

[269] N. Mahoney, "Mexican Tomato Exporters say Their Produce Is Virus Free," *FreightWaves*, January 2, 2020.

[270] Prepared by Renée Johnson, Specialist in Agricultural Policy, CRS.

[271] WTO, "Canada—Measures Governing the Sale of Wine in Grocery Stores (Second Complaint), Request for the Establishment of a Panel by the United States," WT/DS531/7, May 29, 2018. Regulations are available at http://www.bcvqa.ca/regulations/. See also M. LaCombe, "More Problems with Wine Regulation in Canada," *MJIL Online*, vol. 41 (2019).

Québec.[272] The rules allow Quebec small wine producers to bypass the provincial liquor board. Regulations are also in place in Ontario requiring that 50% of the wine on display at a grocery store meet certain requirements that some claim make it difficult for imported products to compete with like domestic products.[273] According to the U.S.-based Wine Institute, Canada is the leading export market for California wine—the leading wine producing state in the United States—accounting for $448 million in sales in 2018.[274]

In January 2017, the Obama Administration initiated trade enforcement action against Canada at the WTO regarding Canada's BC wine measures.[275] Subsequent actions by the Trump Administration, in September 2017, led to the United States requesting formal consultations with Canada regarding BC wine measures.[276] USTR states that "discriminatory regulations implemented by British Columbia are unfairly keeping U.S. wine off of grocery store shelves" and that the measures are inconsistent with Canada's commitments and obligations under the WTO.[277] The United States reiterated its concerns as part of a second complaint issued in this case in July 2018. Argentina, Australia, New Zealand, and the EU joined the consultation. The WTO case remains active.

---

272 Quebec's regulations are at http://legisquebec.gouv.qc.ca/fr/ShowDoc/cr/S-13,%20r.%206. See also USDA, "Canada: The Wine Market in the Province of Quebec," *GAIN* Report CA17013, April 12, 2017.

273 These include conditions related to the size of the winery producing the wine, where the grapes were grown, and whether the wine meets the definition of a *quality assurance wine*. Ontario's regulations are at https://www.ontario.ca/ laws/regulation/160232.

274 Wine Institute, "US Wine Exports Total $1.46 Billion in 2018," April 8, 2019.

275 USTR, "United States Challenges Canadian Trade Measures That Discriminate Against U.S. Wine," January 2017.

276 WTO, "Canada—Measures Governing the Sale of Wine in Grocery Stores (Second Complaint)," WT/DS531/7, May 29, 2018. For more information on Canada's wine regulations in selected provinces, see USTR, *2019 National Trade Estimate Report on Foreign Trade Barriers*, pp. 84-85.

277 USTR "United States Takes Action Against Canadian Trade Measures That Discriminate Against U.S. Wine," press release, May 25, 2018. The U.S. complaint cites Article III:4 of the 1994 General Agreement on Tariffs and Trade.

Status

The USMCA includes a side letter addressing U.S. concerns about Canada's BC wine measures. As outlined in the side letter, Canada would modify certain measures that provide preferential grocery store shelf space to wines produced within the province and "implement any changes no later than November 1, 2019."[278] At this time, it is unclear whether Canada has taken additional action to address U.S. concerns about the status of BC's regulations. The USMCA side letter does not address potential market barriers to U.S. wine in Quebec and Ontario. Canada's wine regulations in certain provinces continues to be a concern to some in Congress.[279]

ISSUES RELATED TO LIVESTOCK AND MEAT TRADE[280]

In 2019, exports of U.S. livestock and poultry products totaled $24.1 billion, and imports totaled $14.2 billion. Foreign demand for U.S. animals and products supports prices of domestic livestock and poultry producers, while imports supplement U.S. consumer demand for a variety of livestock and poultry products. Recent trade agreements with Canada and Mexico, China, and Japan will facilitate increased livestock and poultry product exports to these four markets, which accounted for 65% of the value of total U.S. exports of these products in 2019. The U.S.-Japan agreement lowers tariffs for U.S. beef and pork products, and adjusts beef and pork safeguards. These measures offer U.S. livestock producers benefits that competing exporters have enjoyed under the TPP-11, the successor to Trans-Pacific Partnership agreement—from which the Trump Administration withdrew the United States before its ratification. Under U.S.-China Phase One trade agreement, China agreed to abide by

[278] USMCA side letter from USTR Robert E. Lighthizer to Canada's Minister of Foreign Affairs, Chrystia Freeland, November 30, 2018.

[279] See, for example, Senator Schumer, "Schumer To USTR: Level the Playing Field and Uncork Potential for the Thousand Islands Seaway & St. Lawrence Wine Trails," press release, March 21, 2019.

[280] Prepared by Joel Greene, Analyst in Agricultural Policy, CRS.

international standards and guidelines for trade, while expanding market access for more meat products that the USDA Food Safety and Inspection Service regulates should ease the process for U.S. meat and poultry exporters.

Export Bans on U.S. Meat and Poultry

USDA forecasts that exports of meat and poultry products will represent about 17% of U.S. domestic production in 2020.[281] Periodically, foreign countries impose export bans on U.S. meat products in response to an outbreak of certain animal diseases. The bans are disruptive for livestock producers and meat exporters, are often inconsistent with internationally accepted protocols, and vary in terms of scope and duration. For example, bans were imposed on U.S. beef exports because of the discovery of bovine spongiform encephalopathy (BSE, or mad cow disease) in 2003. An outbreak of highly pathogenic avian influenza (HPAI) at the end of 2014 and early 2015 in U.S. turkey and egg-laying flocks triggered export bans on poultry products by more than 30 countries. The bans were imposed on all U.S. products even though the HPAI outbreaks were not in areas in close proximity to commercial broiler production.[282]

The World Organization for Animal Health (known as OIE) has established trade protocols when disease outbreaks occur in countries that export meat and poultry products.[283] According to OIE, in most cases total export bans are not recommended or needed when there is a BSE or HPAI discovery or outbreak in exporting countries. In 2013, the OIE determined that the United States is at "negligible risk"[284] for BSE, meaning that U.S. surveillance and safeguard systems are adequate. For HPAI, USDA, in

---

[281] USDA, "World Agricultural Supply and Demand Estimates," January 10, 2020, p. 32.

[282] For more information on this issue, see CRS Report R44114, *Update on the Highly-Pathogenic Avian Influenza Outbreak of 2014-2015*.

[283] OIE, "Terrestrial Animal Health Code (2018)," vol. II: Recommendations Applicable to OIE Listed Diseases and Other Diseases of Importance to International Trade, http://www.oie.int/en/international-standard-setting/terrestrial- code/access-online/.

[284] OIE, "Terrestrial Animal Health Code (2018)," Chapter 11.4 Bovine Spongiform Encephalopathy.

collaboration with states, has implemented increased flock biosecurity and has placed a system to rapidly contain and eradicate an outbreak of HPAI.

Over the years, while some foreign markets imposed total bans on U.S. beef exports following the 2003 BSE incident, other export markets for U.S. beef imposed specific conditions for imports of U.S. beef. For example, Japan and South Korea—two major importers of U.S. beef—required that imported U.S. beef be produced from cattle under 30 months of age. China did not lift its ban on U.S. beef exports until 2017 and included an under 30-month age restriction. Regarding poultry, some foreign markets imposed total bans on poultry exports during the HPAI outbreak, while other markets imposed export bans only from the regions affected by the outbreak, consistent with the recommended OIE regionalization protocol that allows for trade from regions that are disease free.[285] As the United States demonstrated that the outbreak was contained and then eliminated, most of these bans were lifted.

Status

China lifted the ban on U.S. beef in 2017 but continued to restrict imports of U.S. beef to cattle under 30 months of age, similar to other countries maintaining age restrictions.[286] However, under the U.S.-China Phase One trade agreement, China agreed to amend import protocols that align with international standards. China agreed to (1) eliminate the cattle age restriction;[287] (2) recognize that the U.S. traceability system meets or exceeds OIE guidelines for maintaining "negligible risk" for bovine disease, and if the U.S. status should change, China would set import regulations that follow OIE guidelines; and (3) adopt MRLs for certain

---

[285] For example, in 2018, the United States and South Korea reached an agreement accepting regionalization in the event of an HPAI outbreak in the United States instead of imposing nationwide bans. USDA, "USDA Announces Regionalization Agreement with South Korea to Help Protect U.S. Trade During HPAI Detections," press release, March 15, 2018.

[286] USDA, Food Safety and Inspection Service, "Export Requirements for The People's Republic of China," updated March 18, 2019, https://www.fsis.usda.gov/wps/portal/fsis/topics/international-affairs/exporting-products/export- library-requirements-by-country/Peoples-Republic-of-China.

[287] USTR, *Economic and Trade Agreement Between the Government of the United States of America and the Government of the People's Republic of China* (hereafter the U.S.-China trade agreement), Chapter 3, Annex 4.2, January 15, 2020.

hormones used in U.S. beef production, and follow Codex MRL guidelines.[288] China continues to require that U.S. beef exporters participate in the USDA Agricultural Marketing Service export verification program,[289] which verifies that U.S. suppliers are meeting importing country requirements. In 2019, the U.S. shipped about 10,507 MT of beef to China, representing about 1% of total U.S. beef exports. U.S. beef exports to China were valued at $85.3 million.

China lifted its ban on the import of U.S. poultry meat in November 2019, allowing U.S. poultry exports from FSIS-approved poultry plants.[290] Under the U.S.-China Phase One trade agreement, the United States and China agreed to finalize a protocol accepting regionalization when there are outbreaks of poultry diseases, and China agreed to follow OIE guidelines on international trade.[291] Poultry industry analysts believe U.S. poultry exports to China could reach $1 billion in a short time, which would exceed record exports of $750 million in 2008.[292]

China's hog industry was hit hard with African Swine Fever in 2019, leaving a large gap in China's pork supplies and increasing demand for pork imports. In 2019, the value of U.S. pork and pork product exports (includes pork offal) to China more than doubled to $1.3 billion. Under the U.S.-China Phase One trade agreement, China is to increase the number of U.S. pork products inspected by FSIS that are eligible for import.[293]

---

[288] U.S.-China Phase One trade agreement, Chapter 3, Annex 4.2, 4.3, and 4.5, January 15, 2020.

[289] Agricultural Marketing Service, "Bovine, Ovine and Caprine Export Verification Programs," https://www.ams.usda.gov/services/imports-exports/bovine-ovine-and-caprine-export-verification-programs.

[290] FAS, *U.S. Poultry and Poultry Products Return to China*, GAIN Report, CH2019-0153, November 25, 2019. Includes an unofficial translation of China's November 14, 2019, General Administration of Customs and Ministry of Agriculture and Rural Affairs Announcement No. 177 that lifted five separate bans on U.S. poultry.

[291] U.S.-China trade agreement, Chapter 3, Annex 3.1 and 3.3.

[292] Austin Alonzo, "U.S. Could Find $1 Billion Poultry Market in China," *Watt Poultry USA*, January 2020, pp. 13-14.

[293] U.S.-China Phase One trade agreement, Chapter 3, Annex 6.2.

U.S. Meat and Poultry Imports

Currently, 33 countries are eligible to export meat and poultry to the United States.[294] Before the United States authorizes imports of meat or poultry, APHIS conducts risk assessments of any foreign animal diseases that could pose a threat to U.S. animal health. APHIS maintains a list of countries and their animal health status for critical diseases.[295] Also, FSIS must determine if foreign meat or poultry inspection systems provide an "equivalent" level of sanitation and protection of public health as the U.S. system.[296] Foreign governments provide documentation on how their inspection systems are regulated, and FSIS conducts onsite audits of foreign facilities. FSIS also conducts equivalency verification and periodic audits of countries already approved to export meat and poultry to the United States.

Imports of Chicken from China

In August 2013, FSIS confirmed that China's poultry processing inspection system was equivalent to the U.S. inspection system. This allowed China to export processed (cooked) poultry meat that is sourced raw from the United States or from countries eligible to export poultry to the United States. In March 2016, FSIS recommended that the process of verifying equivalency for China's poultry slaughter inspection system move forward.[297] In August 2017, FSIS released an audit report confirming that China's poultry processing system remained equivalent.[298]

---

[294] FSIS, "Import Library: Eligible Countries and Products," accessed March 18, 2019, https://www.fsis.usda.gov/wps/portal/fsis/topics/international-affairs/importing-products/eligible-countries-products-foreign-establishments/eligible- countries-and-products.

[295] APHIS, *Animal Health Status of Regions*, updated September 30, 2019, https://www.aphis.usda.gov/aphis/ourfocus/animalhealth/animal-and-animal-product-import-information/animal-health-status-of-regions.

[296] Equivalency is authorized under the Federal Meat Inspection Act (21 U.S.C. §601 et seq.) and the Poultry Products Inspection Act (21 U.S.C. §451 et seq.). Regulations for FSIS equivalency are in 9 C.F.R. 327. See FSIS, "Process for Evaluating the Equivalence of Foreign Meat, Poultry, and Egg Products Food Regulatory Systems," https://www.fsis.usda.gov/wps/portal/fsis/topics/international-affairs/importing-products/equivalence/equivalence- process-apply-for-initial-equivalence.

[297] FSIS, "Evaluating the Food Safety Systems Governing Slaughtered Poultry for Export to the United States," February 17, 2016, https://www.fsis.usda.gov/wps/wcm/connect/bd2f2159-

In November 2019, FSIS issued a final rule that determined that China's poultry slaughter system is equivalent and that China could export domestically slaughtered poultry meat to the United States.[299] China may only export fully cooked—not shelf stable-products.[300] China is not permitted to export raw poultry products due to animal disease risks. The United States did not import poultry meat from China in 2018 and 2019.

These actions were the culmination of a process that began in 2005, when China requested that USDA evaluate its poultry inspection system. Congress halted the process in FY2006, when appropriations provisions prohibited FSIS from expending funds to evaluate China's poultry inspection system. The process resumed in FY2010 on the condition that FSIS provide Congress with regular reports on the equivalency process. The possibility that the United States could import poultry meat from China has alarmed some food safety advocates and some Members of Congress because of concerns about relatively lax food safety enforcement in China for both domestically consumed products and exports. Testimony presented during a Congressional- Executive Commission on China hearing highlighted concerns regarding China's food safety.[301]

Status

In response to concern about China's record on food safety, Section 738 of Division B of the Further Consolidated Appropriations Act, 2020 (P.L. 116-94) prohibits USDA from using any funds to purchase Chinese raw or processed poultry products for feeding programs, including the school lunch and school breakfast programs. Section 741 of Division B of the FY2020 appropriations act prohibits USDA from finalizing the

---

63b2-4846-a738-7983f38f297f/2015-China-Slaughtered-Poultry-FAR.pdf?MOD=AJPERES.

[298] FSIS, "Evaluating the Food Safety Systems Governing Production of Processed Poultry Products Exported to the United States," August 10, 2017.

[299] 84 *Federal Register* 60318, November 8, 2019.

[300] Products that undergo a full lethality heat process (cooking) and require freezing or refrigeration for food safety.

[301] Hearing, Congressional-Executive Commission on China, *Pet Treats and Processed Chicken from China: Concerns for American Consumers and Pets*, June 17, 2014, http://www.cecc.gov/events/hearings/pet-treats-and-processed- chicken-from-china-concerns-for-american-consumers-and-pets.

proposed rule to allow the importation of slaughtered Chinese poultry unless certain conditions are met to ensure the food safety of poultry meat imports from China.[302]

Under the U.S.-China Phase One trade agreement, China may submit a formal request to the United States to evaluate regional avian influenza (AI) status.[303] Within 30 days of receipt of the request, APHIS would initiate an evaluation of conditions in the regions in order to determine if a region or regions could be recognized as AI-free.[304] Such a determination would allow China to export raw poultry meat if FSIS determines that poultry plants in the region(s) met equivalency standards.

Fresh Beef Imports from Brazil and Argentina

The United States restricts or prohibits imports of animals or animal products (including meat) from countries where highly infectious animal diseases exist in order to protect U.S. herds. Fresh beef imports from Brazil and Argentina have been prohibited or restricted because of foot-and-mouth disease (FMD) in the two countries. U.S. beef imports from Brazil and Argentina have mostly been limited to fully cooked/processed product. Argentina was approved to export fresh beef to the United States from 1997-2001, until the United States halted exports after an Argentine FMD outbreak in 2001.

In December 2013, APHIS proposed a rule that would allow fresh beef imports from 13 regions in Brazil.[305] In August 2014, APHIS proposed a separate rule to allow fresh beef imports from Patagonia and northern Argentina.[306] In July 2015, APHIS released final rules to allow the import

---

[302] FSIS issued the final rule in November 2019 before the FY2020 appropriations act was enacted. The final rule addresses the conditions included in Sec. 741 to ensure that imports from China are safe. The final rule requires China to comply with any APHIS animal health requirements and all imports from China will be subject to re-inspection.

[303] U.S.-China Phase One trade agreement; USTR, *Economic and Trade Agreement Between the Government of the United States of America and the Government of the People's Republic of China*, Chapter 3, Annex 3.4, January 15, 2020.

[304] APHIS, "Regionalization," updated November 9, 2017, https://www.aphis.usda.gov/aphis/ourfocus/animalhealth/export/international-standard-setting-activities-oie/regionalization/ct_reg_request.

[305] 78 *Federal Register* 77370, December 23, 2013.

[306] 79 *Federal Register* 51508, August 29, 2014; and 79 *Federal Register* 51528, August 29, 2014.

of fresh beef from these regions of Brazil and Argentina.[307] USDA risk assessments determined that, under certain circumstances, fresh beef could be safely imported from Brazil and Argentina without threatening the FMD-free status of the United States. Some livestock industry stakeholders, such as the National Cattlemen's Beef Association and the National Farmers Union, have expressed opposition to allowing fresh beef from Brazil and Argentina because neither country is considered to be free of FMD.[308] FMD was eradicated in the United States in 1929, and any introduction of the disease back into the United States could be economically devastating for the livestock industry. In 2013, the Department of Homeland Security estimated that the cost of an FMD outbreak in the United States could exceed $50 billion.[309]

In May 2015, FSIS found that Brazil's beef inspection system would provide an equivalent level of food safety as the U.S. system.[310] In August 2016, USDA announced that Brazil was approved to ship fresh beef to the United States, and the first shipments arrived the following month. In June 2017, USDA suspended imports of fresh beef from Brazil after FSIS found problems with re-inspected Brazilian beef at the U.S. port of entry.[311] According to USDA, FSIS was re-inspecting 100% of Brazilian fresh beef imports and refused entry to 11% of shipments, well above the 1% refusal rate for other beef imports.

In November 2018, FSIS announced that the Argentine beef inspection system was equivalent, and the country could export fresh beef to the

---

[307] 80 *Federal Register* 37923, July 2, 2015; and 80 *Federal Register* 37935, July 2, 2015.

[308] National Cattlemen's Beef Association, Docket Number APHIS 2009-0017, Importation of Beef from a Region in Brazil, April 22, 2014, https://www.regulations.gov/ document?D=APHIS-2009-0017-0820; and National Farmers Union, Docket Number APHIS 2009-0017, Importation of Beef from a Region in Brazil, April 21, 2014, https://www.regulations.gov/document?D=APHIS-2009-0017-0755.

[309] U.S. Department of Homeland Security, "A World Free of One of the Most Virulent Animal Diseases," updated October 28, 2013, http://www.dhs.gov/world-free-one-most-virulent-animal-diseases.

[310] The FSIS audit report for Brazil is available at https://www.fsis.usda.gov/wps/ wcm/ connect/d0c646c1-cc80-4540-b3df-01da1d9e9040/Brazil-2015-FAR.pdf?MOD=AJPERES.

[311] USDA, "Perdue: USDA Halting Import of Fresh Brazilian Beef," press release, June 22, 2017.

United States.[312] FSIS also announced that within six months of the November 2018 equivalency determination, the agency would undertake additional onsite audits of Argentina's raw beef inspection system. The United States imported about 1,623 MT of fresh beef from Argentina in 2019. Argentina holds a 20,000 MT ton duty-free TRQ allotment for beef shipments to the United States.

Status

On February 21, 2020, the United States lifted the suspension on imports of raw, intact beef from Brazil.[313] FSIS released a targeted on-site audit report on February 20, 2020 that addressed corrective actions taken by Brazil.[314] Raw beef imports from Brazil will be subject to re-inspection at U.S. points of entry by FSIS.

FSIS released an on-site audit report on Argentina's meat inspection system in September 2019 and noted that further on-site audits would be conducted to ensure that corrective actions undertaken as a result of the audit were implemented.[315]

Meat Exports under U.S.-Japan Trade Agreement (USJTA)

Japan is a leading export market for U.S. beef and pork products. In 2019, U.S. beef and beef product exports to Japan totaled about $2 billion, and pork and pork products amounted to $1.5 billion. Exports of both products were lower than the value of shipments in 2018, partly due to the preferential tariff treatment that competing exporters, such as Australia,

---

[312] FSIS, "Argentina Eligible to Export Raw Beef Products to the U.S.," November 30, 2018, https://www.fsis.usda.gov/wps/portal/fsis/newsroom/meetings/newsletters/constituent-updates/archive/2018/ ConstUpdate113018.

[313] FSIS, *Constituent Update*, February 21, 2019, https://www.fsis.usda.gov/wps/portal/fsis/newsroom/meetings/newsletters/constituent-updates/archive/2020/ConstUpdate022120.

[314] FSIS, *Evaluating the Food Safety Systems Governing Raw and Processed Meat Products Exported to the United States of America*, Final Followup Report of an Audit Conducted in Brazil, January 13-24, 2020, February 20, 2020, https://www.fsis.usda.gov/wps/wcm/connect/8f9c5e4c-2564-47f8-81af-22bac0cc0b4d/brazil-far-2020.pdf?MOD= AJPERES.

[315] FSIS, *Evaluating the Food Safety Inspection Systems Governing Meat Products Exported to the United States of America*, Final Report of an Audit Conducted in Argentina, February 25-March 15, September 25, 2019, https://www.fsis.usda.gov/wps/wcm/connect/1afd4cf5-b578-4d41-854b-fe6d78d974d4/Argentina-2019-FAR.pdf? MOD=AJPERES.

New Zealand, Canada, and Mexico, have with Japan through the TPP-11 agreement.

For example, Japan's beef imports from TPP-11 member nations entered at a 26.6% tariff rate in 2019 (year 2 of the TPP-11 agreement), but U.S. beef entered with a tariff rate of 38.5%. Under USJTA, the tariff on U.S. beef is now aligned with the TPP-11 tariff rates.[316] Under these agreements, Japan's tariff on beef from the TPP-11 countries and the United States is scheduled to decline until it reaches 9% in year 15 of the USJTA (year 16 of TPP-11).

Similarly, Japan's tariffs on imports of U.S. pork are reduced under the agreement, matching the TPP-11 tariff rates. Instead of an ad valorem rate of 4.3% on U.S. pork, the rate is 1.9% in the first year of the agreement, and is phased out in year 9. Japan maintains a variable duty mechanism (gate price),[317] which is set to a fixed value and will gradually decline until year 9.[318]

U.S. beef and pork exports are not subject to Japan's WTO safeguards,[319] but to U.S.-specific safeguards for beef and pork. The U.S. beef safeguard threshold is set at 242,000 MT and increases annually after year 2 of the agreement. Japan will terminate the beef safeguard measure if it does not trigger for four consecutive years after year 14 of the agreement. The U.S. pork safeguard will trigger if imports of U.S. pork exceed 112% of the largest import volume in the previous three years. The pork safeguard will terminate after year 10 of the agreement.[320]

Status

USJTA has been in effect since January 1, 2020, and U.S. meat exports to Japan are expected to increase as a result.

---

[316] USTR, USJTA, Annex 1, Subsection 2, Tariff Elimination or Reduction, 2(bb), October 7, 2019.

[317] Japan sets a minimum price for pork import values (393 yen per kilogram for pork cuts and 524 yen for processed pork). If the customs value of imported pork is below the gate price, the difference between the two values is paid as a variable duty in addition to the ad valorem tariff.

[318] USTR, USJTA, Annex 1, Subsection 2, Tariff Elimination or Reduction, 2(dd) and (ee).

[319] Safeguard measures allow for the imposition of temporary higher tariff rates if import volumes exceed an established threshold.

[320] USTR, USJTA, Annex 1, Subsection 4, Agricultural Safeguard Measures, Part 9 and 10.

Issues in Dairy Product Trade[321]

The United States exported $6.0 billion in dairy products in 2019, and imported $3.1 billion worth of products. Reform of dairy pricing and establishing specific dairy product TRQs in Canada is expected to expand access in that market for U.S. dairy producers. The USJTA lowers tariffs for U.S. dairy products and expands some dairy product TRQs. Like U.S. livestock producers, dairy producers gain benefits that competing exporters have enjoyed under the TPP-11. Under the U.S.-China Phase One trade agreement, China is to streamline the regulatory process to facilitate trade in U.S. dairy and infant formula.

U.S. Dairy Exports to Canada

The Canadian dairy sector limits production, sets prices, and restricts imports. Canadian imports of dairy products are restricted through TRQs, with over-quota tariffs in excess of 200% for some products. Although Canada is the second-largest market for U.S. dairy exports, U.S. exports would likely be higher but for Canadian import restrictions.

In recent years, U.S. milk producers began exporting increased quantities of ultra-filtered (UF) milk to Canada. UF milk is a high-protein liquid product made by separating and concentrating certain milk components (such as protein and fat) for use as ingredients in dairy products, such as cheese, yogurt, and ice cream. U.S. UF milk found a market among Canadian cheese makers in 2008 after Canada revised its compositional standards for cheese. This revision significantly reduced the use of several milk products that U.S. processors had been supplying to Canadian food manufacturers, including milk protein concentrates and dried protein products.

In recent years, growing demand for butterfat in Canada resulted in increased Canadian milk production and, consequently, surplus supplies of

[321] Prepared by Joel Greene, Analyst in Agricultural Policy, CRS.

skim milk. To address the surplus, Canada adopted the Class 7 milk price classification in 2017 (Class 6 in Ontario). Milk classified as Class 7 comprises skim milk components—primarily milk protein concentrates and skim milk powder (SMP)—used to process dairy products. Prices for Class 7 products were set at low levels. Once the Class 7 regime was implemented, Canadian skim milk products became cheaper. Canada expanded global exports of SMP with the consequence that U.S. producers lost exports of high- protein UF milk to Canadian cheese and yogurt processors.

According to USDA, the value of U.S. UF milk exports to Canada peaked at nearly $107 million in 2015 but declined after the Class 7 regime was implemented in 2017 to $49 million in 2017 and $32 million in 2018.[322] At the same time, Canada's exports of SMP more than tripled in 2017 to $133 million, compared with $42 million in 2016 before the Class 7 price regime was implemented.[323] Eliminating Canada's Class 7 pricing regime became a priority for the U.S. dairy industry when NAFTA renegotiations commenced in 2017.

Status

Under USMCA, Canada agreed to eliminate the Class 7 pricing regime six months after USMCA enters into force. Canada also agreed to reclassify Class 7 products according to their end use and base its selling price on a formula that takes into consideration the USDA reported nonfat dry milk price. Also under the agreement, Canada would be required to monitor its exports of milk protein concentrates, SMP, and infant formula and report at the harmonized tariff schedule level monthly.

Although Canada would maintain its milk supply management system under USMCA, it would expand TRQs for U.S. milk, cheese, cream, skim milk powder, condensed milk, yogurt, and several other dairy products. U.S. dairy products within the USMCA TRQs would enter Canada duty

---

[322] FAS, Global Agricultural Trade System Online, Foreign Agricultural Trade of the United States dairy definition, adjusted to include protein concentrate (UF milk), https://apps.fas.usda.gov/gats/default.aspx.

[323] Global Trade Atlas, export data for skim milk powder (harmonized code 040210).

free, while U.S. exports above the TRQ quantities would be subject to the existing over- quota tariffs. In return, the United States agreed to establish TRQs for imports of Canadian dairy products.

In total, under USMCA Canada would grant the United States duty-free access to nearly 17,000 MT of dairy products in the first year of the agreement, 100,000 MT in the sixth year, and 109,000 MT in year 19. The USMCA quota is specific to the United States and would be in addition to the 93,648 MT of WTO global quota, which is open to U.S. dairy products as well as to those from other WTO member countries as was the case under NAFTA.[324]

Dairy Exports under U.S.-Japan Trade Agreement (USJTA)

U.S. exports of dairy products to Japan totaled nearly $283 million in 2019, making Japan the fifth largest dairy export market for the United States. The Japanese dairy sector is protected by high import tariff rates and TRQ. In addition, competing exporters of dairy products to Japan (Australia, New Zealand, Canada, and the EU) have preferential tariffs through free trade agreements. The USJTA is expected to improve the competitive position of U.S. dairy producers through tariff reductions, and eventual tariff elimination in 15 years. Japan also established a country specific TRQ of 5,400 MT for U.S. whey products that is to increase to 9,000 MT in year 10. In-quota exports are to enter duty-free at the beginning of the agreement and tariffs on over- quota exports are to be eliminated in five years. Over-quota tariffs on other dairy products are to be phased out at various times through the agreement.[325]

---

[324] A. Mussell and D. Hedley, "The Canadian Dairy Sector in Relation to the Canada-US-Mexico Agreement and Comprehensive and Progressive Agreement for Trans-Pacific Partnership," *Agri-Food Economic Systems*, February 2019. For more information, also see CRS In Focus IF11149, *Dairy Provisions in USMCA*.

[325] USTR, "Provisions of the U.S.-Japan Traded Agreement Dairy Products," https://ustr.gov/about-us/policy-offices/press-office/fact-sheets/2019/october/provisions-of-the-us-japan-trade-agreement-dairy-products.

Status

"Stage One" of USJTA became effective on January 1, 2020. Unlike the provisions the United States had negotiated with Japan under the Trans-Pacific Partnership (TPP), USJTA does not include TRQs for certain dairy products such as butter and skim milk powder. The U.S. dairy industry has identified that the lack of provisions on non-tariff measures, such as GIs, could prove to be a market access barrier for certain U.S. cheese exports to the Japanese market. Additional negotiations with Japan toward a more comprehensive agreement are expected in 2020 and may address these issues.

U.S.-China Phase One Trade Agreement: Dairy

China was the third-largest market for U.S. dairy exports in 2019 at nearly $374 million, but this total was 25% lower than in 2018 as retaliatory tariffs hindered trade. Under the U.S.-China Phase One trade agreement, China is to streamline the regulatory process to facilitate U.S. exports. China is to accept dairy products manufactured in facilities compiled by FDA and which have a USDA dairy sanitary certificate. China is to accept that the U.S. dairy regulatory system provides the same level of safety as China's system. FDA is to provide China updated lists of dairy facilities under FDA jurisdiction. In addition, China's General Administration of Customs China and the FDA is to hold technical discussions regarding FDA guidance (U.S. Import Alert 99-30) on dairy products and the presence of melamine in imports of Chinese milk products. For infant formula, China is to also streamline its import approval process (such as issuing product registrations, technical reviews, and considering FDA's review, inspections and regulatory determinations).[326]

[326] U.S.-China Phase One trade agreement, Chapter 3, Annex 2.1, 2.2, and 2.3.

Status

The U.S.-China Phase One trade agreement entered into force February 14, 2020.

U.S.-MEXICO SUGAR SUSPENSION AGREEMENTS[327]

In December 2014, DOC signed suspension agreements with the government of Mexico and Mexican sugar producers and exporters that prevented the imposition of CVD and AD on U.S. imports of Mexican sugar. This was a consequence of U.S. government determinations that Mexican sugar was being subsidized by the government of Mexico and was being sold into the U.S. market at less than fair value.

The suspension agreements limit Mexico's sugar exports to the United States to the residual of U.S. needs for domestic human use in a given marketing year after subtracting U.S. production and imports from other countries. The agreements establish minimum reference prices for Mexican sugar that are above U.S. sugar program loan levels for domestically produced sugar. Another provision limits the share of Mexican sugar that can enter the United States as refined sugar.

After the suspension agreements took effect, a number of stakeholders in the U.S. sugar market asserted that the suspension agreements had not worked as intended and had not entirely eliminated the injury caused by the subsidization and dumping of Mexican sugar. One widely held criticism was that cane refiners who were dependent on imports of raw cane from Mexico had received an inadequate share of sugar from Mexico. Another criticism leveled at the agreements was that Mexican exporters were not always adhering to limits on the share of Mexican sugar imports that are refined sugar as compared with raw sugar, nor to the specified minimum reference prices.[328]

---

[327] Prepared by Joel Greene, Analyst in Agricultural Policy, CRS.

[328] For further information on stakeholder views, see CRS In Focus IF10517, *U.S. Stakeholders Critical of U.S.-Mexico Sugar Agreements*.

In November 2016, the American Sugar Coalition—representing sugar cane and sugar beet producers and sugar processors, refiners, and workers—called on DOC to withdraw from the agreements, an action that could have caused AD and CVD duties to be imposed on Mexican sugar.[329] Imperial Sugar Company, a U.S. cane refiner, also advocated for withdrawal. The Sweetener Users Association, which represents sugar-using businesses, recommended renegotiating the agreements to address their shortcomings and warned that terminating them would virtually eliminate Mexican sugar from the U.S. market. In November 2016, DOC issued results of a preliminary administrative review,[330] in which it concluded that the agreements may not have entirely redressed the injury, and that certain import transactions may not have adhered to the terms in the agreements.

In June 2017, the United States and Mexico agreed to amendments to the suspension agreements.[331] Under the amendments, effective October 1, 2017, the price of imported Mexican raw sugar was increased from $0.2225 per pound to $0.23 per pound. The price of imported refined sugar was increased from $0.26 per pound to $0.28 per pound. The maximum share of refined sugar imports was limited to 30%, with raw sugar imports constituting at least 70% of the total, compared with 53% and 47%, respectively, under the 2014 agreement. The agreement also requires that imported raw sugar be loaded in bulk and be free flowing—that is, not packaged. Any raw sugar imports that are packaged would be counted toward the refined sugar allotment. In addition, if USDA determines that the United States requires additional sugar imports to meet its needs, Mexico would be awarded the first opportunity to fill the need.[332]

---

[329] R. Sterk, *Food Business News*, "Imperial Asks DOC to End Mexican Sugar Deal," December 6, 2016.

[330] 81 *Federal Register* 87539, December 5, 2016; and 81 *Federal Register* 87541, December 5, 2016.

[331] 82 *Federal Register* 31942, July 11, 2017; and 82 *Federal Register* 31945, July 11, 2017.

[332] CRS In Focus IF10693, *Amended Sugar Agreements Recast U.S.-Mexico Trade.*

Status

In October 2019, the U.S. Court of International Trade (USCIT) voided the 2017 suspension agreements because DOC failed to follow recordkeeping requirements during the negotiations over the agreement. CSC Sugar LLC, a sugar trader and refiner of liquid sugar sued because the agreement changed the purity definition of refined sugar, harming its business, and it was unable to provide comment on the changes.[333] As a result of the USCIT ruling, the 2014 suspension agreement provisions went back into force.

On January 15, 2020, the DOC and Mexico agreed to new terms for the suspension agreement, specifically limiting imports from Mexico to 1,004,726 short tons from October 2019 through September 2020, with the share of refined sugar limited to 30% of import volume.[334] CSC Sugar LLC again filed suit in the USCIT to block the new agreements between the United States and Mexico.[335]

---

[333] Sugar purity is measured by polarity and in the 2014 agreement refined sugar was defined as 99.5% polarity. The 2017 agreement defined refined sugar at 99.2% polarity.

[334] Khorri Atkinson, *Law360,* "US, Mexico Sign Deal to Settle Sugar Trade Dispute," January 21, 2020. See 85 *Federal Register* 3613 (January 22, 2020).

[335] Sarah Martinson, *Law360,* "US. Sugar Co. Wants Revised US-Mexico Sugar Deal Scrapped," January 24, 2020.

In: COVID 19: Agricultural and Food Issues ISBN: 978-1-53618-561-4
Editor: Kristian Goldschmidt

Chapter 3

AGRICULTURAL DISASTER ASSISTANCE (UPDATED)*

Megan Stubbs

ABSTRACT

The U.S. Department of Agriculture (USDA) offers several programs to help farmers recover financially from natural disasters, including drought and floods. All the programs have permanent authorization, and one requires a federal disaster designation (the emergency loan program). Most programs receive mandatory funding amounts that are "such sums as necessary" and are not subject to annual discretionary appropriations.

The federal crop insurance program offers subsidized policies designed to protect crop producers from risks associated with adverse weather, as well as weather-related plant diseases and insect infestations and declines in commodity prices. Policies must be purchased prior to the planting season. Eligible commodities include most major crops and many specialty crops (including fruit, tree nut, vegetable, and nursery crops), as well as forage and pastureland for livestock producers. Producers who grow a crop that is currently ineligible for the federal crop

* This is an edited, reformatted and augmented version of Congressional Research Service, Publication No. RS21212, dated April 23, 2020.

insurance program may apply for the Noninsured Crop Disaster Assistance Program (NAP). NAP provides a basic level of coverage (when loss exceeds 50% of expected production) as well as options to purchase additional coverage. Similar to crop insurance, policies must be purchased prior to the planting season.

There are four permanently authorized disaster programs for livestock and trees. Producers do not pay a fee to participate, and advanced sign-up is not required. The programs are:

5. the Livestock Indemnity Program (LIP), which provides payments to eligible livestock owners and contract growers at a rate of 75% of market value for livestock deaths in excess of normal mortality or sold at a reduced sale price caused by adverse weather, attacks by reintroduced wild animals, and disease;
6. the Livestock Forage Disaster Program (LFP), which provides payments to eligible livestock producers who have suffered grazing losses on drought-affected pasture or grazing land or on rangeland managed by a federal agency due to a qualifying fire;
7. the Emergency Assistance for Livestock, Honey Bees, and Farm-Raised Fish Program (ELAP), which provides payments to producers of livestock, honey bees, and farm-raised fish as compensation for losses due to animal disease, adverse weather, and feed or water shortages; and
8. the Tree Assistance Program (TAP), which provides payments to orchardists/nursery tree growers for losses in excess of 15% to replant trees, bushes, and vines damaged by natural disasters.

Separately, for all types of farms and ranches, when a county has been declared a disaster area by either the President or the Secretary of Agriculture, producers in that county may become eligible for low-interest emergency disaster loans from USDA.

USDA has several permanent disaster assistance programs designed to help producers repair damaged land following natural disasters. It also has authority to issue disaster payments to farmers with funds from the Commodity Credit Corporation (CCC) or "Section 32." Finally, USDA can use a variety of existing programs to address disaster issues as they arise, such as allowing emergency grazing on land enrolled in the Conservation Reserve Program.

Congressional actions through the Agriculture Improvement Act of 2018 (P.L. 115-334) and the FY2019 supplemental appropriation (P.L. 116-20) have made a number of amendments to the agricultural disaster assistance programs and authorized supplemental appropriations covering crop and livestock losses from natural disasters in recent years. USDA has implemented this ad hoc assistance through two versions of the

Wildfires and Hurricanes Indemnity Program (WHIP). Losses occurring in 2017 were eligible for 2017 WHIP. An expanded set of losses occurring in 2018 and 2019 are eligible for WHIP Plus (WHIP+). In addition to WHIP+, USDA has implemented two subprograms—the On-Farm Storage Loss Program and the Milk Loss Program—as well as block grants with states and payments for acres that were prevented from being planted.

Economic and market losses related to the COVID-19 pandemic are not considered eligible losses under the USDA disaster assistance programs. While USDA has provided some flexibilities through existing programs, most disaster assistance programs require some level of production or physical loss from an unavoidable natural cause in order to trigger eligibility.

INTRODUCTION

The U.S. Department of Agriculture (USDA) offers several programs designed to help farmers and ranchers recover from the financial effects of natural disasters, including (1) federal crop insurance, (2) the Noninsured Crop Disaster Assistance Program (NAP), (3) livestock and fruit tree disaster programs, and (4) emergency disaster loans for both crop and livestock producers. All have permanent authorization, while the emergency loan program is the only one requiring a federal disaster designation (see Table 1). Most programs receive mandatory funding amounts of "such sums as necessary" and are not subject to annual discretionary appropriations.[1] The Agricultural Improvement Act of 2018 (2018 farm bill, P.L. 115-334) made a number of amendments to these programs, generally to expand availability and support.

Prior to the creation of many of the permanently authorized programs at USDA, Congress had historically provided farmers and ranchers with ad hoc disaster assistance payments authorized through supplemental appropriations.

---

[1] For a brief summary of these programs, see CRS In Focus IF10565, *Federal Disaster Assistance for Agriculture*.

Table 1. Select Permanent USDA Agricultural Disaster Assistance Programs (all programs permanently authorized—disaster designation required only for loans)

Program and Agency	Commodity Coverage and Requirements	Payment Trigger	Timing and Funding
Crop insurance indemnifies yield, revenue, or margin losses—RMA.	More than 100 eligible crops; livestock margins and pasture also covered. Producer must purchase a subsidized policy prior to planting.	Yield, revenue, or margin loss greater than the deductible specified in the policy.	Indemnity payment issued after claim is filed, adjusted, and processed; unlimited funding.
Noninsured Crop Disaster Assistance Program (NAP) payments for reduced yield and loss—FSA.	Crops not currently eligible for crop insurance. Producer pays fee prior to planting.	Crop loss in excess of 50%; additional coverage available.	Payment issued after claim is filed and processed; unlimited funding.
Supplemental agricultural disaster assistance programs issue payments for qualifying losses—FSA			
Livestock Indemnity Program (LIP); payments for livestock death losses or livestock sold at a reduced sale price.	Beef and dairy cattle, hogs, chickens, ducks, geese, turkeys, sheep, goats, alpacas, deer, elk, emus, and equine; no fee.	Livestock deaths in excess of normal mortality or sold at reduced sale price caused by adverse weather and disease.	Payment issued after application is processed; unlimited funding.
Livestock Forage Disaster Program (LFP); payments for grazing losses from drought/fire.	Drought-affected pastureland and cropland planted for grazing; rangeland managed by a federal agency due to a qualifying fire; no fee.	Drought intensity level for an individual county, as published in the U.S. Drought Monitor.	Payment issued after application is processed; unlimited funding.
Emergency Assistance for Livestock, Honey Bees, and Farm-Raised Fish Program (ELAP); payments for losses not covered by LIP/LFP.	Livestock, honey bees, and farm-raised fish; losses may be caused by disease, adverse weather, feed or water shortages, or wildfires; no fee.	Bee losses in excess of normal mortality; certain losses and costs for livestock, including feed and water costs.	Payment issued after application is processed; unlimited funding.
Tree Assistance Program (TAP); payments for tree, bush, and vine losses.	Trees, bushes, and vines from which an annual crop is produced; no fee.	Tree/vine mortality loss or damage in excess of 15%.	Payment issued after application is processed; unlimited funding.

Program and Agency	Commodity Coverage and Requirements	Payment Trigger	Timing and Funding
Emergency (EM) disaster loans offer low-interest financing for production or physical losses—FSA.	Damage to crops, livestock, equipment, or farmland when commercial credit is not available.	County disaster declaration by President, Secretary, or FSA; 30% crop loss or a physical loss.	Application due within eight months of disaster; funding subject to appropriations.
Disaster Set-Aside allows deferred loan payment for direct loans—FSA.	One full year's loan payment can be moved to end of loan period.	Disaster declaration by President, Secretary, or FSA.	Upon declaration of disaster.

Source: Compiled by CRS.

Notes: FSA = Farm Service Agency; RMA = Risk Management Agency. USDA also offers other emergency assistance for land rehabilitation and forbearance on rural housing loans and rural water and waste water disposal loans. See CRS Report R42854, *Emergency Assistance for Agricultural Land Rehabilitation* and CRS Report RL31837, *An Overview of USDA Rural Development Programs.*

Subsequently, policies shifted away from the temporary forms of assistance in favor of enacting more permanent forms of support.[2] More recently, policies have shifted to supplement permanent programs with ad hoc assistance for select agriculture losses. Supplemental appropriations in FY2018 and FY2019 amended existing disaster assistance programs and authorized over $5 billion for production losses assistance from natural disasters in 2017, 2018, and 2019.[3] USDA has also used separate discretionary authorities to respond to disasters administratively.

Economic and market-related losses are not generally considered eligible losses under USDA's disaster assistance programs. Most disaster assistance programs are statutorily required to include some level of production or physical loss from an unavoidable natural cause in order to trigger eligibility. As such, this chapter does not discuss USDA payment programs related to the COVID19 pandemic or trade disputes.[4]

---

[2] For example, the Food, Conservation, and Energy Act of 2008 (2008 farm bill; P.L. 110-246) and the Agricultural Act of 2014 (2014 farm bill; P.L. 113-79) authorized the "Supplemental Agricultural Disaster Assistance Programs" discussed below.

[3] Bipartisan Budget Act of 2018 (P.L. 115-123) and Additional Supplemental Appropriations for Disaster Relief Act, 2019 (P.L. 116-20).

[4] For additional information on USDA COVID-19 assistance, see https://www.usda.gov/coronavirus. For additional information on USDA's trade aid assistance, see CRS Report R45865, *Farm Policy: USDA's 2019 Trade Aid Package*.

This chapter provides an overview of permanently authorized federal disaster assistance programs for agricultural losses from natural disaster as well as discretionary authority that USDA may use to provide assistance. It also discusses ad hoc assistance authorized by Congress.

FEDERAL CROP INSURANCE

The federal crop insurance program is permanently authorized by the Federal Crop Insurance Act, as amended (7 U.S.C. 1501 et seq.) and is operated through public and private entities. Private insurance companies, known as Approved Insurance Providers (AIPs), are the primary insurers selling and servicing the insurance policies. USDA's Risk Management Agency (RMA) drafts federal crop insurance policies, sets premium rates, and regulates AIPs. The Federal Crop Insurance Corporation (FCIC) reinsures crop insurance policies and subsidizes the AIPs expenses. The program is designed to protect crop producers from risks associated with adverse weather, as well as weather-related plant diseases and insect infestations and declines in commodity prices.

Crop insurance is available for most major crops and many specialty crops (including fruit, tree nut, vegetable, and nursery crops) as well as forage and pastureland for livestock producers. A producer who chooses to purchase an insurance policy must do so by an administratively determined deadline date, which varies by crop and usually coincides with the beginning of planting season. Insurance products that protect against loss in revenue and profit margins are also available. Policies are typically available in major growing regions.

The federal crop insurance program was instituted in the 1930s and was subject to major legislative reforms in 1980 and again in 1994 and 2000. The Agriculture Risk Protection Act of 2000 (P.L. 106-224) authorized new federal spending for the program primarily through more generous premium subsidies to help make the program more affordable to farmers and enhance farmer participation levels in an effort to preclude the need for ad hoc emergency disaster payments.[5]

Under the current crop insurance program, a producer who grows an insurable crop selects a level of crop yield and price coverage and pays a premium that increases as the levels of yield and price coverage rise. However, all eligible producers can receive catastrophic (CAT) coverage without paying a premium. The premium for CAT coverage is subsidized by the federal government. Under CAT coverage, participating producers can receive a payment equal to 55% of the estimated market price of the commodity on crop losses in excess of 50% of normal yield— referred to as 50/55 coverage.

Although eligible producers do not have to pay a premium for CAT coverage, they are required to pay upon enrollment a $655 administrative fee per covered crop for each county where they grow the crop. USDA can waive the fee for financial hardship cases. In addition to the administrative fee, producers can elect to pay a premium, which is partially subsidized by the government, to increase the 50/55 CAT coverage to any equivalent level of coverage between 50/100 and 85/100 (i.e., 85% of yield and 100% of the estimated market price) in increments of 5%. These higher levels of coverage are known as *buy-up* coverage.

For many insurable commodities, an eligible producer can purchase revenue insurance. Under such a policy, a farmer could receive an indemnity payment when actual farm revenue for a crop falls below the target level of revenue, regardless of whether the shortfall in revenue was caused by poor production or low farm commodity prices. Insured producers are also eligible for reduced coverage if they are late or prevented from planting because of flooding.

The annual agriculture appropriations bill traditionally makes two separate appropriations for the federal crop insurance program. It provides discretionary funding for the salaries and expenses of the RMA. It also provides "such sums as are necessary" for the FCIC, which finances all

[5] H.Rept. 106-639; "Conference Report on H.R. 2559, Agricultural Risk Protection Act of 2000," House debate, *Congressional Record*, vol. 146, part 67 (May 25, 2000), p. H3817-H3828; and "Risk Management for the 21st Century Act," Senate debate, *Congressional Record*, vol. 146, part 34 (March 23, 2000), pp. S1627-S1642.

other expenses of the program, including premium subsidies, indemnity payments, and reimbursements to the AIPs.[6] The total cost of the program varies by year, primarily due to fluctuating levels of indemnity payments from changes in commodity prices, planting decisions, and weather conditions.[7] Across all policies, the average premium subsidy was about 63% of total premiums in 2018.[8] The federal government also subsidizes the costs of selling and servicing the policies (as delivery subsidies to Approved Insurance Providers) and absorbs underwriting losses (indemnities in excess of premiums received) in years when indemnities are high.

Noninsured Crop Disaster Assistance Program (NAP)

Producers who grow a crop that is currently ineligible for crop insurance may apply for NAP. NAP has permanent authority under Section 196 of the Federal Agriculture Improvement and Reform Act of 1996 (7 U.S.C. 7333) and is administered by USDA's Farm Service Agency (FSA). It was first authorized under the Federal Crop Insurance Reform Act of 1994 (P.L. 103 354). NAP is not subject to annual appropriations. Instead, it receives such sums as are necessary through USDA's Commodity Credit Corporation (CCC), which has a line of credit with the U.S. Treasury to fund an array of farm programs.[9]

---

[6] The FCIC is funded through a mandatory indefinite appropriation that draws necessary funds directly from the U.S. Treasury.

[7] The total direct cost of the federal crop insurance program to the federal government for crop years 2007-2016 was about $72 billion, of which about $43 billion (60%) directly benefitted producers, $28 billion (39%) went to approved insurance providers, and $754 million (1%) was for RMA salaries and expenses. See CRS Report R45193, *Federal Crop Insurance: Program Overview for the 115th Congress*, for a more detailed discussion on cost.

[8] In crop year 2018, total premiums were $9.9 billion, and premium subsidies were $6.3 billion. See RMA, *Crop Year Government Cost of Federal Crop Insurance*, March 11, 2020, https://rma.usda.gov/-/media/RMAweb/AboutRMA/Program-Budget/18cygovcost.ashx ?la=en.

[9] For more information on CCC, see CRS Report R44606, *The Commodity Credit Corporation: In Brief*.

Eligible crops under NAP include any commercial crops grown for food, fiber, or livestock consumption for which there is no CAT coverage available under the federal crop insurance program, with limited exceptions. These crops include mushrooms, floriculture, ornamental nursery, Christmas trees, turfgrass sod, aquaculture, honey, maple sap, ginseng, and industrial crops used in manufacturing or grown as a feedstock for energy production, among others. Trees grown for wood, paper, and pulp products are not eligible. To be eligible for a NAP payment, a producer first must apply for coverage by the application closing date, which varies by crop but is generally about 30 days prior to the final planting date for an annual crop. Like CAT coverage under crop insurance, NAP applicants must also pay an administrative fee at the time of application. The NAP service fee is the lesser of $325 per crop or $825 per producer per administrative county, not to exceed a total of $1,950 for farms in multiple counties.[10]

In order to receive a NAP payment, a producer must experience at least a 50% crop loss caused by a natural disaster or be prevented from planting more than 35% of intended crop acreage. For any losses in excess of the minimum loss threshold, a producer can receive 55% of the average market price for the covered commodity. Hence, NAP is similar to CAT coverage in that it pays 55% of the market price for losses in excess of 50% of normal historical production.

Additional coverage (referred to as buy-up coverage) may be purchased at 50% to 65% (in 5% increments) of established yield and 100% of average market price, contract price, or other premium price. The farmer-paid fee for additional coverage is 5.25% times the product of the selected coverage level and value of production (acreage multiplied by

---

[10] For more on NAP, see USDA factsheet at https://www.fsa.usda.gov/Assets/USDA-FSA-Public/usdafiles/FactSheets/noninsured_crop_disaster_assistance_program-nap-fact_sheet.pdf. Also, see NAP crop eligibility, premium, and payments estimator at https://apps.fsa.usda.gov/nap/costtool/login.do.

yield multiplied by the average market price multiplied by the producer's share of the crop).[11] Grazing land is not eligible for buy-up coverage.

A producer of a noninsured crop is subject to an annual payment limit of $125,000 per person for catastrophic coverage and $300,000 for buy-up coverage. A producer is ineligible under NAP if the producer's total adjusted gross income (AGI) exceeds $900,000.[12] The total federal cost of NAP was $165 million in FY2014, $125 million in FY2015, $137 million in FY2016, $157 million in FY2017, $164.3 million in FY2018, and $138.4 million in FY2019.[13]

SUPPLEMENTAL AGRICULTURAL DISASTER ASSISTANCE PROGRAMS

Four agricultural disaster assistance programs are permanently authorized for livestock and fruit trees: (1) the Livestock Indemnity Program (LIP); (2) the Livestock Forage Disaster Program (LFP); (3) the Emergency Assistance for Livestock, Honey Bees, and Farm-Raised Fish Program (ELAP); and (4) the Tree Assistance Program (TAP). They operate nationwide and are administered by FSA. Producers do not pay fees to participate and can apply at their local FSA offices.[14]

All programs receive "such sums as necessary" in mandatory funding via the CCC to reimburse eligible producers for their losses. Total payments vary each year based on eligible loss conditions. For FY2019, LFP payments totaled $287.9 million, LIP payments totaled $49.9 million,

---

[11] For NAP enhancements, see CCC and FSA, "Noninsured Crop Disaster Assistance Program; Final Rule," 85 *Federal Register* 12213-12221, March 2, 2020, https://www.fsa. usda.gov/ Internet/FSA_Federal_Notices/nap_2020.pdf.

[12] For additional information on payment limits and AGI requirements, see CRS Report R44739, *U.S. Farm Program Eligibility and Payment Limits*.

[13] USDA, *2021 Explanatory Notes*, Commodity Credit Corporation, February 2020, http://www.obpa.usda.gov/ 26ccc2021notes.pdf.

[14] For local FSA contact information, see FSA locator at http://offices.sc.egov.usda.gov/ locator/app?agency=fsa.

TAP payments totaled $15.4 million, and ELAP payments totaled $46.0 million.[15] All payments are reduced by sequestration.[16]

For individual producers, payments under LFP may not exceed $125,000 per year. There are no limits on the amount of payments received under LIP, ELAP, and TAP. To be eligible for a payment under any of these programs, a producer's total AGI cannot exceed $900,000.[17]

Livestock Indemnity Program (LIP)

LIP provides payments to eligible livestock owners and contract growers for livestock deaths in excess of normal mortality caused by an eligible loss condition (e.g., adverse weather, disease, or animal attack). Payments may also be made when the animal is injured as a direct result of an eligible loss condition but does not die and is sold at a reduced price. Eligible loss conditions may include (1) extreme or abnormal damaging weather that is not expected to occur during the loss period for which it occurred, (2) disease that is caused or transmitted by a vector and is not susceptible to control by vaccination, and (3) an attack by animals reintroduced into the wild by the federal government or protected by federal law. Eligibility is predicated on not only the occurrence of an eligible loss condition but direct causation to the death or injury of the animal.

Eligible livestock include beef and dairy cattle, bison, hogs, chickens, ducks, geese, turkeys, sheep, goats, alpacas, deer, elk, emus, llamas, reindeer, caribou, and equine. The livestock must have been maintained for commercial use and not produced for reasons other than commercial use as

---

[15] USDA, *2021 Explanatory Notes*, February 2020, https://www.obpa.usda.gov/explan_notes.html.

[16] Sequestration is a process to reduce federal spending through automatic, largely across-the-board reductions that permanently cancel mandatory and/or discretionary budget authority. The supplemental agricultural disaster assistance programs classify as *nonexempt mandatory* accounts and are therefore reduced by sequestration. For additional information on sequestration, see Appendix C in CRS Report R45974, *Agriculture and Related Agencies: FY2020 Appropriations*.

[17] For additional information on payment limits, see CRS Report R46248, *U.S. Farm Programs: Eligibility and Payment Limits*.

part of a farming operation. The program excludes wild free-roaming animals, pets, and animals used for recreational purposes, such as hunting, roping, or for show. Poultry and swine are the only eligible livestock for contract growers.

The LIP payment rate is equal to 75% of the average fair market value of the deceased animal type. USDA publishes a payment rate for each type of livestock for each year (e.g., $889.89 per adult beef cow and $3.70 per duck as of February 2020).[18] For eligible livestock contract growers, the payment rate is based on 75% of the national average input costs for the applicable livestock. For livestock sold at a reduced sale price, payments are calculated by multiplying the national payment rate for the livestock category minus the amount the owner received at sale multiplied by the owner's share.

Livestock Forage Disaster Program (LFP)

LFP makes payments to eligible livestock producers who have suffered grazing losses on drought-affected pastureland (including cropland planted specifically for grazing), or on rangeland managed by a federal agency due to a qualifying fire.

U.S. Drought Monitor

The U.S. Drought Monitor is a collaboration between the USDA, National Oceanic and Atmospheric Administration, and the National Drought Mitigation Center at the University of Nebraska-Lincoln. Weekly maps are released based on measurements of climatic, hydrologic, and soil conditions and are combined with local impacts and observations across the country from over 400 participants (primarily from government and academic backgrounds). Eleven climatologists rotate as lead author to produce the weekly maps. The drought monitor is used to determine drought relief for USDA programs (e.g., LFP and the Non-Fat Dry Milk Program) and by the Internal Revenue Service (IRS) to determine the replacement period for livestock sold because of drought. Congressional interest in the U.S. Drought Monitor has grown in recent years as additional program benefits are tied to it. For additional information, see http://droughtmonitor.unl.edu/.

[18] Payment rates are available in the USDA fact sheet at https://www.fsa.usda.gov/Assets/USDA-FSA-Public/usdafiles/FactSheets/livestock_indemnity_program_lip-fact_sheet.pdf. For more information, see 7 C.F.R. §1416 Subpart D— Livestock Indemnity Program and the FSA LIP handbook at https://www.fsa.usda.gov/Internet/FSA_File/1-lip_r00_a03.pdf.

Table 2. Livestock Forage Program (LFP) (drought intensity and time period determine the number of monthly payments)

Drought Monitor Intensity	Qualifying Time Period	No. of Monthly Payments
D2 (severe drought)	For at least eight consecutive weeks during the normal grazing period	one monthly payment
D3 (extreme drought)	At any time during the normal grazing period	three monthly payments
D3 (extreme drought)	For at least four weeks during the normal grazing period	four monthly payments
D4 (exceptional drought)	At any time during the normal grazing period	four monthly payments
D4 (exceptional drought)	For four weeks (not necessarily consecutive) during the normal grazing period	five monthly payments

Source: 7 U.S.C. §9081(c).

Notes: Drought intensity level can apply to any area of a county. The LFP monthly payment rate for drought is equal to 60% of the lesser of the monthly feed cost based on either (a) corn prices, specified feeding requirements, and number of animals or (b) the normal carrying capacity of the land. For details on monthly feed costs and examples, see FSA LFP handbook at http://www.fsa.usda.gov/Internet/FSA_File/1-lfp_r00_a01.pdf. In the case of a producer who sold livestock because of drought conditions, the payment rate is equal to 80% of the monthly feed cost. For fire on federally managed rangeland, the payment rate is 50% of the monthly feed cost, adjusted for the number of days the producer is prohibited from grazing (not to exceed 180 days).

Eligible producers must own, cash or share lease, or be a contract grower of covered livestock during the 60 calendar days before the beginning date of a qualifying drought or fire. They must also provide pastureland or grazing land for covered livestock that is either (a) physically located in a county affected by a qualifying drought during the normal grazing period for the county or (b) managed by a federal agency where grazing is not permitted due to fire.

Eligible livestock types are livestock that have been grazing on eligible grazing land or pastureland or would have been had a disaster not struck. These include alpacas, beef cattle, buffalo, beefalo, dairy cattle, sheep, deer, elk, emus, equine, goats, llamas, poultry, reindeer, and swine. Livestock must be maintained for commercial use as part of a farming operation.

Livestock owned for noncommercial uses, or livestock that is in (or would have been in) feedlots, are excluded.

Payments are generally triggered by the drought intensity level for an individual county as published in the U.S. Drought Monitor, a federal report published each week. The number of monthly payments depends on the drought severity and length of time the county has been designated as such (Table 2). For drought, the payment amount is equal to the number of monthly payments multiplied by 60% of estimated monthly feed cost. For producers who sold livestock because of drought conditions, the payment rate is equal to 80% of the estimated monthly feed cost.[19]

Emergency Assistance for Livestock, Honey Bees, and Farm-Raised Fish Program (ELAP)

ELAP provides payments to producers of livestock, honey bees, and farm-raised fish as compensation for losses due to disease, adverse weather, feed or water shortages, or other conditions (such as wildfires) that are not covered under LIP or LFP.[20]

ELAP specifically provides assistance for the loss of honey bee colonies in excess of normal mortality. In order to meet the eligibility requirements for honey bee colony losses, they must be the direct result of an eligible adverse weather or loss condition such as colony collapse disorder, eligible winter storm, excessive wind, and flood. For livestock losses, ELAP covers three categories: (1) livestock feed and grazing losses

---

[19] For more information on LFP, see 7 C.F.R. §1416 Subpart C—Livestock Forage Disaster Program; the USDA fact sheet at https://www.fsa.usda.gov/Assets/USDA-FSA-Public/usdafiles/FactSheets/2019/livestock_forage_disaster_program-fact_%20sheet_july-2019.pdf; and the FSA LFP handbook at http://www.fsa.usda.gov/Internet/FSA_File/1-lfp_r00_a01.pdf.

[20] For more information on ELAP, see 7 C.F.R. §1416 Subpart B—Emergency Assistance for Livestock, Honeybees, and Farm-Raised Fish Program; general ELAP fact sheet at https://www.fsa.usda.gov/Assets/USDA-FSA-Public/usdafiles/FactSheets/elap_general_factsheet.pdf; ELAP honey bee fact sheet at https://www.fsa.usda.gov/Assets/ USDA-FSA-Public/usdafiles/FactSheets/elap_honeybee_factsheet.pdf ; ELAP livestock fact sheet at https://www.fsa.usda.gov/Assets/USDA-FSA-Public/usdafiles/FactSheets/2019/elap_livestock_fact_sheet.pdf; ELAP farm-raised fish fact sheet at https://www.fsa.usda.gov/Assets/USDA-FSA-Public/usdafiles/FactSheets/2019/elap_farm-raised_fish_fact_sheet.pdf; and the FSA ELAP handbook at https://www.fsa.usda.gov/Internet/FSA_File/1-elap_r00_a03.pdf.

that are not due to drought or wildfires on federally managed lands, (2) losses resulting from the additional cost of transporting water to livestock due to an eligible drought, and (3) losses resulting from the additional cost associated with gathering livestock for inspection and treatment related to cattle tick fever. Applicants who are defined as a *socially disadvantaged*, *limited resource*, *beginning*, or *veteran* farmer or rancher are eligible for up to 90% of cost of loss.[21]

Tree Assistance Program (TAP)

TAP makes payments to qualifying orchardists and nursery tree growers to replant or rehabilitate trees, bushes, and vines damaged by natural disasters. Losses in crop production—as opposed to the tree, bush, or vine itself—are generally covered by federal crop insurance or NAP. Eligible trees, bushes, and vines are those from which an annual crop is produced for commercial purposes. Nursery trees include ornamental, fruit, nut, and Christmas trees produced for commercial sale. Trees used for pulp or timber are ineligible.

To be considered an eligible loss, the individual stand must have sustained a mortality loss or damage loss in excess of 15% after adjustment for normal mortality or damage.[22] Normal mortality or damage is determined based on (a) each eligible disaster event, except for losses due to plant disease, or (b) for plant disease, the time period for which the stand is infected. Also, the loss could not have been prevented through reasonable and available measures.

For replacement, replanting, and/or rehabilitation of trees, bushes, or vines, the payment calculation is the lesser of (a) 65% of the actual cost of replanting (in excess of 15% mortality) and/or 50% of the actual cost of

[21] For additional information and definitions for *socially disadvantaged*, *limited resource*, *beginning*, or *veteran* farmer or rancher, see https://www.nrcs.usda.gov/wps/portal/nrcs/main/national/people/outreach/slbfr/.

[22] Note that the FY2019 supplemental appropriation (P.L. 116-20) extended TAP payments to pecan tree losses in excess of 7.5% (adjusted for normal mortality) during calendar year 2018.

rehabilitation (in excess of 15% damage), or (b) the maximum eligible amount established for the practice by FSA.[23] Applicants who are defined as *beginning* or *veteran* farmer or ranchers are eligible for higher payments, including the lesser of (a) 75% of the actual cost of replanting (in excess of 15% mortality) and/or 50% of the actual cost of rehabilitation (in excess of 15% damage) or (b) the maximum eligible amount established for the practice by FSA.[24] The total quantity of acres planted to trees, bushes, or vines for which a producer can receive TAP payments cannot exceed 1,000 acres annually.[25]

Emergency Disaster Loans

When either the President or the Secretary of Agriculture declares a county a disaster area or quarantine area, agricultural producers in that county may become eligible for low-interest emergency disaster (EM) loans available through FSA.[26] Producers in counties that are contiguous to a county with a disaster designation also become eligible for EM loans. EM loan funds may be used to help eligible farmers, ranchers, and aquaculture producers recover from *production* losses (when the producer

---

[23] Higher payment rates of up to 75% for replanting and rehabilitation are available for beginning and veteran farmers and ranchers.

[24] For additional information on beginning and veteran farmer and rancher provision in farm bill programs, see CRS In Focus IF11227, *2018 Farm Bill Primer: Beginning Farmers and Ranchers*; and CRS In Focus IF11093, *2018 Farm Bill Primer: Veteran Farmers and Ranchers*.

[25] For more information on TAP, see 7 C.F.R. §1416 Subpart E—Tree Assistance Program; the general TAP fact sheet at https://www.fsa.usda.gov/Assets/USDA-FSA-Public/usdafiles/FactSheets/tree_assistance_program-tapfact_sheet.pdf; the 2017 TAP pecan loss fact sheet at https://www.fsa.usda.gov/Assets/USDA-FSA-Public/usdafiles/FactSheets/2018/tap_pecan_tree_losses_fact_sheet_august_2018.pdf; and the TAP Florida citrus greening fact sheet at https://www.fsa.usda.gov/Assets/USDA-FSA-Public/usdafiles/FactSheets/2018/tap_florida_citrus_greening_fact_sheet_may_2018.pdf.

[26] The Secretary of Agriculture may designate quarantine areas under the Plant Protection Act (Title VI of P.L. 106 224) or animal quarantine laws. For an overview of the USDA emergency disaster designation and declaration process, see https://www.fsa.usda.gov/Assets/USDA-FSA-Public/usdafiles/FactSheets/2017/emergency_disaster_designation_and_declaration_process_oct2017.pdf. For the USDA fact sheet on EM loans, see https://www.fsa.usda.gov/Assets/USDA-FSA-Public/usdafiles/FactSheets/2019/emergency-loan-program.pdf .

suffers a significant loss of an annual crop) or from *physical* losses (such as repairing or replacing damaged or destroyed structures or equipment or for the replanting of permanent crops such as orchards). A qualified applicant can then borrow up to 100% of actual production or physical losses (not to exceed a loan total of $500,000).

Once a county is declared eligible, an individual producer within the county (or a contiguous county) must also meet the following requirements for an EM loan: A producer must (1) be an established family farmer and a citizen or permanent resident of the United States; (2) experience a crop loss of more than 30% or a physical loss of livestock, livestock products, real estate, or property; and (3) be unable to obtain credit from a commercial lender but still show the potential to repay the loan, including having acceptable credit history and collateral to secure the loan. Applications must be received within eight months of the county's disaster designation date. Loans for nonreal-estate purposes must generally be repaid within seven years. Loans for physical losses to real estate have terms up to 20 years. Depending on the repayment ability of the producer and other circumstances, these terms can be extended to 20 years for nonreal-estate losses and up to 40 years for real estate losses.

The EM loan program is permanently authorized by Title III of the Consolidated Farm and Rural Development Act (P.L. 87-128), as amended, and is subject to annual appropriations. In FY2020, the program received $37.7 million of new loan authority.[27] Unused funds are carried over and available in the next fiscal year. Therefore, the total loan authority can vary greatly depending on appropriated levels, annual use, and total carryover. In FY2020, the total loan authority is $122.4 million.[28]

Also in counties with disaster designations, producers who have existing direct loans with FSA may be eligible for Disaster Set-Aside. If, as a result of disaster, a borrower is unable to pay all expenses and make loan

---

[27] Division B, Title II of P.L. 116-94.

[28] For the most recent funding levels available, see emergency loan funding totals at http://www.fsa.usda.gov/programs-and-services/farm-loan-programs/funding/index.

payments that are coming due, up to one full year's payment can be moved to the end of the loan.[29]

Other USDA Assistance

USDA also has several permanent disaster assistance programs that help producers repair damaged land following natural disasters. It also has authority to issue disaster payments to farmers with "Section 32" or "CCC" funds and can use a variety of existing programs to address disaster issues as they arise.

Emergency Agricultural Land Assistance Programs

Several USDA programs offer financial and technical assistance to producers to repair, restore, and mitigate damage by a natural disaster on private land.

- The Emergency Conservation Program (ECP) and the Emergency Forest Restoration Program (EFRP) are administered by FSA. For both programs, FSA pays participants a percentage of the cost to restore the land to a productive state. ECP also funds water for livestock and installing water conserving measures during times of drought. EFRP was created to assist private forestland owners with losses caused by a natural disaster on nonindustrial private forest land.
- The Emergency Watershed Protection (EWP) program and the EWP floodplain easement program are administered by USDA's Natural Resources Conservation Service and the U.S. Forest Service. EWP assists sponsors, landowners, and operators in

---

[29] No loan may receive more than one disaster set-aside unless it is later restructured. For more information, see USDA fact sheet at https://www.fsa.usda.gov/Assets/USDA-FSA-Public/usdafiles/FactSheets/2017/disaster_set_aside_program_oct2017.pdf.

implementing emergency recovery measures for runoff retardation and erosion prevention to relieve imminent hazards to life and property created by a natural disaster. The EWP floodplain easement program is a mitigation program that pays for permanent easements on private land meant to safeguard lives and property from future floods and drought and the products of erosion.

"Section 32" and "CCC" Funds for Farm Disaster Payments

USDA has discretionary authority to distribute emergency payments to farmers with Funds for Strengthening Markets, Income and Supply (referred to as "Section 32") and CCC funds.[30] While both Section 32 and CCC have broad authority to support U.S. agriculture, many of their activities are required under various statutes, such as omnibus farm bills.[31] Beginning in FY2012, annual appropriations acts limited select discretionary uses of CCC and Section 32. The FY2018 omnibus appropriation removed this limitation for CCC and allowed for limited carryover funding under Section 32 to be used pending congressional notification.

- USDA's Section 32 program is funded by a permanent appropriation of 30% of the previous year's customs receipts. Funds are used for a variety of activities, including child nutrition programs, the purchase of commodities for domestic food programs, and farm disaster assistance. The statutory authority is Section 32 of the Agricultural Adjustment Act Amendment of 1935 (P.L. 74-320, 7 U.S.C. 612c). The authority to provide disaster assistance is attributed to clause 3 of Section 32, which provides that funds "shall be used to re-establish farmers'

---

[30] For additional information on Section 32, see CRS Report RL34081, *Farm and Food Support under USDA's Section 32 Program*. For additional information on the CCC, see CRS Report R44606, *The Commodity Credit Corporation: In Brief.*

[31] Most CCC- and Section 32-funded programs are classified as mandatory spending programs and therefore do not require annual appropriations.

purchasing power by making payments in connections with the normal production." The FY2020 appropriation (§714, P.L. 116-94) limits use of clause 3 to carryover funding of no more than $350 million following a two-week advance notice to Congress.

- The CCC serves as the funding institution for carrying out federal farm support programs, such as commodity price support and production programs, conservation programs, disaster assistance, agricultural research, and bioenergy development. It is federally chartered by the CCC Charter Act of 1948 (P.L. 80- 806), as amended. The authority to provide disaster assistance is attributed to Section 5 of the act (15 U.S.C. 714c), which, among other activities, authorizes the CCC to support the prices of agricultural commodities through loans, purchases, payments, and other operations.

Adjustments to Existing USDA Programs

In addition to implementing the disaster assistance programs discussed above, USDA can use authority under other existing programs to help producers recover from natural disasters. Examples of USDA adjustments to existing programs in recent years include:

- extending emergency grazing and haying on Conservation Reserve Program (CRP) acres;[32]
- reducing the emergency loan interest rate and making emergency loans available earlier in the season;

[32] Emergency grazing and haying on CRP acres are generally restricted during primary nesting season for certain wildlife. Authorization is provided either at the national FSA level or by a state FSA committee utilizing the U.S. Drought Monitor. For more information on CRP emergency grazing and haying requirements, see FSA, "Emergency Haying and Grazing," https://www.fsa.usda.gov/programs-and-services/conservation-programs/conservation-reserveprogram/emergency-haying-and-grazing/index.

- targeting conservation assistance through the Environmental Quality Incentives Program (EQIP), including providing additional funding for farmers and ranchers to implement conservation practices that prevent excessive soil erosion, restore streambanks, and respond to animal mortality (farmers and ranchers contribute about half the cost of implementing the practices);[33] and
- directing Emergency Community Water Assistance Grants for rural water systems experiencing emergencies resulting from a significant decline in quantity or quality of drinking water.[34]

Assistance to Prevent Spread of Animal Diseases

Under the Animal Health Protection Act (7 U.S.C. 8301, et seq.), USDA is authorized to take protective actions against the spread of livestock disease, including seizing, treating, or destroying animals if USDA determines that an extraordinary emergency exists because of the presence of a pest or disease of livestock. As part of its animal health program, USDA's Animal and Plant Health Inspection Service compensates producers for animals that must be euthanized, for their disposition, and for infected materials that must also be destroyed. Funding is provided by annual appropriations or through the CCC for larger amounts. The most recent example of a large-scale outbreak that resulted in payments to producers was in 2015 and 2016 during outbreaks of highly pathogenic avian influenza affecting the U.S. poultry industry.[35]

---

[33] For additional information on EQIP Disaster Recovery Assistance, see https://www.nrcs.usda.gov/Internet/FSE_MEDIA/nrcseprd1429025.pdf.

[34] USDA's Rural Utilities Service provides grants and loans for rural water systems in communities with fewer than 10,000 inhabitants. Its programs are for domestic water service, not water for agricultural purposes. For more information, see CRS Report R43408, *Emergency Water Assistance during Drought: Federal Non-Agricultural Programs*.

[35] For more information, see CRS Report R44114, *Update on the Highly-Pathogenic Avian Influenza Outbreak of 2014-2015*.

AD HOC ASSISTANCE

The U.S. farm policy mix that provides assistance to agricultural producers for damage and loss following a natural disaster continues to include both permanent and temporary authorized support. The authorization of permanent disaster assistance programs in the 2008 and 2014 farm bills,[36] as well as expanded crop insurance and NAP policies, were intended to reduce the need for ad hoc disaster assistance. Following enactment of the 2008 farm bill, Congress appropriated little in the way of supplemental disaster assistance for agriculture for several years, most of which was for land rehabilitation efforts under EWP and ECP. This changed in 2018, when an active hurricane and wildfire season in 2017 resulted in Congress authorizing supplemental assistance in the Bipartisan Budget Act of 2018 (P.L. 115-123) and again in 2019 in the FY2019 supplemental appropriation (P.L. 116-20). USDA has implemented this ad hoc assistance through two versions of the Wildfires and Hurricanes Indemnity Program (WHIP). USDA has also further implemented two subprograms—the On-Farm Storage Loss Program and the Milk Loss Program—as well as block grants to states and direct payments for acres prevented from planting.

Appropriations

The Bipartisan Budget Act of 2018 authorized $2.36 billion for production losses not covered under NAP or crop insurance in 2017.[37] The majority of this funding was made available through WHIP, referred to as *2017 WHIP*.[38]

---

[36] See the "Supplemental Agricultural Disaster Assistance Programs" section discussed above.
[37] CRS In Focus IF10829, *Agriculture Funding in the Bipartisan Budget Act of 2018*.
[38] USDA, "Nearly $2 Billion Now Available for Eligible Producers Affected by 2017 Hurricanes and Wildfires," press release, July 16, 2018, https://www.usda.gov/media/press-releases/2018/07/16/nearly-2-billion-now-available-eligibleproducers-affected-2017.

Only crop, tree, bush, and vine losses from a wildfire or hurricane in 2017 were directed by Congress as eligible for assistance under 2017 WHIP.[39] A portion of the funding was also used for a $340 million block grant to the State of Florida to assist the citrus industry.[40]

An FY2019 supplemental appropriation was enacted on June 6, 2019, which authorized an additional $3 billion to USDA to cover crop, tree, bush, and vine losses from 2018 and 2019 natural disasters that were not covered under crop insurance or NAP.[41] The majority of the FY2019 supplemental appropriation funding was announced by USDA as an expanded version of WHIP, referred to as *WHIP+*. Congress also required payments for additional losses not previously covered, including crops that were prevented from being planted in 2019, on-farm stored commodities, and milk losses. The FY2019 supplemental appropriation authorized, and USDA made available, block grants to select states to cover livestock, poultry, and forestry losses.[42]

Approximately 60% of 2017 WHIP went unobligated before funds expired on December 31, 2019. The FY2020 Further Consolidated Appropriation Act (P.L. 116-94, Division B, §791) repurposed those funds to WHIP+ (that is, rescinded the 2017 WHIP unobligated balance, estimated to be $1.5 billion, and reallocated it to WHIP+) with expanded eligibility and additional program requirements.

---

[39] For additional information, see FSA, *2017 Wildfires and Hurricanes Indemnity Program (WHIP)*, https://www.fsa.usda.gov/programs-and-services/disaster-assistance-program/wildfires-and-hurricanes-indemnityprogram/index. The FY2019 supplemental expanded eligibility under 2017 WHIP to include losses from Tropical Storm Cindy, peach and blueberry losses from extreme cold in calendar year 2017, and blueberry productivity losses from extreme cold and hurricanes in calendar year 2018.

[40] An additional $3.2 million was available for administrative expenses. FSA, "Perdue Announces Florida Citrus Hurricane Recovery Details," press release, May 1, 2018, https://www.fsa.usda.gov/news-room/news-releases/2018/nr_20180501_rel_0068.

[41] CRS in Focus IF11245, *FY2019 Supplemental Appropriations for Agriculture*.

[42] FSA, "USDA Announces Block Grants for Three States as Part of Broader Disaster Relief Package," press release, November 8, 2019, https://www.fsa.usda.gov/news-room/news-releases/2019/usda-announces-block-grants-for-threestates-as-part-of-broader-disaster-relief-package.

Wildfires and Hurricanes Indemnity Program Plus (WHIP+)

WHIP+ provides assistance payments to cover a portion of production losses in calendar years 2018 and 2019 from Hurricanes Michael or Florence, other hurricanes, floods, tornadoes, typhoons, volcanic activity, snowstorms, wildfires, drought, excessive moisture, and other natural disasters.[43] Losses may include crops, trees, bushes, and vines. Payments for WHIP+ are based on several factors, including (1) the expected value of the crop, (2) the level of insurance or NAP coverage, (3) a WHIP+ payment factor, (4) the value of the crop harvested, and (5) insurance payments received. Payment calculations vary based on production loss; value of crop loss; and tree, bush, and vine loss.[44] An additional payment factor varies by state and commodity and is set by USDA to reflect the decreased costs incurred by producers when the crop is not harvested or is prevented from being planted. Table 3 outlines the USDA-created WHIP+ payment calculation. All WHIP+ participants are required to purchase crop insurance or NAP for the next two crop years.

Payment limits for WHIP+ are based on an individual's or entity's average AGI over a three-year period—2015, 2016, and 2017—depending on how much of that income is derived from farming.[45] Generally, payments are limited to $125,000 per person or legal entity. If more than

---

[43] The FY2020 appropriation expanded losses for WHIP+ to include quality loss, drought (D3 U.S. Drought Monitor intensity or higher), and excessive moisture. USDA opened WHIP+ sign up for the expanded set of losses on March 23, 2020. USDA, "USDA Opens Wildfires and Hurricane Indemnity Program Plus (WHIP+)," press release, February 28, 2020, https://www.usda.gov/media/press-releases/2020/02/28/usda-opens-wildfires-and-hurricane-indemnity-programplus-whip. The FY2020 appropriation also expanded WHIP+ to losses previously eligible under 2017 WHIP, including losses from Tropical Storm Cindy, peach and blueberry losses due to freeze in 2017 and blueberry productivity losses in 2018. USDA reopened 2017 WHIP sign up for these losses on February 10, 2020. FSA, *2017 WHIP Policy Update*, Notice WHIP-14, February 10, 2020, https://www.fsa.usda.gov/Internet/FSA_Notice/whip_14.pdf.

[44] For additional information, see *FSA, Wildfire and Hurricane Indemnity Program-Plus (WHIP+),* October 2019, https://www.fsa.usda.gov/Assets/USDA-FSA-Public/usdafiles/FactSheets/2019/wildfire-and-hurricane-indemnityprogram-plus_whip.pdf.

[45] Producers are assumed to be in the lowest payment limit category unless an exception to the payment limit is filed with USDA and documentation from a certified public accountant or attorney states that at least 75% of the person's or legal entity's average AGI was from adjusted gross farm income. Farm income includes income received or obtained from farming, ranching, and forestry operations.

75% of the participant's AGI is from farming, then payments are limited to $250,000 for each crop year with an overall payment limit of $500,000 for all crop years. The AGI eligibility threshold that apply to farm commodity support programs and permanent disaster assistance programs, such as the "Supplemental Agricultural Disaster Assistance Programs" described above, do not apply in determining WHIP+ eligibility.

Table 3. WHIP+ Payment Calculation and Definition of Terms

Payment Calculation		Definition of Terms
Expected Value		Revenue the producer would have received if the crop had been harvested based on historical yield.
X	WHIP+ Payment Factor	Based on crop insurance or NAP coverage. Higher coverage = higher payment factor. Ranges from 75%-95% for participates that purchased crop insurance or NAP and 70% if uninsured.
–	Actual Value	Revenue received for the crop unit's production for the intended use and coverage period.
–	Salvage Value	Secondary use and salvage value payments received for the crop, by type, and planting period.
X	Payment Factor	Applied to payments when significant and variable harvesting expenses are not incurred because the crop acreage was either prevented from being planted or planted but not harvested. Does not apply to tree, bush, and vine loss.
–	Indemnity Payment	RMA indemnities and NAP payments received for the crop, by type, intended use, and planting period.
= WHIP+ Payment		

Source: CRS using 7 C.F.R. §760; *FSA, Wildfire and Hurricane Indemnity Program-Plus (WHIP+)*, October 2019; and FSA WHIP+ handbook (2-WHIP), Part 10, http://www.fsa.usda.gov/Internet/FSA_File/2-whip_r00_a04.pdf.

Notes: All payments are adjusted based on the participant's percentage share of the crop. Some areas and crops may have altered payment calculations from the one shown in this table. All WHIP+ participants should inquire about eligibility and payment calculations with their local FSA offices. The calculation does not reflect payments made under 2017 WHIP.

State Block Grants

Both the Bipartisan Budget Act of 2018 and the FY2019 supplemental appropriation provided USDA with the flexibility to provide assistance through block grants. One block grant was issued under 2017 WHIP to the State of Florida to assist the citrus industry with the cost of buying and planting replacement trees—including resetting and grove rehabilitation—and for repairing damage to irrigation systems, among other activities.[46]

Under WHIP+, USDA announced over $800 million for block grants with Alabama, Florida, and Georgia for losses of timber, cattle, and poultry as well as necessary expenses related to losses of horticulture crops and pecan production.[47]

The FY2020 appropriation required that USDA use an additional $400 million of the repurposed funds for WHIP+ through block grants with eligible states.[48]

On-Farm Storage Loss Program

The FY2019 supplemental appropriation specifically covered the loss of commodities stored on farm due to a qualifying natural disaster in 2018 and 2019.[49] USDA administered this through an On-Farm Storage Loss Program (OFSLP).[50] The program provides payments for the uncompensated losses of harvested commodities including grains, oilseeds, and hay stored in on-farm structures.[51] Only losses related to eligible disaster events (e.g., floods, tornadoes, snowstorms, and wildfires) that occurred during 2018 or 2019 may receive payments. Payments are calculated using a national payment rate per commodity based on 75% of the market or harvest price. A person or legal entity is limited to no more than $125,000 per loss year. OFSLP payments are for the loss of the stored commodity, not the structure itself.

---

[46] USDA, "Perdue Announces Florida Citrus Hurricane Recovery Details," press release, May 1, 2018, https://www.usda.gov/media/press-releases/2018/05/01/perdue-announces-florida-citrus-hurricane-recovery-details.

[47] USDA, "USDA Announces Block Grants for Three States as Part of Broader Disaster Relief Package."

[48] P.L. 116-94, Division B, §791(c).

[49] Most crop insurance policies do not cover the loss of a crop once harvested unless a supplemental policy is purchased. Trade disruptions in 2018 and 2019 resulted in higher than normal commodity stocks stored on-farm. Heavy rains and flooding in 2018 and 2019, combined with these large commodity stocks, resulted in numerous losses of on-farm stored commodities that were ineligible for indemnity payments.

[50] 7 C.F.R. §760.

[51] Examples of eligible OFSLP commodities include barley, canola, chickpeas (large and small), corn, cotton, crambe, dry peas, flaxseed, grain sorghum, hay (alfalfa and all-hay), lentils, mustard seed, oats, peanuts, rapeseed, rice, safflower seed, sesame seed, soybeans, sunflower seed, and wheat.

Milk Loss

The FY2019 supplemental appropriation specifically covered the loss of milk due to a qualifying natural disaster during 2018 and 2019. USDA administered this assistance through a Milk Loss program (WHIP-ML).[52] The WHIP-ML program provides indemnity payments to eligible dairy operators for milk that was dumped or removed without compensation from the commercial milk market due to qualifying natural disasters (e.g., hurricanes, floods, tornadoes, typhoons, and wildfires).[53] In some cases, dairy operations received a partial payment for milk that was dumped through a marketing organization, insurance, or other source. Only the portion of milk that was not compensated for is eligible for WHIP-ML. Payments are limited to $125,000 per year per person or legal entity.

Prevented Planting

The FY2019 supplemental appropriation specially covered losses for crops in 2019 that were either prevented from being planted or delayed in begin planted. USDA implemented this provision by providing a *top-up* payment to producers who claimed prevent plant losses (PPL) in 2019 through their federal crop insurance policy.[54] In 2019, payments were automatically issued to producers with an approved non-drought-related PPL indemnity due to excessive moisture, flooding, or other non-drought causes. Payments varied based on the type of crop insurance policy purchased, and payments ranged from 10% to 15% of the PPL indemnity received.

Producers who received top-up payment are required to purchase crop insurance for the next two crop years, similar to WHIP+ requirements. As of November 2019, USDA has paid over $570 million in top-up payments.[55]

---

[52] 7 C.F.R. §760.

[53] 7 C.F.R. §760.

[54] For additional information on PPL, see CRS Report R46180, *Federal Crop Insurance: Record Prevent Plant (PPL) Acres and Payments in 2019*.

[55] USDA, "USDA Announces Block Grants for Three States as Part of Broader Disaster Relief Package."

Sugar Beets

The FY2020 appropriation required that all sugar beet losses in 2018 and 2019 be paid through cooperative processors.[56] USDA announced that $285 million is to be provided through sugar beet processing cooperatives to member growers.[57] Cooperatives are to enter into an agreement with USDA to administer assistance to their members.[58] Sugar beet producers who are not members of a cooperative processor may apply for WHIP+ through FSA.

---

[56] P.L. 116-94, Division B, §791(c).

[57] USDA, "USDA Opens Wildfires and Hurricane Indemnity Program Plus (WHIP+)."

[58] As of April 14, 2020, approved sugar beet processors include American Crystal Sugar Company, Michigan Sugar Company, Minn-Dak Farmers Cooperative, Snake River/Amalgamated Sugar Company, Southern Minnesota Beet Sugar Cooperative, Western Sugar Cooperative, and Wyoming Sugar Company.

In: COVID 19: Agricultural and Food Issues ISBN: 978-1-53618-561-4
Editor: Kristian Goldschmidt

Chapter 4

FOOD SAFETY AND COVID-19*

Renée Johnson

Coronavirus Disease 2019 (COVID-19) is thought to spread mainly from person to person through respiratory droplets produced when an infected person sneezes, coughs, or speaks. Currently there is no evidence indicating the transmission of COVID-19 through food, according to U.S. and international public health agencies. However, China—a major U.S. agricultural export market—has reportedly started to institute requirements to address COVID-19-related concerns about the safety of imported food, which could impact global agricultural trade.

EVIDENCE OF FOOD SAFETY RISK

In the United States, the Centers for Disease Control and Prevention (CDC) states "there is no evidence to suggest that handling food or consuming food is associated with COVID-19." The U.S. Department of

---

* This is an edited, reformatted and augmented version of Congressional Research Service Publication No. IN11453, dated July 9, 2020.

Agriculture (USDA) states that the agency is "not aware of any reports at this time of human illnesses that suggest COVID-19 can be transmitted by food or food packaging." USDA further indicates there is no evidence to support transmission of COVID-19 associated with food imported to the United States from other countries. While there may be the potential for transmission through contaminated surfaces—for example, if someone with an active infection coughs on packaging later touched by a noninfected person who then touches their eyes, nose, or mouth—the risk of such transmission is low, according to the CDC. To avoid transfer through contaminated surfaces, it is necessary to follow good hygiene practices.

At food production facilities, the Food and Drug Administration (FDA), an agency of the Department of Health and Human Services, asserts the need to take necessary precautions to protect workers and minimize the disease transmission, including practicing social distancing and disinfecting surfaces, as well as ensuring workers have access to personal protective equipment (PPE) such as masks and gloves.

While some labeling requirements have been temporarily relaxed to provide regulatory flexibility to food producers related to COVID-19 (for example, certain food nutrition labeling requirements, packaging and labeling requirements for shell eggs, and menu labeling requirements), U.S. food safety standards and requirements have otherwise remained in effect. Because of the ongoing crisis, however, some facility inspections have been delayed or rescheduled. For the meat and poultry industries, CDC and the Occupational Safety and Health Administration have issued guidance outlining steps that meatpacking and meat processing workers and employers should follow to reduce the risk of exposure to COVID-19.

There is also a consensus among international organizations that COVID-19 transmission is not associated with food and food consumption. The World Health Organization (WHO), an agency of the United Nations, states there is "currently no evidence that people can catch COVID-19 from food or food packaging." WHO further claims coronaviruses cannot multiply in food but require a live animal or human host to multiply and survive. WHO considers food packaging, preparation, and delivery safe if proper hygiene protocols are followed such as hand washing and sanitization. WHO has issued guidance for governmental authorities responsible for national food safety control systems to contain widespread food safety risks and to reduce serious disruption to national food safety programs. The guidance also emphasizes the need for PPE to help reduce the spread of disease among workers within the food sector.

The Food and Agriculture Organization (FAO) of the United Nations has also issued guidance for additional food safety measures related to COVID-19 for food businesses and wholesale food markets.

WHO, FAO, and the World Trade Organization (WTO) have acknowledged the role of food safety standards in promoting fair practices in food trade. They have called on national governments to minimize the impact of COVID-19-related border restrictions on food trade to ensure food safety and food security.

POTENTIAL TRADE IMPLICATIONS

Despite lack of evidence that COVID-19 is transmitted through food, concerns that restrictions related to COVID-19 are affecting trade have been reported by U.S. agricultural producers regarding shipments to certain countries. Specifically, in China, there have been reports of COVID-related product testing for meat, seafood, fresh fruit, and bulk grains. According to U.S. fruit and vegetable exporters, China has also instituted administrative measures requiring registration of foreign manufacturers and a Letter of Commitment from foreign suppliers certifying that food shipments are not contaminated with coronavirus. Such additional regulatory requirements have the potential to impact global trade and could reduce or block U.S. food and agricultural exports.

In response, USDA and FDA issued a joint statement claiming the United States is "taking every necessary precaution to prioritize food safety especially during these challenging times" and reiterating that there is "no evidence that people can contract COVID-19 from food or from food packaging." The European Union has also stated "there has been no report of transmission of COVID-19 via consumption of food to date," which is further supported by the findings of its European Food Safety Authority.

Both the United States and the European Union have issued statements to the WTO's Committee on Sanitary and Phytosanitary Measures arguing that these types of border restrictions are not based on risk. Under WTO rules, global food safety regulations and standards are to operate within the framework of internationally accepted trade rules and norms, including safety and public health protections addressed in the multilateral Agreement on the Application of Sanitary and Phytosanitary Measures (SPS Agreement) and the Agreement on Technical Barriers to Trade. Under the SPS Agreement, for example, countries may adopt and enforce public health and safety measures "necessary to protect human, animal or plant life or health," so long as such measures are not applied in an arbitrary or discriminatory manner. Yet trade restrictions citing food safety concerns are often implemented through various SPS measures and other technical regulations such as testing, registration, certification, and labeling requirements.

Congress may examine the extent to which U.S. food and agricultural exports might be affected by such trade restrictions as the United States and its trading partners continue to grapple with the ongoing COVID-19 crisis.

In: COVID 19: Agricultural and Food Issues ISBN: 978-1-53618-561-4
Editor: Kristian Goldschmidt

Chapter 5

COVID-19 AND GLOBAL FOOD SECURITY: ISSUES FOR CONGRESS*

Alyssa R. Casey and Emily M. Morgenstern

Some Members of Congress have demonstrated an ongoing interest in global food security, including how the Coronavirus Disease 2019 (COVID-19) pandemic may affect global food security needs in 2020. Some policymakers are examining whether existing U.S. global food security programs are sufficient to meet potentially increased needs. They are also considering what legislative changes, if any, might help address the emerging global food assistance challenges caused by COVID-19.

GLOBAL FOOD SECURITY OUTLOOK

Food security generally refers to individuals' ability to access food that meets their food preferences and dietary needs to lead a safe and healthy

* This is an edited, reformatted and augmented version of Congressional Research Service Publication No. IF11575, dated June 15, 2020.

life. Global food security worsened in 2019, according to the Global Network Against Food Crises and Food Security Information Network's Global Report on Food Crises 2020, which relied on data collected prior to the COVID-19 pandemic. The estimated number of acutely food-insecure people totaled 135 million in 2019, an increase of 22 million from 2018. (Acute food insecurity occurs when food deficits immediately endanger individuals' lives or livelihoods.) This deterioration reflects continued widespread conflict and health crises, and an increase in severe natural disasters like drought, hurricanes, and pest infestations. In addition to the acutely food insecure, the United Nations (U.N.) Food and Agriculture Organization (FAO) estimated that millions more experienced hunger, including chronic food insecurity, undernutrition, and malnutrition (821 million people experienced hunger in 2018; complete data for 2019 are not yet available).

EFFECTS OF COVID-19

Research suggests the COVID-19 pandemic is worsening the lack of consistent access to enough food both in already vulnerable populations and in previously food secure populations. The U.N. World Food Program (WFP) estimates that up to 265 million people may face acute food insecurity by the end of 2020, a 96% increase from its 2019 estimate. Research suggests that COVID-19 is constraining access to food for some populations, primarily through supply chain disruptions and declining household purchasing power. The FAO projects that global supplies of staple commodities, such as wheat and rice, will remain stable. However, social distancing and movement restrictions have resulted in supply chain disruptions, such as delays at ports and border crossings, and migrant workers being unable to travel for harvesting and food- processing work. Commodities such as fruits, vegetables, and fish are more perishable and labor-intensive than staple commodities, and thus more susceptible to transportation and labor force disruptions.

COVID-19 is also causing widespread job and income losses, largely due to movement restrictions and slowingeconomies. As incomes decline, remittances, a key source of income for many poor households in some countries, are also expected to fall. Increased health care and sanitation costs related to COVID-19 may further strain household food budgets. Displaced populations and populations in conflict areas that were already struggling to access food before COVID-19 will be more vulnerable to supply chain disruptions and movement restrictions. Experts anticipate that these and other factors will negatively affect the ability of many households to afford food, driving up acute food insecurity in 2020.

COUNTRY AND MULTILATERAL RESPONSES

Some foreign governments and multilateral institutions are providing financial or other assistance to mitigate the food insecurity impacts of COVID-19. For example, some governments, such as those of Ethiopia, Kyrgyzstan, and Nigeria, have instituted cash transfer and food aid programs for their citizens. In addition, the World Bank and WFP are providing cash transfers and food aid to vulnerable populations. WFP has also transitioned many of its school meal programs to provide take home rations, and it is reinforcing its stocks of pre-positioned food at storage hubs around the world to prepare for a potential need to further scale up food aid.

U.S. POLICY

Congress funds a range of international food assistance programs that may be used to address global food security needs through annual Agriculture and Department of State, Foreign Operations, and Related Programs appropriations. The U.S. Department of Agriculture (USDA) and the U.S. Agency for International Development (USAID) implement international food assistance programs. USAID provides emergency food assistance through the Food for Peace Title II Program (Title II) and the

Emergency Food Security Program (EFSP). USDA administers the Bill Emerson Humanitarian Trust (BEHT), a reserve of funds that can supplement emergency assistance in years with unanticipated food assistance needs. USDA also administers nonemergency food assistance through the McGovern-Dole Food for Education and Child Nutrition (McGovern-Dole), and Food for Progress (FFPr) programs. In FY2020, Congress provided more than $4 billion for these and other related food assistance programs, compared with $4.6 billion in FY2019.

Both USAID and USDA have sought to adapt their international food assistance programs in response to growing food insecurity amidst COVID-19. USAID has targeted its emergency food assistance not only to countries already receiving food assistance that saw their needs increase, but also countries that previously did not receiveU.S. food assistance but developed food security needs as a result of the pandemic. USAID is also modifying existing development programs to address longer-term food needs primarily through agricultural development and livelihood programs. For example, according to USAID, the agency's Fish Right program in the Philippines, which predominantly focuses on biodiversity and marine conservation, launched an online platform in an effort to safely connect fisherfolk with consumers during COVID-19 movement restrictions and to help "protect fisherfolk income and prevent a food crisis."

USDA has also sought to adapt its programming due to COVID-19. For example, in the wake of school closures due to COVID-19, some McGovern-Dole projects have pivoted from providing school meals to providing take- home rations. FFPr provides *in-kind* food assistance—U.S. food shipped to the recipient country—and technical assistance to local farmers. In April 2020, USDA published notices of supplemental funding opportunities for McGovern-Dole and FFPr in response to reports of increased transportation or distribution costs due to COVID-19-related movement restrictions.

SELECTED ISSUES FOR CONGRESS

Over the years, Congress has granted USDA and USAID increased programming flexibilities in response to changing global needs and to agencies' advocacy for a "right tool, right time" approach. These include the limited use of market-based approaches (e.g., cash transfers for food, food vouchers, and locally and regionally procured food [LRP]) in Title II, the establishment and expansion of the market- based EFSP, and the use of LRP in McGovern-Dole.

However, COVID-19 presents a unique set of challenges, particularly those related to the distribution of food assistance. For example, in-kind food aid requires distribution sites, which can become overcrowded and potentially increase the risk of spreading COVID-19 among beneficiaries. As Congress considers increasing funds for food aid programs, Members may explore whether or not existing programs have the authorities to meet current needs while addressing COVID-19-related challenges.

FFP TITLE II FUNDING STRUCTURE

To address the anticipated global food crisis, and to support U.S. farmers, some in Congress have considered providing supplemental funding for Title II. When programming Title II commodities, USAID partly relies on funding for associated costs authorized in Section 202(e) of the Food for Peace Act, as amended (7 U.S.C. §1722(e)). Section 202(e) states that not less than 7.5% and not more than 20% of Title II funds shall be made available for certain associated costs. Examples of these costs include staff salaries, milling or fortifying commodities, and distribution site maintenance.

The Section 202(e) statutory cap may present a challenge to, and possibly prevent, the purchase and programming of Title II commodities. Some organizations implementing Title II projects have reported increased associated costs due to COVID-19-related project modifications (e.g., additional food distribution sites to reduce crowding and/or adding soap and other hygiene materials to food baskets). These modifications, which would increase Section 202(e) costs, have presented a challenge to USAID. If Section202(e) costs risk exceeding the 20% cap, USAID may need to either reject the project modifications or decrease the overall amount of Title II assistance.

To address this challenge, some in the food aid community have advocated for Congress to lift the 20% cap on Section 202(e). This has prompted considerable debate. Advocates of lifting the cap assert that the funds are necessary to safely program Title II commodities during the pandemic; opponents have expressed concern that lifting the caps could reduce the portion of Title II funds used to purchase commodities, to the detriment of U.S. farmers.

Potential Increase in EFSP Funds

Some Members and stakeholders have advocated for increased funds for EFSP, arguing that its market-based interventions offer more flexibility than in-kind aid through Title II or BEHT and would be particularly helpful in adapting to COVID-19-related challenges (e.g., a mobile cash transfer for food means that an individual would not need to interact with another to receive assistance). Others suggest that there are situations in which in-kind food assistance remains the most appropriate intervention, particularly in areas in which food is sparse. Further, International Disaster Assistance (IDA), the account through which Congress funds EFSP, also provides for nonfood assistance, such as hygiene and medical supplies. Some assert that directing more funds toward EFSP would reduce available IDA funding for necessary health supplies.

As of this writing, two COVID-19 supplemental acts, P.L. 116-123 and P.L. 116-136, provided a combined $558 million for IDA. USAID has indicated that $100 million will be used for EFSP and the remainder will fund nonfood assistance, such as hygiene and medical supplies.

POSSIBLE USE OF RESERVE FUNDS

Some stakeholders have suggested using BEHT to bolster U.S. international food assistance in the wake of COVID-19. BEHT funds are subject to the same constraints on associated costs as Title II. While BEHT would allow for providing additional in-kind assistance, the program would face the same associated cost challenges as Title II.

Currently, Congress has authorized BEHT to finance only standalone in-kind assistance projects. BEHT funds cannot be used to cover increased associated costs for projects funded under other food assistance programs, such as Title II. However, some in Congress have considered amending BEHT authorities to allow funds to be used for the associated costs of existing Title II projects. This could allow USAID to cover additional Title II costs without requiring increased flexibilities in the Title II program (such as increasing the Section 202(e) cap). The potential impact of this change on the total amount of in-kind food assistance is difficult to estimate. Less BEHT funding might be used to purchase commodities. However, using BEHT to cover some Title II associated costs would allow USAID to purchase more commodities under Title II.

In: COVID 19: Agricultural and Food Issues ISBN: 978-1-53618-561-4
Editor: Kristian Goldschmidt

Chapter 6

FOOD BANKS AND OTHER EMERGENCY FEEDING ORGANIZATIONS: FEDERAL AID AND THE RESPONSE TO COVID-19*

Kara Clifford Billings, Erica A. Lee and Elizabeth M. Webster

ABSTRACT

During the Coronavirus Disease 2019 (COVID-19) pandemic, rising unemployment rates and other factors led an increasing number of Americans to seek food assistance. Food banks, food pantries, and other emergency feeding organizations—sometimes referred to as the *charitable food assistance network*—are often the first place people go when they become food insecure, and these organizations have experienced an influx in demand during the pandemic. At the same time, they have had to address safety concerns resulting from COVID-19, often

---

* This is an edited, reformatted and augmented version of Congressional Research Service, Publication No. R46432, dated June 25, 2020.

operating with fewer staff or utilizing the National Guard to distribute food, and setting up drive-through and mobile food distributions.

While the charitable food assistance network relies heavily on private donations, federal programs also provide a significant source of food and revenue for emergency feeding organizations. The primary federal programs that support food banks and other emergency feeding organizations are administered by the U.S. Department of Agriculture (USDA) and the Department of Homeland Security's (DHS's) Federal Emergency Management Agency (FEMA). Ongoing programs, including The Emergency Food Assistance Program (TEFAP) and Emergency Food and Shelter Program (ESFP), have received supplemental appropriations during the pandemic. Other programs have been activated (USDA's Disaster Household Distribution program and FEMA's Public Assistance (PA) program) or created (USDA's Farmers to Families Food Box program) in response to the pandemic. Recent legislation has proposed additional funding for existing programs as well as the creation of new programs and funding streams for emergency feeding organizations.

INTRODUCTION

In April 2020, news reports of farmers dumping milk and plowing under crops corresponded with footage of cars lining up for miles outside of some food banks in the United States as a result of the COVID-19 pandemic.[1] Many of these stories raised concerns about farmers' livelihoods as well as the rising number of unemployed and food insecure Americans. They also generated increased interest in federal aid for food banks, food pantries, and other similar organizations feeding Americans, including the newly unemployed.

Such organizations are called emergency feeding organizations, and they comprise the emergency food assistance network (also known as the charitable food assistance network). The term *emergency feeding*

---

[1] For example, see D. Yaffe-Bellany and M. Corkery, "Dumped Milk, Smashed Eggs, Plowed Vegetables: Food Waste of the Pandemic," *New York Times,* April 11, 2020, https://www.nytimes.com/2020/04/11/business/coronavirusdestroying-food.html; J. Burnett, "Thousands Of Cars Line Up At One Texas Food Bank As Job Losses Hit Hard," *NPR,* April 17, 2020, https://www.npr.org/2020/04/17/837141457/thousands-of-cars-line-up-at-one-texas-food-bankas-job-losses-hit-hard; and AP News, "1,000 cars line up for Nebraska food handout," April 12, 2020, https://apnews.com/56decced810f1d737eae0d8201df9909.

organization, as used in this chapter, refers to institutions—mostly private nonprofit institutions—that facilitate the provision of food to households and individuals in times of need or emergency (*emergency*, in this sense, refers to personal or household emergencies rather than large-scale disasters or emergencies).[2] Emergency feeding organizations include food banks, food pantries, shelters, soup kitchens, and hunger relief agencies. *Food banks* are typically regional warehouses that store and deliver food to smaller, client-facing organizations like food pantries, shelters, and soup kitchens. In some cases, food banks may also provide food directly to individuals and households.

In recent years, new types of organizations have arguably become part of the emergency food assistance network, such as restaurants and chef-led organizations like World Central Kitchen, which provides emergency food assistance in the United States and internationally.[3] During the COVID-19 pandemic, many schools have also become emergency food providers. While schools qualify for some of the programs discussed in this chapter, the primary sources of federal aid for schools' feeding efforts—the federal child nutrition programs—are not discussed here.[4]

The emergency food assistance network has historically relied heavily on private donations and state and local funding in addition to federal aid. For example, Feeding America (a nonprofit membership and advocacy organization)[5] reported that federal commodities represented one-quarter of

---

[2] This definition is similar to the characterization of the charitable food assistance system in D. O'Brien, E. Staley, S. Uchima, E. Thompson, and H.T. Aldeen, *The Charitable Food Assistance System: The Sector's Role in Ending Hunger in America*, The UPS Foundation and the Congressional Hunger Center, 2004. A definition of *emergency feeding organization* can also be found in Section 201A of the Emergency Food Assistance Act (7 U.S.C. §7501).

[3] World Central Kitchen, "Our Mission," https://wck.org/mission.

[4] For more information on child nutrition program operations during the pandemic, see U.S. Department of Agriculture (USDA), Food and Nutrition Service (FNS), "FNS Response to COVID-19: Child Nutrition Programs," https://www.fns.usda.gov/disaster/pandemic/covid-19. For background on the child nutrition programs, see CRS Report R46234, *School Meals and Other Child Nutrition Programs: Background and Funding*.

[5] The Feeding America network represents a large segment of emergency feeding organizations nationwide—80% of such organizations were affiliated with Feeding America in 2000, according to a study published in 2002 (J.C. Ohls et al., "The Emergency Food Assistance System—Findings From the Provider Survey," prepared by Mathematica Policy Research,

the meals served by its network of 200 food banks and 60,000 food pantries in FY2019.[6]

This chapter discusses the sources of federal aid available to emergency feeding organizations, including new federal programs created during the COVID -19 pandemic. The primary federal programs that support food banks and other emergency feeding organizations are administered by the U.S. Department of Agriculture (USDA) and the Department of Homeland Security's (DHS's) Federal Emergency Management Agency (FEMA). The focus of this chapter is on federal programs that have a specific emphasis on emergency feeding organizations. However, there are other programs that are not discussed in this chapter that may provide funding for emergency feeding organizations among other grant recipients; these include the Commodity Supplemental Food Program (CSFP), Milk Donation Reimbursement Program (MDRP), and Food Distribution Program on Indian Reservations (FDPIR).[7]

The emergency food assistance programs discussed in this chapter can be distinguished from other federal nutrition assistance programs for households such as the Supplemental Nutrition Assistance Program (SNAP) in that they provide in-kind (food) aid rather than near-cash benefits to households. Food banks and other emergency feeding organizations serve households that do not participate in SNAP, such as households with incomes that are too high to qualify and ineligible noncitizen households. They also provide supplemental foods for SNAP households; an analysis by USDA's Economic Research Service (ERS) found that roughly half of low-income households that used food pantries and soup kitchens in 2018 also received SNAP benefits.[8] Policy proposals

---

Inc. for USDA, Economic Research Service, Food and Rural Economics Division, October 2002, p. 2, https://www.ers.usda.gov/publications/pub-details/?pubid=46507).

[6] CRS communication with Feeding America on June 2, 2020.

[7] For more information on CSFP and FDPIR, see CRS Report R42353, *Domestic Food Assistance: Summary of Programs*. For more information on MDRP, see CRS In Focus IF11188, *2018 Farm Bill Primer: Dairy Programs*.

[8] Among households with annual incomes below 185% of the poverty guidelines, 52.7% that used food pantries and 57.8% that used emergency kitchens in 2018 also received SNAP benefits in the previous month, according to A. Coleman-Jensen, M.P. Rabbitt, C.A. Gregory, and A. Singh, *Statistical Supplement to Household Food Security in the United*

to address increased food insecurity during the COVID-19 pandemic may focus on distributing foods through the emergency food assistance network, providing near-cash benefits (e.g., SNAP or unemployment benefits) to help households purchase foods, or other approaches.

IMPACT OF COVID-19 ON EMERGENCY FEEDING ORGANIZATIONS

The COVID-19 pandemic has caused an influx in demand for food banks and other emergency feeding organizations.[9] At the same time, it has posed unique challenges for many of these organizations, including a decline in private donations from grocery stores, restaurants, and individuals, fewer or more expensive foods available for purchase, and reduced ability to utilize volunteers for safety reasons.[10]

Prior to the pandemic, analyses of national survey data found that 5% to 10% of households utilized food pantries, food banks, soup kitchens, or other charitable food assistance.[11]

States in 2018, AP-081, USDA, Economic Research Service, September 2019, pp. 20-21, https://www.ers.usda.gov/publications/pub-details/?pubid=94869.

[9] For example, see N. Kulish, "'Never Seen Anything Like It': Cars Line Up for Miles at Food Banks," *New York Times,* April 8, 2020, https://www.nytimes.com/2020/04/08/business/economy/coronavirus-food-banks.html and A. Vesoulis, "'It's a Bucket Brigade on a 5-Alarm Fire.' Food Banks Struggle to Keep Up With Skyrocketing Demand," *Time,* April 24, 2020, https://time.com/5825944/food-banks-coronavirus/.

[10] For example, see L. Reiley, "Food banks are seeing volunteers disappear and supplies evaporate as coronavirus fears mount," *Washington Post,* March 16, 2020, https://www.washingtonpost.com/business/2020/03/16/food-banks-are- seeing-volunteers-disappear-food-supply-evaporate-coronavirus-fears-mount/; and Feeding America, "Feeding America Food Bank Network Projects $1.4 Billion Shortfall Due T o The COVID-19 Crisis," April 1, 2020, https://www.feedingamerica.org/about-us/press-room/feeding-america-food-bank-network-projects-14-billionshortfall- due-covid-19.

[11] According to an analysis of Current Population Survey (CPS) data by USDA ERS, an estimated 5.7 million households (4.4%) utilized food pantries and at least 657,000 households (0.5%) utilized soup kitchens in 2018 (A. Coleman-Jensen, M.P. Rabbitt, C.A. Gregory, and A. Singh, *Statistical Supplement to Household Food Security in the United States in 2018*, AP-081, USDA, ERS, September 2019). However, this is likely an underestimate of the population using emergency food assistance because the sample did not include certain households with incomes over 185% of the poverty guidelines and the CPS does not fully capture households who are homeless or in tenuous housing arrangements. For comparison, an analysis of a survey by the Urban Institute (E. Waxman,

During the pandemic, Feeding America has administered biweekly surveys to its 200 food bank members.[12] Based on one of these surveys, conducted from May 4, 2020, to May 6, 2020, Feeding America reported that 98% of the network's food banks experienced an increase in demand compared to the same time in 2019. The reported increase in clients ranged from 10% to 200%, with an average increase of 59%.[13] The surveyed food banks estimated that nearly 4 out of 10 clients were new to charitable food assistance.

In addition to the challenges of increasing demand, news outlets reported that many food banks and pantries have turned away volunteers during the pandemic out of safety concerns, particularly for high-risk populations such as older volunteers. Several states have mobilized the National Guard to help fill the gap and meet increasing demand.[14] The Feeding America survey found that "30 percent of food banks are still accepting and in need of volunteers, while 53 percent of food banks or agency partners have deployed the National Guard for support."[15]

Whether because of safety, financial, or other factors, the Feeding America survey found that approximately 18% of the food pantries and meal programs that partner with its food banks have closed or suspended operations during the pandemic.[16]

N. Joo, and D. Gonzalez, "Who Is Accessing Charitable Food in America?," *Urban Institute,* December 2019, https://www.urban.org/research/ publication/who-accessing-charitable-food-america) found that 10.3% of adults reported that they or someone in their household used charitable food (defined as receiving free groceries or a free meal in the 30 days before the survey).

[12] CRS communication with Feeding America on June 2, 2020.

[13] Feeding America, "Feeding America Network Stays Resilient During COVID-19 Crisis," May 12, 2020, https://www.feedingamerica.org/about-us/press-room/feeding-america-network-stays-resilient-during-covid-19-crisis.

[14] For example, see A. Quigley, "National Guard to distribute food at sites across Vermont," *VTDigger,* April 21 2020, https://vtdigger.org/2020/04/21/national-guard-to-distribute-food-at-sites-across-vermont/; E. Wilkinson, "National Guard deployed to help Washington food banks," *King5,* April 3, 2020, https://www.king5.com/article/news/health/ coronavirus/ national-guard-deployed-t o-help-washington-food-banks; J. Ruwitch, "California Leans On National Guard T o Help With Food Banks," *NPR,* March 29, 2020, https://www.npr.org/2020/03/29/823592895/californialeans-on-national-guard-to-help-with-food-banks.

[15] Feeding America, "Feeding America Network Stays Resilient During COVID-19 Crisis," May 12, 2020, https://www.feedingamerica.org/about-us/press-room/feeding-america-network-stays-resilient-during-covid-19-crisis.

[16] Ibid.

Food Insecurity During the COVID-19 Pandemic

Recent analyses have shown an increase in food insecurity rates during the pandemic. *Food insecurity* refers to reduced food intake and disrupted eating patterns due to a lack of resources, and is a way that researchers quantify hunger (*food security* is the absence of such conditions).[17] In a survey administered between March 25, 2020, and April 10, 2020, the Urban Institute found that 21.9% of adults were food insecure.[18] This was nearly double the household food insecurity rate of 11.1% in 2018 (the most recent official federal-level estimate).[19] The Urban Institute found that food insecurity rates were significantly higher among adults whose families lost jobs, work hours, or work-related income (29.6%) compared to those whose families did not (16.3%). Using data from the Urban Institute survey and another survey, the Hamilton Project (an initiative of the Brookings Institution), found a similar increase in food insecurity among households with children.[20]

An analysis by Feeding America published in May 2020 projected future increases in food insecurity under different scenarios of poverty and unemployment. It projected that 3.3 million to 17.1 million Americans could become newly food insecure should poverty and unemployment continue to increase.[21] Feeding America's research also projected variations in food insecurity rates by state and county.[22]

---

[17] For the definition of food insecurity, see A. Coleman-Jensen, M.P. Rabbitt, C.A. Gregory, and A. Singh, *Household Food Security in the United States in 2018*, ERR-270, USDA, ERS, 2019. For further background on the measurement of food insecurity, see CRS Report R42353, *Domestic Food Assistance: Summary of Programs*.

[18] M. Karpman, S. Zuckerman, D. Gonzalez, and G.M. Kenney, *The COVID-19 Pandemic Is Straining Families' Abilities to Afford Basic Needs*, Urban Institute, April 2020, https://www.urban.org/research/publication/covid-19- pandemic-straining-families-abilities-afford-basic-needs.

[19] A. Coleman-Jensen, M.P. Rabbitt, C.A. Gregory, and A. Singh, *Household Food Security in the United States in 2018*, ERR-270, U.S. Department of Agriculture, Economic Research Service, 2019.

[20] L. Bauer, *The COVID-19 Crisis Has Already Left Too Many Children Hungry in America*, Hamilton Project, Brookings Institution, May 6, 2020, https://www.hamiltonproject.org/blog/the_covid_19_crisis_has_already_left_too_many_children_hungry_in_america.

[21] Feeding America, *The Impact of the Coronavirus on Food Insecurity*, brief, April 2020, https://www.feedingamerica.org/research/coronavirus-hunger-research.

[22] Feeding America, *The Impact of the Coronavirus on Local Food Insecurity*, brief, May 2020, https://www.feedingamerica.org/research/coronavirus-hunger-research.

In addition, the U.S. Census Bureau has administered Household Pulse Surveys during the COVID-19 pandemic, which include questions on *food insufficiency*—a narrower measure than food insecurity that examines whether households have enough to eat. The most recent Household Pulse Survey as of the date of this chapter, conducted between June 11, 2020, and June 16, 2020, found that 9.8% of adults (24.5 million individuals) reported food insufficiency in the last seven days: 8.1% said that they sometimes did not have enough to eat and 1.7% said that they often did not have enough to eat.[23] For comparison, 8.1% of adults (20.1 million individuals) reported food insufficiency prior to the pandemic (which the Census Bureau classified as March 13, 2020).[24] There was a larger increase in the proportion of adults who reported that they had enough food "but not always the kinds of food [they] wanted to eat": 19.8% of adults (49.4 million individuals) reported this was the case before the pandemic, and 31.1% of adults (77.5 million individuals) reported this was the case between June 11 and June 16.[25]

Of adults who reported food insufficiency in the most recent Household Pulse Survey, 19.3% reported receiving free groceries or a free meal, 7.8% reported receiving a free meal from a food pantry or food bank, 4.2% reported receiving a free meal from a religious organization, and 0.8% reported receiving a free meal from a shelter or soup kitchen in the last seven days (respondents could select more than one category).[26]

Despite challenges, many food banks and other emergency feeding organizations have continued providing meals during the pandemic by setting up drive-through food distributions, delivering food directly to households, and partnering with other meal providers such as schools, among other strategies.[27] While private donations from grocery stores and individuals have declined in some places, new private donations have

[23] U.S. Census Bureau, Household Pulse Survey: June 11-June 16, "Table 2b. Food Sufficiency for Households, in the Last 7 Days, by Select Characteristics," https://www.census.gov/data/tables/2020/demo/hhp/hhp7.html.

[24] 6.4% of adults said that they sometimes did not have enough to eat and 1.7% said that they often did not have enough to eat prior to the pandemic. U.S. Census Bureau, Household Pulse Survey: April 23-May 5, "Table 2a. Food Sufficiency for Households, Prior to COVID-19 Pandemic, by Select Characteristics," https://www.census.gov/data/tables/2020/demo/hhp/hhp5.html.

[25] Ibid.

[26] Ibid.

[27] For example, see A. Tadayon, "California food banks partner with schools to serve families of students," *EdSource,* March 30, 2020, https://edsource.org/2020/california-food-banks-partner-with-schools-to-serve-families-of-students/ 627322.

occurred as a result of the pandemic.[28] In addition, some states may provide emergency funding for emergency feeding organizations. For example, on May 3, 2020, New York Governor Andrew Cuomo announced the Nourish New York initiative, which will provide $25 million to food banks and emergency feeding providers, primarily to purchase surplus milk and dairy products.[29] Also, as discussed in the sections to follow, emergency feeding organizations have received increased federal aid in FY2020, largely as a result of emergency funding and programs.

FEDERAL PROGRAMS THAT SUPPORT EMERGENCY FEEDING ORGANIZATIONS DURING THE COVID-19 PANDEMIC

The following sections discuss a variety of federal programs, most of which existed prior to the pandemic. Others have been established specifically to respond to demands resulting from the pandemic. Of the pre-existing programs, some provide food assistance on an ongoing basis, while others are emergency authorities triggered by presidentially declared major disasters and/or emergencies under the Robert T. Stafford Disaster Relief and Emergency Assistance Act (hereinafter, the Stafford Act).[30] These emergency response programs (USDA's Disaster Household Distribution program and FEMA's Public Assistance program) have been

---

[28] For example, see Feeding America, "COVID-19 Response Partners," https://www.feedingamerica.org/about-us/ partners/covid-19-response-partners; and Feeding America, "Feeding America Responds To $100 Million Gift from Jeff Bezos In Support of Food Banks During COVID-19 Pandemic," April 2, 2020, https://www.feedingamerica.org/about-us/press-room/jeff-bezos-support-food-banks.

[29] New York State Department of Agriculture and Markets, "New York State Provides Additional Details on $25 Million in Funding for Food Banks Through Nourish New York Initiative," May 3, 2020, https://agriculture.ny.gov/ news/new-york-state-provides-additional-details-25-million-funding-food-banks-through-nourish-new.

[30] P.L. 93-288 as amended (42 U.S.C. §§5121 et seq.).

activated in response to the President's emergency declaration and subsequent major disaster declarations in response to COVID-19.[31]

Table 1 summarizes the federal programs that have provided aid to emergency feeding organizations during the pandemic.

Table 1. Federal programs providing aid to emergency feeding organizations during the COVID-19 pandemic

Programs	Operates on	Operates	Established	Established
USDA Programs				
The Emergency Food Assistance Program	X			
Section 32 bonus commodity purchases	X			
Commodity Credit Corporation (CCC)				X
Disaster Household Distribution		X		
Farmers to Families Food Box Program			X	
FEMA Programs				
Emergency Food and Shelter Program	X			
Public Assistance (PA) program		X		

Source: Congressional Research Service (CRS), based on current law.

[a] Some of these programs received supplemental funding in COVID-19 response legislation.

USDA Programs

The Emergency Food Assistance Program (TEFAP)

Administered by USDA's Food and Nutrition Service (FNS), TEFAP provides federally purchased foods (hereinafter, USDA Foods) and administrative funds to states for distribution to emergency feeding organizations.[32] While certain other charitable institutions are eligible for TEFAP aid, priority is given to emergency feeding organizations that

---

[31] Letter from Donald J. Trump, President of the United States, to Acting Secretary Wolf, Secretary Mnuchin, Secretary Azar, and Administrator Gaynor, March 13, 2020, https://www.whitehouse.gov/wp-content/uploads/2020/03/ LetterFromThePresident.pdf. The Stafford Act may be found at 42 U.S.C. §§5121 et seq; Stafford Act Declarations as of June 2, 2020. FEMA, "COVID-19 Disaster Declarations," https://www.fema.gov/ coronavirus/disaster-declarations.

[32] Further detail on TEFAP can be found in CRS Report R45408, The Emergency Food Assistance Program (TEFAP): Background and Funding.

"relieve situations of emergency and distress through the provision of food to needy persons, including low-income and unemployed persons."[33]

There are two types of TEFAP commodities: *entitlement* commodities and *bonus* commodities. Both types of commodities are domestically grown, but entitlement commodities are selected from a list of USDA Foods by states and emergency feeding organizations, whereas bonus commodities are selected by USDA based on agricultural surpluses and made available to states and emergency feeding organizations (discussed further in the "USDA Food Purchasing and Delivery Process" section).

In FY2020, legislation responding to COVID-19 has more than tripled the amount of funding available for TEFAP entitlement commodities compared to prior years. The annual appropriation for TEFAP entitlement commodities, provided in the Further Consolidated Appropriations Act for FY2020 (P.L. 116-94), enacted on December 20, 2019, was $322 million.[34] Subsequently, the Families First Coronavirus Response Act (FFCRA; P.L. 116-127), enacted on March 18, 2020, provided an additional $400 million for TEFAP entitlement commodities, up to $100 million of which could be used for food distribution costs. The Coronavirus Aid, Relief, and Economic Security Act (CARES Act, P.L. 116-136), enacted on March 27, 2020, provided another $450 million for TEFAP entitlement commodities, up to $150 million of which could be used for food distribution costs. In total, at least $972 million[35] is available for TEFAP entitlement commodities in FY2020, compared to approximately $295 million in FY2019.

There is also a higher-than-usual amount of bonus commodities available for TEFAP in FY2020, due in part to a food purchasing and distribution component of the Administration's second round of aid for producers affected by trade disputes. From FY2014 to FY2018, USDA purchased roughly $300 million annually in bonus commodities for

---

[33] 7 C.F.R. §251.4; Section 203B of the Emergency Food Assistance Act of 1983 (codified at 7 U.S.C. §7505).

[34] The annual appropriation for TEFAP's entitlement commodities is based on a level of funding specified in Section 27 of the Food and Nutrition Act of 2008, which creates a mandatory level of spending that requires annual appropriation.

[35] This assumes that states opt to use the maximum funding available for food distribution costs under FFCRA (up to $100 million) and the CARES Act (up to $150 million).

TEFAP.[36] In FY2019, bonus purchases increased to $1.46 billion with the first round of trade aid purchases. FY2020 bonus purchases are likely to exceed the FY2019 amount (discussed in the next section).

TEFAP statute also authorizes discretionary administrative funds that participating state agencies and emergency feeding organizations can use for food storage, preparation, and distribution; and equipment, personnel, and related costs.[37] The FY2020 annual appropriation for TEFAP administrative funds was $80 million (P.L. 116-94). As noted above, FFCRA subsequently provided up to $100 million and the CARES Act provided up to $150 million "for food distribution costs." FNS guidance indicates that these funds will be treated as administrative funds.[38]

Table 2 displays FY2020 funding for TEFAP and the other programs discussed in this chapter.

Table 2. Sources of federal funding for food banks and other emergency feeding organizations, FY2020

	Authorizing Law	**FY2020 Available Funding ($ in millions)**
U.S. Department of Agriculture (USDA) Programs		
Farmers to Families Food Box program	Families First Coronavirus Response Act (FFCRA; P.L. 116-127)	3,000
Disaster Household Distribution	Section 32 (7 U.S.C. §612c) and Section 4(a) of the Agricultural and Consumer Protection Act of 1973 (7 U.S.C. §612c note)	Not available
Commodity Credit Corporation (CCC) Food Purchase and Distribution Program (FPDP)	CCC Charter Act (15 U.S.C. §714)	1,460[a]
Section 32 bonus commodity purchases	Section 32 of the act of August 24, 1935 (7 U.S.C. §612c)	873[b]

[36] USDA, Office of Budget and Program Analysis, "2021 Explanatory Notes – Food and Nutrition Service," p. 34-104, https://www.usda.gov/sites/default/files/documents/32fns 2021notes.pdf.

[37] Section 204 of the Emergency Food Assistance Act of 1983 (7 U.S.C. §7508); 7 C.F.R. §251.8; USDA FNS, "The Emergency Food Assistance Program (TEFAP): Administrative Costs," FNS Instruction 716-3 (Revision 1), August 9, 2002, https://www.fns.usda.gov/tefap/emergency-food-assistance-program-tefap-administrative-costs.

[38] USDA, FNS, "The Emergency Food Assistance Program (TEFAP): Allocation of Coronavirus Aid, Relief, and Economic Security Act Supplemental Appropriations," April 24, 2020.

	Authorizing Law	FY2020 Available Funding ($ in millions)
The Emergency Food Assistance Program (TEFAP)		
Entitlement commodities	Further Consolidated Appropriations Act of FY2020 (P.L. 116-94); Division A of FFCRA (P.L. 116-127); Division B of CARES Act (P.L. 116-136)	972[c]
Administrative and food distribution costs	Further Consolidated Appropriations Act of FY2020 (P.L. 116-94); Division A of FFCRA (P.L.116-127); Division B of CARES Act (P.L. 116-136)	330[c]
Farm to Food Bank Projects	Section 203D of the Emergency Food Assistance Act of 1983 (7 U.S.C. §7507)	4
Federal Emergency Management Agency (FEMA) Programs		
Emergency Food and Shelter Program (EFSP)	McKinney-Vento Homeless Assistance Act (P.L.100-77, reauthorized under P.L. 100-628; 42 U.S.C. §11331 et seq.)	445[d]
Public Assistance (PA) program	Sections 402, 403, 418, 419, and 502 of them Stafford Act (P.L. 93-288, as amended; (42 U.S.C. §§5121 et seq.)	No predetermined limit[e]
Total		More than 7,084[f]

Source: Congressional Research Service (CRS), based on current law. Notes: Amounts reflect budget authority unless noted.

[a]Includes $1.4 billion for food purchases and $60 million for administrative costs.

[b]This is the amount available for the remainder of FY2020 as of April 17, 2020. Actual amount may be higher.

[c]Assumes that states opt to use the maximum funding available for food distribution costs under FFCRA (up to $100 million) and the CARES Act (up to $150 million). TEFAP entitlement commodities funds include $322 million provided in FY2020 annual appropriations, $300 million provided by FFCRA, and $350 million provided by the CARES Act. TEFAP administrative and food distribution funds include $80 million provided in FY2020 annual appropriations, $100 million provided by FFCRA, and $150 million provided by the CARES Act.

[d]The $445 million in total funding awarded to the Emergency Food and Shelter Program (EFSP) National Board by the Federal Emergency Management Agency (FEMA) for implementation in calendar year 2020 includes the $200 million appropriated through the CARES Act, $125 million appropriated in FY2020 annual appropriations, and $120 million appropriated in FY2019 annual appropriations.

[e]Public Assistance is funded through the Disaster Relief Fund (DRF), the primary source of funding for the federal government's domestic general disaster relief programs. DRF appropriations are not allocated to specific emergencies, disasters, or specific forms of assistance, including food assistance. Division B of the CARES Act (P.L. 116-136) included $45 billion for the DRF in March 2020, bringing its available balance for the costs of major disasters to more than $87 billion. As of April 30, 2020, the balance of the fund is approximately $80 billion.

[f]Total does not include funding for the USDA Disaster Household Distribution program or FEMA Public Assistance (PA) program.

Bonus Commodity Purchasing Authorities: Section 32 and Commodity Credit Corporation

TEFAP has the capability to receive bonus commodities from two USDA market support programs: Section 32 and the Commodity Credit Corporation (CCC).

Section 32 of the Act of August 24, 1935, is a permanent appropriation that sets aside 30% of annual customs receipts ($15.1 billion in FY2020) to support the farm sector through the purchase of surplus commodities and other activities.[39] USDA's Agricultural Marketing Service (AMS) uses a portion of these funds to purchase surplus commodities for distribution through TEFAP and other nutrition assistance programs. In FY2019, AMS purchased $395.7 million in bonus commodities for nutrition assistance programs using Section 32 authority.[40]

An April 17, 2020, USDA press release on the COVID-19 response noted:

> USDA has up to an additional $873.3 million available in Section 32 funding to purchase a variety of agricultural products for distribution to food banks. The use of these funds will be determined by industry requests, USDA agricultural market analysis, and food bank needs.

On May 4, 2020, USDA announced $470 million in purchases using Section 32 authority, including fruits, vegetables, dairy, poultry, fish, and pork, for distribution through TEFAP and possibly other nutrition assistance programs.[41]

The CCC, a government-owned entity that finances authorized programs that support U.S. agriculture,[42] may provide budget authority for

---

[39] For more information, see CRS Report RL34081, *Farm and Food Support Under USDA's Section 32 Program*.

[40] USDA, Office of Budget and Program Analysis, "2021 Explanatory Notes – Agricultural Marketing Service," p. 23- 126, https://www.usda.gov/sites/default/files/documents/21ams2021notes.pdf.

[41] USDA, AMS, "USDA Announces Additional Food Purchase Plans," May 4, 2020, https://www.ams.usda.gov/pressrelease/usda-announces-additional-food-purchase-plans.

[42] For CCC purchasing authorities, see Section 5 of the CCC Charter Act. The Secretary of Agriculture's authority to donate such commodities to TEFAP is established by Section 17 of the Commodity Distribution Reform and WIC Amendments Act Of 1987.

bonus purchases for delivery through TEFAP and additional avenues, among its other purposes. It has permanent authority to borrow up to $30 billion from the U.S. Treasury to finance its programs, which is reimbursed as needed by annual Agriculture appropriations acts.[43]

In recent years, USDA has used the CCC to fund a Food Purchase and Distribution Program (FPDP), administered by AMS, to purchase surplus commodities affected by trade retaliation.[44] USDA's 2018 trade aid package, announced in August 2018, included $1.2 billion in bonus purchases for distribution through TEFAP and other nutrition assistance programs.[45] The vast majority of these foods—$1.1 billion worth—were distributed through TEFAP in FY2019.[46] USDA's second trade aid package, announced in May 2019, provided another $1.4 billion for bonus purchases, largely for distribution through TEFAP in FY2020.[47] Planned purchases include "fruits, vegetables, some processed foods, beef, pork, lamb, poultry, and milk."[48]

Using CCC authority, USDA also provided more than $50 million in administrative funds to participating TEFAP organizations for costs

---

[43] For more information, see CRS Report R44606, *The Commodity Credit Corporation: In Brief.*

[44] For more information, see CRS Report R45310, *Farm Policy: USDA's 2018 Trade Aid Package*; and CRS Report R45865, *Farm Policy: USDA's 2019 Trade Aid Package.*

[45] USDA, "USDA Announces Details of Assistance for Farmers Impacted by Unjustified Retaliation," press release, August 27, 2018, https://www.usda.gov/media/press-releases/2018/08/27/usda-announces-details-assistance-farmersimpacted-unjustified. The largest purchases announced include pork, sweet cherries, apples, pistachios, dairy, and almonds. For more information, see CRS Report R45310, Farm Policy: USDA's 2018 Trade Aid Package; and CRS Report R45865, Farm Policy: USDA's 2019 Trade Aid Package.

[46] USDA, Office of Budget and Program Analysis, "2021 Explanatory Notes – Agricultural Marketing Service," p. 23- 126, https://www.usda.gov/sites/default/files/documents/21ams2021notes.pdf.

[47] USDA, "USDA Announces Support for Farmers Impacted by Unjustified Retaliation and Trade Disruption," press release, May 23, 2019, https://www.usda.gov/media/press-releases/2019/05/23/usda-announces-support-farmersimpacted-unjustified-retaliation-and. Historically, CCC surplus commodities have been distributed through TEFAP and other FNS nutrition assistance programs. However, in July 2019 USDA issued regulations to establish its authority to distribute these foods to outlets outside of FNS nutrition assistance programs. Therefore, emergency feeding organizations may not be the only recipients of these commodities. See USDA, CCC, "Trade Mitigation Program," 84 *Federal Register* 36456, July 29, 2019, https://www.federalregister.gov/documents/2019/07/29/2019-15700/trademitigation-program.

[48] USDA, AMS, "USDA Announces Additional Food Purchase Plans," May 4, 2020, https://www.ams.usda.gov/pressrelease/usda-announces-additional-food-purchase-plans.

associated with the receipt, storage, and distribution of FPDP foods in FY2019.[49] In FY2020, approximately $60 million is available for this purpose.[50]

USDA Food Purchasing and Delivery Process

The processes for procuring and delivering USDA Foods are based on federal law, regulations, and agency policy. The process described in this section is specific to TEFAP, but similar processes are used for other federal nutrition assistance programs that use USDA Foods.

The process starts with determining what foods will be available. In the case of entitlement commodities, FNS and AMS work together to determine a list of foods for the fiscal year that will be procurable and usable by emergency feeding organizations.[51] In the case of bonus commodities, AMS determines which foods it will purchase on an ongoing basis throughout the year based on the needs of producers.

State agencies, in consultation with emergency feeding organizations, place orders via the USDA Foods catalog (in some cases, state agencies may delegate this responsibility to food banks).[52] They use their allocated entitlement funds to choose foods from a list that includes meats, eggs, grains, fruits, vegetables, beans, nuts, and milk. They can also opt to receive preselected bonus foods, if available.[53] Most foods show up in the

---

[49] USDA, Office of Budget and Program Analysis, "2021 Explanatory Notes – Food and Nutrition Service," p. 34-104, https://www.usda.gov/sites/default/files/documents/32fns2021notes.pdf.

[50] USDA, FNS, "Terms and Conditions for Funds Available for Storage and Distribution of Food Purchase Distribution Program Foods," December 5, 2019, https://www.fns.usda.gov/tefap/terms-and-conditions-funds-available-storageand-distribution-food-purchase-distribution.

[51] For the FY2020 list, see USDA, FNS, "USDA Foods Available List for The Emergency Food Assistance Program (TEFAP) 2020," August 2019, https://fns-prod.azureedge.net/sites/default/files/resource-files/tefap-foods-available.pdf.

[52] 7 C.F.R. §250.10; 7 C.F.R. §251.2(i).

[53] In FY2019, bonus foods included apricots, beans, cheese, cherries, chicken, eggs, fig pieces, milk, oranges, orange juice, peaches, pears, plums, pollock, pumpkin, raisins, salmon, strawberries, sweet potatoes, and walnuts. USDA, Office of Budget and Program Analysis, "2021 Explanatory Notes – Food and Nutrition Service," p. 34-103, https://www.usda.gov/sites/default/files/documents/32fns2021notes.pdf.

catalog on a monthly or quarterly basis; others are only available in the catalog at certain times of the year (e.g., fresh whole apples).[54]

Once organizations make their selections, AMS publishes a solicitation via an online system, which "describes USDA's needs in terms of the product, volume, delivery destination (city/State) and delivery window, and invites approved vendors to submit offers to fill the demand."[55] Pre-approved vendors may respond to the solicitation and then AMS awards contracts. Per the terms of the contract, a vendor may deliver food to a state warehouse, directly to a food bank or emergency feeding organization, or to a food processor (if the state or food bank opts to use a portion of their entitlement funds for this purpose).[56] Once the product has been delivered, AMS pays the vendor.[57] Depending on the product, it takes roughly two to five months from solicitation through delivery.[58]

Disaster Household Distribution

During a presidentially declared disaster or emergency, there are program options through which states may be able to repurpose existing inventories of USDA Foods intended for other nutrition assistance programs for disaster/emergency feeding efforts. Specifically, during a presidentially declared disaster or emergency, state agencies may (1) use USDA Foods for prepared meals in congregate settings (mass feeding) and/or (2) request FNS approval to use USDA Foods for household distribution (pick up or delivery).[59] For congregate feeding, USDA Foods are typically repurposed from the National School Lunch Program (NSLP),

---

[54] See "AMS CPP Procurement Schedule for January 2020 to June 2021 (Schedule) (xlsx)" at https://www.ams.usda.gov/selling-food/solicitations.

[55] USDA, AMS, "How the Process Works," https://www.ams.usda.gov/selling-food/how-process-works.

[56] 7 C.F.R. §250.11.

[57] USDA, AMS, "How the Process Works," https://www.ams.usda.gov/selling-food/how-process-works.

[58] See "AMS CPP Procurement Schedule for January 2020 to June 2021 (Schedule) (xlsx) at https://www.ams.usda.gov/selling-food/solicitations.

[59] 7 C.F.R. §250.69.

given their institutional sizing. For Disaster Household Distribution, foods are typically repurposed from TEFAP due to their smaller packaging.[60]

Mass feeding is not a viable option during the COVID-19 pandemic, and several states have requested USDA FNS approval to operate Disaster Household Distributions. According to the Disaster Household Distribution COVID-19 response website, "FNS may approve state requests for Disaster Household Distributions for targeted areas to meet specific needs when traditional channels of food are unavailable and not being replenished on a regular basis."[61] As of May 11, 2020, USDA had approved Disaster Household Distribution programs in 18 states, Guam, and 33 tribal nations in response to COVID-19.[62]

Under an approved Disaster Household Distribution program, USDA Foods are provided to disaster feeding organizations, which are selected by the state and may or may not include TEFAP-participating organizations. These organizations distribute food boxes to households in need of food assistance because of the disaster or emergency.[63] According to program regulations, "distribution may continue for the period that FNS has determined to be necessary to meet the needs of such households."[64]

USDA replenishes or reimburses federal nutrition assistance programs (e.g., TEFAP and NSLP) for USDA Foods reprogrammed for disaster/emergency feeding during a presidentially declared disaster or emergency.[65] According to agency guidance, replacement or reimbursement of USDA Foods may be funded under Section 4(a) of the Agriculture and Consumer Protection Act of 1973 (7 U.S.C. §612c note) and Section 32 (7 U.S.C §612c).[66] USDA also reimburses state agencies

---

[60] USDA, FNS, Food Distribution Division, "USDA Foods Program Disaster Manual," revised September 2017, https://fns-prod.azureedge.net/sites/default/files/fdd/disaster-manual.pdf.

[61] USDA, FNS, "Disaster Household Distribution," May 11, 2020, https://www.fns.usda.gov/usda-foods/covid-19- disaster-household-distribution.

[62] Ibid.

[63] 7 C.F.R. §250.69(c). Households receiving Disaster Supplemental Nutrition Assistance Program (D-SNAP) benefits are not eligible for Disaster Household Distribution foods.

[64] 7 C.F.R. §250.69(b).

[65] 7 C.F.R. §250.69(g).

[66] USDA, FNS, Food Distribution Division, "USDA Foods Program Disaster Manual," revised September 2017, pp. 8- 9, https://fns-prod.azureedge.net/sites/default/files/fdd/disaster-manual.pdf. Sections 412 and 413 of the Robert T. Stafford Disaster Relief and Emergency Assistance Act authorize USDA to use Section 32 funds "to purchase food commodities

for costs associated with transporting USDA Foods (both within a state and from one state to another) during presidentially declared disasters and emergencies.[67]

As noted, Disaster Household Distribution is separate from TEFAP. Some state agencies may see an advantage in operating a Disaster Household Distribution program because of more flexible administrative requirements (e.g., loosening eligibility rules), while others may see a disadvantage in fewer foods being available for current TEFAP recipients and a delay in replenishing those foods. For example, a FNS letter dated March 24, 2020, approving New York's Disaster Household Distribution program stated that orders to replenish TEFAP inventory "may not be available for delivery prior to July 2020."[68] FNS guidance emphasizes that "during disasters, emergencies, or situations of distress, States should continue to operate TEFAP if possible."[69] FNS has also publicized "flexibilities" within TEFAP that, for example, encourage states to consider expanding eligibility requirements.[70] Recent letters from FNS approving extensions of states' Disaster Household Distribution programs have included the following statement:[71]

---

necessary to provide adequate supplies for use in any area of the United States in the event of a major disaster or emergency in such area" (42 U.S.C. §5180).

[67] 7 C.F.R. §250.69(h).

[68] USDA, FNS, "Approval of Disaster Household Distribution in Response to the National Emergency Declaration due to COVID-19," March 24, 2020, https://fns-prod.azureedge.net/sites/default/files/resource-files/NY-COV-19-DHHDFinal-Approval.pdf.

[69] USDA, FNS, Food Distribution Division, "USDA Foods Program Disaster Manual," revised September 2017, p. 26, https://fns-prod.azureedge.net/sites/default/files/fdd/disaster-manual.pdf.

[70] USDA, FNS, "The Emergency Food Assistance Program (TEFAP) Flexibilities for State Agencies," March 30, 2020, https://www.fns.usda.gov/tefap/flexibilities-state-agencies.

[71] For example, see New York's extension: USDA, FNS, "Extension Approval of Disaster Household Distribution in Response to the National Emergency Declaration Due to COVID-19," May 28, 2020, https://fns-prod.azureedge.net/ sites/default/files/resource-files/NERO-NYWestchester-COV-DHD-Extension2-Final-Approval.pdf; and Pennsylvania's extension: USDA, FNS, "Extension Approval of Disaster Household Distribution in Response to the National Emergency Declaration Due to COVID-19," May 23, 2020, https://fns-prod.azureedge.net/sites/default/files/ resource-files/MARO-PDA-COV-DHD-Extension2-Final-Approval.pdf.

> Consideration of any further requests to extend disaster household distribution will be dependent on the targeting of the request to specific areas most in need and justification as to why the availability of other programs and resources are not meeting the State's food assistance needs.

Farmers to Families Food Box Program

On April 17, USDA announced a Coronavirus Food Assistance Program[72] that includes $3 billion for a new Farmers to Families Food Box initiative to facilitate the distribution of foods—specifically, fresh fruits and vegetables, dairy, and meat products—from local and regional suppliers to public and nonprofit organizations, including food banks, schools, tribal organizations, and faith-based organizations.[73] The stated goals of the program are to expedite federally sponsored food deliveries to food banks and other feeding organizations and to "sell food previously destined for restaurants and bulk purchasers to distributors, preventing waste."[74] USDA is operating the program using authority provided by Section 1101(g) of FFCRA, which gives the Secretary of Agriculture the authority to "purchase commodities for emergency distribution in any area of the United States during a public health emergency designation" during FY2020.[75]

---

[72] For more information on the Coronavirus Food Assistance Program, see CRS Report R46347, COVID-19, U.S. Agriculture, and USDA's Coronavirus Food Assistance Program (CFAP); and CRS Report R46348, COVID-19. *Supply Chain Disruptions in the U.S. Fruit and Vegetable Industry: In Brief.*

[73] USDA, "USDA Announces Coronavirus Food Assistance Program," April 17, 2020, https://www.usda.gov/media/ press-releases/2020/04/17/usda-announces-coronavirus-food-assistance-program. Nonprofits must have 501(c)(3) tax exempt status and all recipient organizations must "demonstrate that they have the operational and financial capability to receive, store and distribute requested food items." USDA, AMS, "Farmers to Families Food Box Program FAQs," April 27, 2020, https://www.ams.usda.gov/publications/ content/farmers-families-food-box-program-faqs.

[74] USDA, AMS, "Farmers to Families Food Box Infographic," https://www. ams.usda.gov/sites/default/files/media/ FarmerstoFamiliesFoodBox.pdf; USDA AMS webinar on April 21, 2020, recording available at https://www. ams.usda.gov/selling-food-to-usda/farmers-to-families-food-box.

[75] USDA, AMS, "Notice of Funds Availability (NOFA); Purchase of Fruit, Vegetable, Dairy, and Meat Products Due to COVID-19 National Emergency-USDA Food Box Distribution Program," 85 *Federal Register* 23325, April 27, 2020, https://www.federalregister.gov/

Unlike TEFAP, state agencies do not play a role in the administration of the Farmers to Families Food Box program. Instead, AMS awards contracts to businesses through a solicitation process. According to the terms of the contracts, suppliers must package products into "family-sized boxes" and distribute the boxes to food banks and other recipient organizations selected by the supplier (which may or may not include TEFAP-participating organizations).[76] Boxes must contain fresh fruit and vegetables, dairy products, fluid milk, precooked meats, or a combination of those items.[77]

On May 8, 2020, AMS awarded $1.2 billion for the first round of contracts. The initial food distribution was scheduled to take three weeks to two months from solicitation (April 24) to delivery (May 15 to June 30).[78] Following the awards, some news outlets, industry organizations, and lawmakers raised concerns about the supplier selection process.[79] Some food banks also raised concerns that the program does not include funding for their food handling and distribution costs.[80] Others have argued that the program has sped up federal aid and expanded the foods available to those

---

documents/2020/04/27/2020-08979/notice-of-funds-availability-nofa-purchaseof-fruit-vegetable-dairy-and-meat-products-due-to.

[76] USDA, AMS, "Farmers to Families Food Box," https://www.ams.usda.gov/selling-food-to-usda/farmers-to-familiesfood-box, accessed on June 8, 2020.

[77] USDA, AMS, Solicitation AG-12-3J14-20-R-0377, April 24, 2020, p. 5, https://www.ams.usda.gov/sites/default/ files/media/RFP_ERAcquisition.pdf.

[78] USDA, AMS, Solicitation AG-12-3J14-20-R-0377, April 24, 2020, p. 6, https://www.ams.usda.gov/sites/default/ files/media/RFP_ERAcquisition.pdf.

[79] U.S. Congress, House Agriculture Committee, "Fudge, Costa, and Plaskett Push USDA for Answers on Food Box Contracts," May 22, 2020; J. Hagstrom, " Stabenow leads Dem to question Perdue about food box program," *Hagstrom Report*, June 8, 2020; R. McCrimmon, "USDA food boxes start flowing, direct aid on deck," *Politico Morning Agriculture,* May 18, 2020, https://www.politico.com/newsletters/morning-agriculture/2020/05/18/usda-food-boxesstart-flowing-direct-aid-on-deck-787676; and R. McCrimmon, "Filling the safety gap for farmworkers," *Politico Morning Agriculture,* May 12, 2020, https://www.politico.com/newsletters/morning-agriculture/2020/05/12/filling-thesafety-gap-for-farmworkers-787524.

[80] N. Rasul, "'Last Mile' Concerns Complicate USDA Food Box Program," *Food Bank News,* June 2, 2020, https://foodbanknews.com/last-mile-concerns-complicate-food-box-program/; T. Orsborn, "San Antonio Food Bank footing costly bill for USDA program, adds delay in getting food to those in need," *San Antonio Express-News,* June 13, 2020, https://www.expressnews.com/news/local/article/San-Antonio-Food-Bank-footing-costly-bill-for15337758.php.

in-need of food as sistance.[81] Solicitation and awards for the remaining portions of the program are expected through the summer and fall of 2020.

FEMA Programs

Emergency Food and Shelter Program

The Emergency Food and Shelter Program (EFSP) provides grants for ongoing programs by private nonprofit organizations and local governments to provide shelter, food, and supportive services for individuals and families who are homeless or experiencing economic emergencies.[82] The EFSP is governed by a National Board[83] that is chaired by a director (by statute, the FEMA Administrator).[84] The National Board also includes representatives from private nonprofit organizations.[85] When Congress appropriates funding to the EFSP, the director is responsible for awarding a grant to the National Board,[86] and the National Board is

---

[81] For example, see K. Whitcomb, "Free food program also helps farmers," *Rutland Herald,* June 1, 2020, https://www.rutlandherald.com/news/free-food-program-also-helps-farmers/article_14479ae4-6706-5a06-85cb897500559729.html and L. Moore, "Farmers' bounty distributed for free in ongoing Muskegon area relief effort," *MLive,* June 5, 2020, https://www.mlive.com/news/muskegon/2020/06/farmers-bounty-distributed-for-free-in-muskegon-heights.html.

[82] EFSP was first authorized under the Stewart B. McKinney Homeless Assistance Act of 1987 (P.L. 100-77), later renamed the McKinney-Vento Homeless Assistance Act and reauthorized under the Stewart B. McKinney Homeless Assistance Amendments Act of 1988 (P.L. 100-628). 42 U.S.C. §11343; FEMA, "Emergency Food and Shelter Program," last updated May 7, 2020, https://www.fema.gov/pt-br/media-library/assets/documents/24422.

[83] 42 U.S.C. §11331.

[84] 42 U.S.C. §11331(c) and 42 U.S.C. §11351(1).

[85] 42 U.S.C. §11331(b). The nonprofit organizations that comprise the EFSP National Board include the American Red Cross, Catholic Charities U.S.A., the Jewish Federations of North America (referred to in statute as the Council of Jewish Federations, Inc.), the National Council of Churches of Christ in the U.S.A., the Salvation Army, and United Way Worldwide (referred to in statute as the United Way of America). The United Way serves as the EFSP National Board's Secretariat and Fiscal Agent, and administers the program nationally on a day -to-day basis (see FEMA, "Emergency Food and Shelter Program," fact sheet, April 29, 2020, https://www.fema.gov/media-library-data/ 1588188618188-ef2cf79e8202bfb49bfe3c7c877e8542/FACTSHEETEmergencyFoodandShelterProgramMay2020compliant.pdf (hereinafter, FEMA, "EFSP Fact Sheet")).

[86] 42 U.S.C. §11341.

responsible for establishing program guidelines[87] and disbursing the funding.[88] Localities (i.e., cities or counties) are designated for funding based on a formula that includes population, poverty, and unemployment data.[89] Each locality designated for funding must establish a Local Board, which is responsible for advertising funding availability and determining how their allotment of local funding will be distributed, including selecting the grant recipients (i.e., private nonprofit organizations or public organizations of the local government in the individual locality).[90] The National Board disburses funds directly to local recipient organizations (LROs) selected by the Local Boards.[91] Additionally, localities that do not qualify under the formula may receive funds through the State Set-Aside process, which can also be used to allow localities that are designated for funding to receive additional funding.[92]

Eligible applicants for EFSP grants through the Phase 37, Phase 38, and Phase CARES funds (described in the following section) include private nonprofit organizations[93] that provide food and shelter services.[94] Thus, food banks and other emergency feeding organizations that meet the

---

[87] 42 U.S.C. §11346.

[88] 42 U.S.C. §11345.

[89] Emergency Food and Shelter National Board Program (EFSP National Board), "How Areas Qualify," https://www.efsp.unitedway.org/efsp/website/websiteContents/index.cfm?template=qualify.cfm (last accessed May 29, 2020) (hereinafter, EFSP National Board, "How Areas Qualify"). Per the EFSP website, "For fiscal years 2019 and 2020 funding, jurisdictions qualified for funding if they met one of the following criteria: [1] Number of unemployed: 300 or more with a 3.7% rate of unemployment [and] [2] Number of unemployed: 300 or more with a 14.1% rate of poverty."

[90] 42 U.S.C. §11332.

[91] FEMA, "EFSP Fact Sheet."

[92] EFSP National Board, "How Areas Qualify." Per the EFSP website, "[j]urisdictions that do not qualify under the formula and thus do not receive funding directly from the National Board may receive funds through the State Set - Aside process, and jurisdictions that do qualify may receive additional funding the same way."

[93] 42 U.S.C. §11351(6) defines *private nonprofit organization* as "an organization—(A) no part of the net earnings of which inures to the benefit of any member, founder, contributor, or individual; (B) that has a voluntary board; (C) that has an accounting system, or has designated a fiscal agent in accordance with requirements established by the Director; and (D) that practices nondiscrimination in the provision of assistance."

[94] EFSP National Board Program, "Preliminary Funding Notice: Phases 37, 38 and CARES Funding," https://www.efsp.unitedway.org/efsp/website/websiteContents/pdfs/PreliminaryFundingNoticesP37_38_CARES.pdf (last accessed June 3, 2020) (hereinafter, EFSP, "Preliminary Funding Notice").

eligibility requirements may apply to receive a grant; however, it is a competitive grant process.[95]

Available EFSP Funding

In calendar year 2020, $445 million in EFSP funding is set for implementation.[96] Congress appropriated $120 million for the EFSP in FY2019,[97] and $125 million in FY2020.[98] Congress appropriated an additional $200 million for the EFSP in FY2020 through the CARES Act.[99] The EFSP refers to the grant cycles as "Phases." Thus, Phase 37 refers to the FY2019 annual funding, Phase 38 refers to the FY2020 annual funding, and Phase CARES refers to the FY2020 CARES Act supplemental funding.[100]

---

[95] In order to receive an EFSP grant, potential applicants must submit an application to the Local Board in the jurisdiction where they are providing services. EFSP, "Preliminary Funding Notice." See also EFSP National Board Program, "Overview and Guidance: Phases 37, 38 and CARES Calendar Year 2020," presentation, slide 3, https://www.efsp.unitedway.org/efsp/website/websiteContents/pdfs/OverviewandGuidance(Phases%2037%2038%20and%20CARES).pdf (hereinafter, EFSP, "Overview and Guidance Presentation"). Additionally, food banks are past EFSP grant recipients. For past funded organizations, see the "Funded Organizations" tab of the EFSP National Board website, available at https://www.efsp.unitedway.org/efsp/website/index.cfm.

[96] An additional $30 million was appropriated through Title III of the Emergency Supplemental Appropriations for Humanitarian Assistance and Security at the Southern Border Act, 2019 (Border Supplemental, P.L. 116-26). The EFSP funding through the Border Supplemental is being administered in a different way than the annually appropriated funding and CARES Act funding and is not relevant to the discussion about programs for COVID-19 assistance. More information regarding the Border Supplemental funding for the EFSP is available through the EFSP National Board's website, available at https://www.efsp.unitedway.org/efsp/website/index.cfm (see the "Supplemental Funding Info" tab).

[97] Title III of Division A of the Consolidated Appropriations Act, 2019 (P.L. 116-6). The EFSP National Board published the list of FY2019 Phase 37 allocations (appropriated in P.L. 116-6), available at https://www.efsp.unitedway.org/efsp/website/websiteContents/pdfs/Phase%2037%20Allocations.pdf.

[98] Title III of Division D of the Consolidated Appropriations Act, 2020 (P.L. 116-93). EFSP, "Preliminary Funding Notice"; EFSP, "Overview and Guidance Presentation," slide 5.

[99] Title VI of Division B of the CARES Act (P.L. 116-136). The EFSP National Board published the list of FY2020 Phase CARES allocations (appropriated in P.L. 116-136), available at https://www.efsp.unitedway.org/efsp/website/websiteContents/pdfs/Phase%20CARES%20Allocations.pdf.

[100] EFSP, "Preliminary Funding Notice."

EFSP Services and CARES Act Supplemental Funding for COVID-19 Feeding Support

The following services are eligible for EFSP funding:

- food (i.e., served meals or groceries);
- lodging (i.e., mass shelter or hotel/motel, limited to 30 days per individual or household);
- one month's rent or mortgage payment;
- one month's utility bill; and
- equipment necessary to feed or shelter people (up to $300 per item).[101]

EFSP funding may also be used for expenses related to repairing necessary equipment (not to exceed $300 per item), as well as repairs to mass feeding or sheltering facilities for building code compliance (not to exceed $2,500), and expenses related to making mass feeding or sheltering facilities accessible for individuals with disabilities (not to exceed $2,500).[102]

Although the EFSP is administered by FEMA, it does not require a major disaster or emergency declaration for implementation.[103]

Consistent with normal program administration, according to the Preliminary Funding Notice for Phase CARES (which also includes the Phase 37 and Phase 38 funding), the EFSP funds "must be used to supplement food and shelter services and may not be used as seed money for new programs."[104] Eligible use of funds mirror the above-listed eligible services. The Preliminary Notice of Funding and the EFSP National Board's presentation providing an overview and guidance for the Phase 37, Phase 38, and Phase CARES funding also provide details on the use of funds relevant to providing services during the COVID-19 pandemic. For

---

[101] EFSP National Board, "About the Emergency Food and Shelter Program," https://www.efsp.unitedway.org/efsp/website/websiteContents/index.cfm?template=about.cfm (last accessed May 29, 2020).

[102] EFSP, "Preliminary Funding Notice."

[103] 42 U.S.C. §5122(1) and (2).

[104] EFSP, "Preliminary Funding Notice."

example, "food services" includes home delivery meals (e.g., Meals on Wheels),[105] "lodging" includes lodging in a non-congregate setting such as a hotel/motel,[106] and "supplies essential to feeding or sheltering people" include cleaning supplies.[107] Additionally, administrative funds (which are limited to 2% of the jurisdiction's award and require Local Board approval) "may be used to offset costs to administer the program, including staff salary."[108]

Stafford Act Public Assistance Program

The Public Assistance (PA) program authorized in the Stafford Act provides grant assistance to state, tribal, territorial, and local governments as well as eligible nonprofits for the costs of urgent response and long-term recovery work following an emergency or major disaster declared under the act.[109] The President issued an emergency declaration and subsequent

---

[105] EFSP, "Overview and Guidance Presentation," slide 10.

[106] EFSP, "Preliminary Funding Notice."

[107] EFSP, "Preliminary Funding Notice"; and EFSP, "Overview and Guidance Presentation," slide 10.

[108] EFSP, "Preliminary Funding Notice." More information regarding the Phase 37 and Phase CARES funding for the EFSP is available through the EFSP National Board's website, available at https://www.efsp.unitedway.org/efsp/ website/index.cfm (see the "FY2019/ 2020/CARES Funding" tab).

[109] Public Assistance is authorized under Stafford Act Sections 402, 403, 406, 407, 418, 419, 428, and 502; 42 U.S.C. §§5170a-5170b, 5172; 5173, 5185-86, 5189f; 5192. See 44 C.F.R. §206.220 for general eligibility for public assistance. Per 44 C.F.R. Section 206.225(c) and (d), PA "emergency work" includes emergency communications and emergency transportation authorized under Stafford Act Sections 418 and 419. FEMA's Public Assistance guidance explains: "[a] State, Territorial, Tribal, or local government may provide emergency communication services and public transportation when existing systems are damaged to the extent vital functions of community life or incident response are disrupted. The costs of these services are not eligible for reimbursement. However, FEMA may provide short -term DFA [Direct Federal Assistance] for these services," FEMA, *Public Assistance Program and Policy Guide (PAPPG)*, FP 104-009-2, effective June 1, 2020, p. 114, https://www.fema.gov/media-library-data/1589895203535-0badc3ece1c1d4efe5e599566654860a/PAPPG_V4_Final_6-1-2020_508.pdf (hereinafter, FEMA, *PAPPG*).

major disaster declarations for all 50 states, five territories, the District of Columbia and one tribe authorizing PA for COVID-19 response efforts.[110]

The Stafford Act declarations for the COVID-19 pandemic specifically authorized assistance for PA Category B: Emergency Protective Measures, defined as work undertaken to save lives and protect property or public health and safety, or avert the threat of a catastrophe.[111] This includes the provision of food to achieve these ends.[112] For the pandemic, FEMA reimburses state, tribal, territorial, and local governments as well as eligible nonprofits (PA Applicants) for 75% of eligible costs incurred while performing eligible work.[113] There is no predetermined limit on the amount of funding available through the PA program.

For the pandemic, FEMA provides PA for food assistance required as a lifesaving or life-sustaining commodity given stay-at-home orders that reduce food access, atypical demand for food, or disruptions to food supply chains.[114] State, tribal, territorial, or local governments may receive reimbursement for such assistance, or may contract with other entities including food banks to provide food assistance and receive reimbursement for eligible costs incurred through those contracts.[115] FEMA provided a

---

[110] Stafford Act Declarations as of June 2, 2020. FEMA, "COVID-19 Disaster Declarations," https://www.fema.gov/ coronavirus/disaster-declarations.

[111] FEMA released guidance specific to COVID-19 including a non-exclusive list of emergency protective measures. FEMA, "Coronavirus (COVID-19) Pandemic: Eligible Emergency Protective Measures," fact sheet, March 19, 2020, https://www.fema.gov/news-release/2020/03/19/coronavirus-covid-19-pandemic-eligible-emergency-protective-measures. Assistance for emergency protective measures is authorized under Stafford Act Sections 402, 403, 407, 418, and 419 for major disasters and 502 for emergencies; 42 U.S.C. §§5170a-5170b, 5173, 5185-86, 5192.

[112] The provision of emergency food is authorized under Stafford Act Section 403(a)(3)(B); 42 U.S.C. §5170b(a)(3)(B).

[113] See, for example, 42 U.S.C. Section 5170b(b), which states "[t]he Federal share of assistance under this section [essential assistance] shall be not less than 75 percent of the eligible cost of such assistance." See also 44 C.F.R. Section 206.65. On general public assistance eligibility, see 44 C.F.R. Section 206.220 and FEMA, *PAPPG,* pp. 42-47, 51-59, 65.

[114] FEMA, "Coronavirus (COVID-19) Food Eligible for PA."

[115] On reasonable costs, see 2 C.F.R. Section 200.404, 44 C.F.R. Section 206.228, and FEMA, "Public Assistance: Reasonable Cost Evaluation," Job Aid, October 2018, https://www.fema.gov/media-library-data/1539879525279-d00ae1c43f9765c5a4b415e1a31202c5/PA_Reasonable_Cost_Evaluation_Job_Aid_508_FINAL_10-16-2018.pdf. Additionally, PA Applicants must follow procurement procedures to ensure that costs are eligible. FEMA released a memorandum for COVID-19 procurement; FEMA, "Procurement Under Grants Conducted Under Emergency or Exigent

non-exclusive list of populations that may require food assistance due to the pandemic, including those diagnosed with or exposed to COVID-19 who do not require hospitalization, high-risk individuals, and other populations as identified by appropriate health officials.[116] FEMA may reimburse a range of costs through PA, including the purchase, packaging, preparation, storage, and delivery of food to individuals or points of distribution.[117]

When PA Applicants do not have capacity to perform eligible work under the PA program, FEMA may provide direct assistance (referred to as Direct Federal Assistance or DFA).[118] In these cases, FEMA may assign work to FEMA personnel or other federal personnel. Under this authority, FEMA tasked USDA with food supply assistance in certain instances.[119]

There are certain limitations to food assistance that may be provided through PA for the pandemic. FEMA stated that it will provide PA for food assistance for an initial 30-day period. Thereafter, PA Applicants may request extensions from their respective FEMA Regional Administrators.[120] Additionally, FEMA only reimburses for food assistance that is directly required as a result of the pandemic. For example, FEMA may not reimburse PA Applicants for ongoing costs predating the pandemic of operating food banks, though FEMA may reimburse PA Applicants for the costs of food storage, purchase, preparation, and distribution at food banks in response to the pandemic.[121] FEMA does not

---

Circumstances for COVID-19," memorandum, March 17, 2020, https://www.fema.gov/media-library-data/1584457999950-7186ffa29ace3e6faf2ca2f764357013/Procurement_Under_EE_Circumstances_Memo_final_508AB.pdf. See also FEMA, "Procurement Under Grants: Under Emergency or Exigent Circumstances," fact sheet, March 20, 2020, https://www.fema.gov/news-release/2020/ 03/20/procurement-under-grants-under-exigent-or-emergency-circumstances.

[116] FEMA, "Coronavirus (COVID-19) Food Eligible for PA."

[117] FEMA, "Coronavirus (COVID-19) Food Eligible for PA."

[118] Authorized in Stafford Act Sections 402, 418, 419, and 502; 42 U.S.C. §§5170a, 5185-86, 5192. See also 44 C.F.R. Section 206.208.

[119] FEMA, "Region 7 COVID-19 Fact Sheet," April 20, 2020, https://www.fema.gov/news-release/2020/04/20/region7-covid-19-fact-sheet.

[120] FEMA, "Coronavirus (COVID-19) Food Eligible for PA."

[121] 44 C.F.R. §206.223(a)(1).

reimburse for costs funded by other means, including assistance provided by federal agencies like USDA.[122]

FEMA has obligated funds specifically for food assistance for states including Oregon, California, Texas, Kansas, and Arizona through the PA program.[123] Work descriptions include reimbursement for National Guard food missions, food bank purchases, expanded Meals on Wheels funding, and California's Great Plates program, a public program that delivers restaurant-prepared meals to populations vulnerable to COVID-19.[124] Additional funds may have been obligated for food assistance but not designated as such in publicly available data.

Public Assistance Funding

PA is funded through the Disaster Relief Fund, the primary source of funding for the federal government's domestic general disaster relief programs. DRF appropriations are not allocated for specific emergencies, disasters, or forms of assistance, including PA. There is no statutory limit on the amount of funding that may be available to any PA recipient, PA Applicant, declaration, or project—including food distribution—for the COVID-19 pandemic or other Stafford Act declarations. The terms of assistance provided pursuant to each disaster are set by an agreement negotiated between FEMA and the declaration recipient.

Division B of the CARES Act (P.L. 116-136) included $45 billion for the DRF in March 2020, bringing its available balance for the costs of major disasters to more than $87 billion at the end of the month.[125] This is the first time FEMA has exercised Stafford Act authorities as a primary

---

122 Stafford Act §312; 42 U.S.C. §5155. See also 44 C.F.R. Section 206.250(c).

123 Based on application titles for COVID-19 declarations including the search terms *food*, *plates*, and *meals* in OpenFEMA, "Public Assistance Funded Projects – Details," available at https://www.fema.gov/openfema-datasetpublic-assistance-funded-projects-details-v1. The most recent data in the dataset were last updated on May 30, 2020.

124 California Governor's Office of Emergency Services, "Great Plates Delivered Program Guidance," May 7, 2020, p. 1, https://covid19.ca.gov/img/wp/great-plates-delivered-program-guidance.pdf.

125 FEMA, "Disaster Relief Fund: Monthly Report, as of March 31, 2020," April 8, 2020, p. 7, https://www.fema.gov/media-library-data/1586553202915-6ad14cfec022590b24 85cba9039190b0/ Apr2020DisasterReliefFundReport.pdf.

response to a pandemic, so FEMA has not provided projections for future spending from the DRF—including PA—on the COVID-19 response.[126]

SELECTED LEGISLATIVE PROPOSALS

In recent months, federal policymakers have introduced bills to increase support for food banks and agricultural producers. Proposals often differ in terms of which entity they provide funding to (e.g., states and/or local organizations) and allowable uses of funds (e.g., food purchases, transportation costs, and/or infrastructure costs). Some of these proposals seek to involve restaurants in emergency feeding efforts.

The House-passed Heroes Act (H.R. 6800) would make changes to existing programs. It would provide an additional $150 million for TEFAP and $200 million for ESFP. It would also waive a state matching requirement that applies when states retain TEFAP administrative funds rather than distribute them to or expend them on behalf of emergency feeding organizations. The bill would also amend and provide an additional $25 million for TEFAP's Farm to Food Bank Projects, which facilitate the harvesting, processing, packaging, and transporting of donated commodities from agricultural producers, processors, and distributors to emergency feeding organizations. H.R. 6800 would also waive a 50% state matching requirement in the Farm to Food Bank Projects and encourage the inclusion of specific types of agricultural products.[127]

Other proposals would create new programs. The Food Supply Protection Act (S. 3840), introduced by Senator Debbie Stabenow, Ranking Member of the Senate Committee on Agriculture, Nutrition, and Forestry, would authorize new temporary funding streams for emergency

---

126 For more information, see CRS Report R45484, *The Disaster Relief Fund: Overview and Issues*; and CRS Report R46326, *Stafford Act Declarations for COVID-19 FAQ*.

127 Specifically, "a dairy, meat, or poultry product, or a specialty crop—(A) packaged or marketed for sale to commercial or food service industries; (B) for which decreased demand exists for such a product due to the COVID–19 outbreak; and (C) the repurposing of which would be impractical for grocery or retail sale."

feeding organizations.[128] Specifically, the act would authorize USDA to reimburse food banks and other eligible organizations[129] for costs associated with the distribution and storage of perishable commodities, providing at least $500 million in CCC funding for this purpose. Additionally, S. 3840 would provide $500 million in CCC funding for USDA to award grants to such eligible organizations for "infrastructure, equipment, and supplies that support the distribution of surplus perishable food or meals prepared with that food."[130] Finally, the act would provide at least $1 billion in formula funding to states and Indian tribes to purchase foods that "would otherwise be unharvested or surplus" for distribution to eligible feeding organizations (including restaurants).[131] It would also authorize USDA to reimburse eligible feeding organizations and eligible food organizations[132] for costs associated with the distribution of these donated foods.[133]

Other proposals would fund state and federal food purchases for distribution to emergency feeding organizations. The Food Bank Access to Fresh Farm Produce Act of 2020 (S. 3605) would provide $8 billion for grants to states through USDA's Specialty Crop Block Grant Program. States would distribute the funds to food banks and food access networks

---

[128] It would authorize such funding streams until December 31, 2020 or "a date after December 31, 2020, if there are remaining food surpluses relating to the COVID-19 emergency, as determined by the Secretary [of Agriculture]." Bill text available at U.S. Congress, U.S. Senate Committee on Agriculture, Nutrition, and Forestry, "Ranking Member Stabenow Introduces Legislation to Protect America's Food Supply," May 27, 2020, https://www.agriculture.senate.gov/newsroom/dem/press/release/ranking-member-stabenow-introduces-legislation-toprotect-americas-food-supply.

[129] *Eligible organizations* would include nonprofits (e.g., food banks, schools, and emergency feeding organizations), faith-based organizations, tribal organizations, child and adult care centers, and "similar entities as determined by USDA."

[130] Ibid, Section 3. The bill would also authorize such sums as necessary from the CCC for this grant program.

[131] Ibid, Section 4. The bill would also authorize such sums as necessary from the CCC for these formula funds. *Eligible feeding organizations* are defined as public and nonprofit organizations, including tribal organizations, and restaurants operating an emergency feeding or food relief program.

[132] *Eligible food organizations* are defined as farming cooperatives; farmers organizations, food hubs, food distributors, majority-controlled producer-based business ventures, or other food aggregators; and food processors.

[133] Such expenses could include harvesting, processing, packaging, transportation, storage, and meal preparation costs.

to purchase and distribute specialty crops.[134] Relatedly, the Farmers Feeding Families Coronavirus Response Act (H.R. 6725 and S. 3655) would provide $200 million in Section 32 funds to states to purchase specialty crops, dairy, and meat and poultry products for distribution through state food assistance programs and emergency feeding organizations. The bills would also allow the use of funds for infrastructure, transportation, food storage, and preparation costs. Additionally, they would provide $300 million for federal food purchases of specialty crops, dairy, and meat and poultry products for distribution through federal nutrition assistance programs including TEFAP.

Other bills would establish partnerships between emergency feeding organizations and restaurants. The FEMA Empowering Essential Deliveries Act (FEED Act) (S. 3614 and H.R. 6700) would authorize FEMA to approve state and local partnerships with qualifying restaurants and nonprofits to provide meals to individuals in need for the duration of the COVID-19 Stafford Act declarations. Through PA, FEMA currently reimburses state, tribal, territorial, and local governments for up to 75% of eligible costs incurred while providing meals prepared by private entities, including restaurants and nonprofits, to respond to elevated food demand as a result of COVID-19.[135] S. 3614 may provide statutory enforcement to this policy. Additionally, S. 3614 would eliminate the 25% nonfederal cost share for all PA authorized pursuant to Stafford Act declarations for COVID-19.[136] The Community Meals Fund (H.R. 6384) would authorize appropriations of $20 million for USDA to award grants to nonprofit feeding organizations137 to partner with small and mid-sized restaurants

---

[134] Specialty crops include "fruits, vegetables, tree nuts, dried fruits, horticulture, and nursery crops (including floriculture)"; Section 101 of the Specialty Crops Competitiveness Act of 2004 (7 U.S.C. 1621 note). For more information, see CRS In Focus IF11317, *2018 Farm Bill Primer: Specialty Crops and Organic Agriculture*.

[135] FEMA, "Coronavirus (COVID-19) Food Eligible for PA."

[136] The Stafford Act authorizes FEMA to reimburse not less than 75% of the eligible costs of specific types of disaster response and recovery work undertaken by eligible PA Applicants. See, for example, Stafford Act Section 403(b), 42 U.S.C. §5170b(b). FEMA may recommend that the President increase the federal cost share in extraordinary cases at the request of a state, territory, or tribe, per 44 C.F.R. Section 206.47.

[137] *Nonprofit feeding organizations* include anti-hunger organizations, food banks, food pantries, soup kitchens, food rescue groups, or community food security organizations.

or food contractors "to expand meal access and delivery" during a pandemic and under a major disaster declaration.

In: COVID 19: Agricultural and Food Issues ISBN: 978-1-53618-561-4
Editor: Kristian Goldschmidt

Chapter 7

COVID-19, U.S. AGRICULTURE, AND USDA'S CORONAVIRUS FOOD ASSISTANCE PROGRAM (CFAP)*

Randy Schnepf and Jim Monke

ABSTRACT

As COVID-19 has spread throughout the United States, it has reduced domestic economic activity and disrupted domestic and international supply chains for goods and services, including food and agricultural products. These disruptions have produced an immediate and very strong demand shock on the U.S. food supply chain that has sent many commodity prices sharply lower.

The food supply chain refers to the path that raw agricultural commodities take from the farm where they are produced, through the food processing and distribution network to the consumer where they are used. Supply chain disruptions have been primarily due to two factors: widespread shutdowns of all but essential businesses; and uncertainty about the availability of labor for the food distribution network—whether

* This is an edited, reformatted and augmented version of Congressional Research Service, Publication No. R46347, dated May 8, 2020.

from illness, fear of illness, or immigration status. The food supply chain has been deemed essential; however, many institutional purchasers (including restaurants, hotels, schools, and entertainment venues) have been closed. According to the U.S. Department of Agriculture (USDA), U.S. consumers normally spend 54% of their food and drink dollars on away-from-home food purchases. Thus, prior to the COVID-19 pandemic a large share of U.S. food products traveling through the food supply chain was going to institutional buyers, often in bulk or vendor-ready form, for away-from-home consumption. In order to redirect this food product flow towards retail outlets and at-home consumption, much of this food would require processing into more consumer-usable forms, repackaging, and relabeling. This requires some level of retooling by food packagers and processors. In addition, several plants in the food processing industry, including meat processing plants, have experienced severe COVID infection outbreaks among workers and been forced to shut (at least temporarily).

Several industry groups from the U.S. agricultural sector have released estimates of the economic damage experienced by producers and ranchers. Most of these early assessments are limited to evaluating the effect of the price decline on any unsold production of crops or livestock remaining under farmer's control, and the unexpected marketing costs of unsold products due to the shutdown of most institutional buying of agricultural products. Cumulatively, industry estimates of COVID-related losses approach $40 billion (this would represent over 10% of annual cash receipts). The effect on farm net income is expected to be similarly negative.

In response to the COVID-19 pandemic, Congress has passed and the President has signed four supplemental appropriations acts (P.L. 116-123, P.L. 116-127, P.L. 116-136, and P.L. 116-139) that have included both direct and indirect funding for the U.S. agricultural sector. On April 17, 2020, Secretary of Agriculture Sonny Perdue announced the Coronavirus Food Assistance Program (CFAP), valued at $19 billion, to provide immediate financial relief to farmers, ranchers, and consumers in response to the COVID-19 national emergency. According to Secretary Perdue, CFAP will include $16 billion in direct payments to producers that have been impacted by the sudden drop in commodity prices and the disruption in food supply chains that has occurred since the outbreak, and $3 billion in commodity purchases for distribution through food banks, faith- based organizations, and other nonprofit organizations.

CFAP direct payments are intended to partially offset the loss of market revenue from the price decline, and the unexpected carry-cost of unsold commodities for producers and ranchers of products that have been negatively affected by COVID-19. USDA has released limited information on the specifics of how CFAP's direct payment program will be implemented. Many of the program specifics are expected to be delineated in an expedited rulemaking process. CFAP direct payments are expected to go out to producers by the end of May or early June.

CFAP's commodity purchase and distribution program serves the dual roles of supporting commodity market prices by temporarily increasing demand, and of expanding the availability of food distribution to consumers that have lost their jobs or have limited resources. Expectations are that it will be operated differently than, and separate from, existing USDA commodity purchase programs such as Section 32 or Food and Nutrition Service (FNS) food distribution programs in two major ways. First, USDA plans to purchase about $100 million per month of fresh produce, $100 million per month of dairy products, and $100 million per month of meat products (chicken and pork). Second, the CFAP purchase program intends for vendors to deliver household-ready boxes—potentially a diversified mix of the previously mentioned produce, meat and dairy products—which are ready for more convenient and immediate distribution ("off the truck and into the trunk") that is consistent with social distancing. Initial program delivery may be as early as May 15. The program is expected to operate through the end of 2020. Potential congressional concerns include how to channel assistance to industries affected by COVID-19, long-term effects of the pandemic, and the capacity of rural banks to help with recovery. Several issues related to CFAP and the U.S. agricultural sector in a post-COVID economy that could be of potential interest to Congress are presented at the end of this chapter.

INTRODUCTION

Since the novel coronavirus—designated Coronavirus Disease 2019, or COVID-19—first appeared in the United States in mid-January, it has contributed to substantial economic upheaval across the U.S. economy, including the agricultural sector.

In response to the COVID-19 pandemic, Congress has passed and the President has signed four supplemental appropriations acts (P.L. 116-123, P.L. 116-127, P.L. 116-136, and P.L. 116-139) that have included both direct and indirect funding for the U.S. agricultural sector.[1] Using funds

[1] For a brief discussion of the first three acts and their assistance for agriculture, see CRS In Focus IF11491, *Supplemental Appropriations for Agriculture and Related Agencies Due to COVID-19*. For a brief summary of the fourth act, see CRS Report R46325, *Fourth COVID-19 Relief Package (P.L. 116-139): In Brief*.

from these acts and from other authorities, the U.S. Department of Agriculture (USDA) announced, on April 17, 2020, the Coronavirus Food Assistance Program (CFAP), valued at $19 billion, to provide immediate financial relief to farmers, ranchers, and consumers in response to the COVID-19 national emergency.[2] USDA could provide additional financial assistance for the U.S. agricultural sector beyond CFAP later in the summer when a replenishment payment of $14 billion for the Commodity Credit Corporation becomes available. Congress is also considering providing additional support.[3]

This chapter describes some of the actions that USDA has taken in response to the COVID-19 emergency, including CFAP—in particular, how CFAP is funded and how USDA intends to use the funds. The description of USDA COVID-19 response efforts is preceded by: first, a brief review of food supply chain issues where the U.S. agricultural sector has experienced economic harm or is potentially vulnerable to the effects of the COVID-19 pandemic; and second, a review of current assessments of the economic harm to U.S. farm income, as well as to individual commodity sectors, resulting from COVID-19.

The report then describes the emergency funds that have been allocated to address the U.S. agricultural sector, and how USDA plans on using those funds, including a detailed description of CFAP producer payments, and USDA's food purchase and distribution efforts. This is followed by a review of the announced positions for selected U.S. agricultural stakeholders in regard to how the COVID-19 pandemic has affected their industries, what their anticipated losses might be, and what their expectations are vis-à-vis congressional funding and USDA's announced program response.

Finally, several issues related to CFAP and the U.S. agricultural sector in a post-COVID economy that could be of potential interest to Congress are presented at the end of this chapter.

---

[2] USDA, "USDA Announces Coronavirus Food Assistance Program," Release No. 0222.20, April 17, 2020.

[3] See the discussion later in this chapter under "Next Steps."

The Coronavirus Disease 2019 (COVID-19)

COVID-19's Global Spread

The novel coronavirus—designated Coronavirus Disease 2019, or COVID-19—began sickening patients in Wuhan, China, in early December 2019. By mid-January the virus had spread to Europe and the United States. The World Health Organization (WHO) declared the outbreak a "pandemic" on March 11, 2020.[4] By mid-April 2020, it had spread across the globe.[5]

The first known case of COVID-19 on U.S. soil was confirmed on January 20, 2020.[6] The Trump Administration announced a national emergency declaration on March 13.[7] By May 1, nearly 3.3 million people worldwide had been infected, resulting in a death toll of over 234,000; the United States had nearly 1.1 million confirmed infections and over 63,000 confirmed deaths.[8]

CRS Coverage: COVID-19

In order to cover the many aspects of the COVID-19 pandemic, the Congressional Research Service has established a Coronavirus Resources web page that displays current CRS reports on COVID-19, the federal response, and its effect across various subsectors of the U.S. economy.[9] In addition, CRS has established an "experts list" to help identify coverage for specific subject areas.[10] This chapter focuses on the potential effects of COVID-19 on the U.S. agricultural sector. As such, a list of CRS experts for major agricultural issue areas is included here.

CRS Coverage of COVID-19 Agriculture-related Issues

- Ad Hoc Direct Payment programs including CFAP & Market Facilitation Program (MFP); crop insurance; traditional farm revenue support

---

[4] CRS In Focus IF11421, *COVID-19: Global Implications and Responses*.

[5] CRS Resource, "Coronavirus Disease 2019 (COVID-19)," https://www.crs.gov/resources/coronavirus-disease-2019.

[6] Michelle L. Holshue, M.P.H. et al., "First Case of 2019 Novel Coronavirus in the United States," *The New England Journal of Medicine,* March 5, 2020.

[7] CRS Insight IN11253, *Domestic Public Health Response to COVID-19: Current Status*.

[8] Johns Hopkins University of Medicine, "Corona-19 Global Map," Corona Virus Resource Center, accessed May 1, 2020, https://coronavirus.jhu.edu/map.html.

[9] CRS Coronavirus Resources web page, https://www.crs.gov/resources/coronavirus-disease-2019.

[10] CRS Report R46209, *Coronavirus Disease 2019 (COVID-19) Outbreak: CRS Experts*.

programs (marketing assistance loans, ARC and PLC): Randy Schnepf

- USDA's Farm Service Agency (FSA) and Farm Credit Service (FCS) farm loans (and flexibilities); Section 32 and CFAP food purchases and distribution: Jim Monke
- Rural development issues; business loans and grants; community programs; agricultural labor; off-farm employment; rural broadband; international food aid: Alyssa Casey and Tadlock Cowan
- Specialty crops; local food systems; organic agriculture; fresh produce issues; food safety; and urban agriculture: Renee Johnson
- Dairy, poultry, and livestock issues; USDA's Food Safety and Inspection Service (FSIS) meat and poultry inspection: Joel Greene
- Commodity Credit Corporation funding and program authority; permanent disaster programs; emergency disaster loans; Noninsured Disaster Assistance Program (NAP); conservation programs: Megan Stubbs
- Agricultural research; extension; biotechnology; climate change and agriculture: Genevieve Croft
- Agricultural trade; U.S. agriculture in international trade agreements including the U.S.-China Phase One Agreement and World Trade Organization (WTO) commitments and dispute settlement: Anita Regmi

An appendix at the end of the report includes a Table (Table A-1) that summarizes estimates— from selected studies—of the economic damage to several major agricultural sectors of the United States due to the COVID-19 emergency.

U.S. Agricultural Sector Vulnerability to COVID-19

As COVID-19 has spread throughout the United States, it has reduced domestic economic activity and disrupted domestic and international supply chains for goods and services, including food and agricultural products. These disruptions have produced an immediate and very strong demand shock on the U.S. food supply chain. In particular, the abrupt shutdown of much food service and institutional buying has affected commodity prices throughout the food supply chain.

On the supply side, there is no food shortage in the United States. Supplies for most commodities remain relatively abundant; however, inflexibilities in the food supply chain (from food product specialization targeting food service and institutional markets that have closed, to pandemic outbreaks at processing plants) have resulted in bottlenecks that have left many farmers with unmarketable surpluses, while some retail outlets have experienced temporary shortages of various food and agricultural products.

Temporarily Empty Grocery Store Shelves

During the early weeks of shutdown in March and April, many grocery stores had empty shelves for basic staples such as pasta, rice, sugar, and flour, as well as frozen ready-to-eat foods, household cleaning products, and toilet paper. This was the result of a temporary surge in consumer food stockpiling—much of it in the form of panic buying and potential hoarding—that occurred in March when consumers worried that they might be locked down in their houses for weeks or months.[11] This type of shortage is temporary. Eventually the existing supplies of foodstuffs and other household products work their way through the food supply chain and restock the grocery store shelves.

The temporary shortage occurred due to two primary factors. First, consumers had been getting a significant share (54%) of their food from restaurants and other away-from-home venues.[12] All of this demand was suddenly diverted to grocery stores and retail outlets almost overnight.[13] This switch is not simply a matter of redirecting truckloads of products

---

[11] Millie Munshi and Isis Almeida, "There's plenty of food in the world, just not where it's needed," *Bloomberg News*, March 21, 2020.

[12] USDA, Economic Research Service (ERS). "Food Expenditure Series," accessed on April 22, 2020, www.ers.usda.gov/data-products/food-expenditure-series/.

[13] Many restaurants remain open for takeout or delivery services, but are closed for dine-in service.

away from institutional buyers towards retail outlets, but also of transforming many products from bulk or vendor-ready forms to consumer-ready forms in smaller packages with new product labelling. Furthermore, truckers would now be making many stops for smaller deliveries to retail outlets along their routes, rather than a few stops to large buyers. All of this would require re-engineering of the food supply networks that ultimately may be temporary once the pandemic eases.

Second, much of the food supply chain operates on a "just-in-time" principle to minimize costs associated with storage and waste of products. Under this principle, many processing plants and transportation routes are designed to operate near full capacity on a routine basis in expectation of slow, steady demand and to avoid the cost of idled resources—this is particularly true for food products and other items like toilet paper that normally do not experience large fluctuations in demand over time.[14]

Thus, not all production processes can ramp up on short notice to meet unexpected surges in demand for their products, or to suddenly alter the product form, packaging, labeling, and shipping methods. As a result, shortages due to consumer stockpiling create bulges in the supply chain that may take weeks or months to eliminate as the food supply chain re-engineers itself in response.

The U.S. Food Supply Is Ample

U.S. and global supplies of major agricultural commodities are in a state of relative abundance— thanks to large harvests over the past several years—which has kept market prices at relatively low levels for the past five years. The surplus supply situation has been compounded by the ongoing trade dispute between the United States and China, which disrupted traditional trade patterns and contributed to lower trade levels

[14] Laura Reily, "Why Shelves are Empty at the Grocery Store," *Post Reports,* Washington Post news podcast, April 20, 2020.

(and greater than expected U.S. stock levels) for several major commodities in the second half of 2018 and 2019.[15]

Globally, the Food and Agriculture Organization (FAO) of the United Nations reports that stocks for many major agricultural commodities—including wheat, rice, corn, barley, coffee, sugar, butter, cheese, palm oil, soybean oil, poultry, and pork—are at or near 20-year highs.[16] A similar situation exists within the United States. On April 9, 2020, USDA reported that the major U.S. grain and oilseed crops, as well as upland cotton, all had relatively large end-of-year stocks levels last fall (2019) to carry into this year.[17] On April 22, 2020, USDA reported abundant supplies of meat, dairy, fruit, vegetable, and tree nut products in cold storage.[18] Similarly, USDA reports that hog and poultry populations in the United States are historically large, while the cattle population has rebounded from its low point in 2014.[19] In addition, USDA reported that for 2020, U.S. farmers intended to increase corn and soybean planted acres by 8% and 10%, respectively, while adjusting wheat (1%) and cotton acres (<1%) down slightly from 2019.[20] If such large planted acres are realized, under normal weather conditions, they could produce bumper crops and further expand domestic supplies.

According to news reports, Secretary of Agriculture Sonny Perdue has said that consumers should not worry about a shortage of food, saying the supply chain is just mismatched.[21] Although demand has surged at the retail level in response to the pandemic, demand from restaurants, cafeterias, sports venues, and tourism has plummeted.

---

[15] CRS Report R45903, *Retaliatory Tariffs and U.S. Agriculture*.

[16] Naveen Thukral et al., "Panic Buying, Lockdowns May Drive World Food Inflation: FAO, Analysts," *Reuters*, March 21, 2020.

[17] USDA, World Agricultural Outlook Board (WAOB), World Agricultural Supply and Demand Estimates (WASDE), April 9, 2020. See also Figure 3 in CRS Report R46249, *U.S. Farm Income Outlook: February 2020 Forecast*.

[18] National Agricultural Statistics Service (NASS), *Cold Storage,* April 22, 2020.

[19] USDA, NASS, *Cattle*, January 31, 2020; USDA, NASS, *Quarterly Hogs and Pigs*, March 26, 2020; USDA, NASS, *Poultry Slaughter*, March 24, 2020; and USDA, NASS, *Poultry-Production and Value, 2018 Summary*, May 23, 2019.

[20] USDA, NASS, *Prospective Plantings*, March 30, 2020.

[21] Helena Bottemiller Evich, "USDA to buy $2B in ag goods as part of coronavirus aid," *Politico*, April 16, 2020.

Disruption of Food Supply Chains Impact Commodity Prices

The food supply chain refers to the path that raw agricultural commodities take from the farm where they are produced, through the food processing and distribution network to the consumer where they are used. The domestic food supply chain has the potential to break down at any of a number of different points: availability of inputs and labor for agriculture production; trucks and truck drivers for transporting raw and finished products; food processing plants, plant workers, and food safety inspectors; packaging, warehousing, and storage capacity; and wholesale and retail outlets and their workers. For exported products, the supply chain includes containers, ships, crew, and port workers.

There is a finite supply of trucks, railcars, and shipping containers, and they may not be situated where they are needed when a temporary surge in demand occurs. Labor shortages at any point along the supply chain can lead to bottlenecks, delays, and regional shortages. Similarly, there is a finite supply of warehousing and cold storage (i.e., refrigerated and frozen) space, which may contribute to temporary regional shortages.

With respect to COVID-19's impact, supply chain disruptions have been primarily due to two factors: widespread shutdowns of all but essential businesses; and uncertainty about the availability of labor for the food distribution network, from farms to retail outlets—whether from illness, fear of illness, or immigration status.

Loss of Institutional Buyers

The first effect of the COVID-19 pandemic on agricultural producers occurred when many states and municipalities closed schools and instituted economy-wide shutdowns of all but essential businesses. The wholesale and retail food distribution network has been deemed essential; however, many institutional purchasers of agricultural products (often referred to as the food services sector, including restaurants, hotels, schools, and entertainment venues) have been closed.[22]

---

[22] Many restaurants remain open for takeout or delivery service, but with no dine-in services.

According to USDA, U.S. consumers normally spend 54% of their food and drink dollars on away-from-home food purchases.[23] Thus, a large share of U.S. food products traveling through the food supply chain was going to the food services sector, often in bulk or vendor-ready form, for away-from-home consumption. In order to redirect this food product flow towards retail outlets and at-home consumption, much of it would require processing into consumer-usable quantities and forms, requiring repackaging, and relabeling. This requires some level of retooling by food packagers and processors.

As a result, the near total stoppage of institutional food purchases contributed to sharp declines in the prices of affected commodities (Figure 1), and led to unanticipated conditions of oversupply from commodities that could no longer move through the food supply chain and were, instead, backing up to the farms that produced them. This left many agricultural producers with excess supplies of perishable products—including fruit, vegetables, milk, and market-ready livestock— that are not easily diverted to alternate uses or retail outlets.[24] In the interim, the temporary glut of perishable products with nowhere to go has led to news reports of producers dumping fresh milk, burying truckloads of raw onions, plowing fields of ripe vegetables back into the ground, and more disturbingly, depopulating millions of market-ready hogs and poultry.[25]

The surplus of perishable, unsold commodities worsened starting in mid-April, when a surge in COVID-19 infections among workers in meat packing plants and other food processing plants led to multiple plant closures, and contributed to both animal surpluses on farms and public concerns about the reliability of the nation's food supply.[26]

---

[23] USDA, ERS. "Food Expenditure Series," accessed on April 22, 20202, ww.ers.usda.gov/data-products/food- expenditure-series/.

[24] Mark Heller, "Farmers fret as pandemic creates animal surplus," E&E News, April 20, 2020.

[25] Aime Williams, "US restaurant closings spur farmers to destroy food," *Financial Times*, April 22, 2020.

[26] Kyle Bagenstose, Sky Chadde and Matt Wynn, "Coronavirus at meatpacking plants worse than first thought, USA TODAY investigation finds," *USA TODAY*, Apr. 22, 2020.

Uncertainty over Labor Availability

Agriculture has been classified by the federal government as a critical industry that must remain operating, even as much of the rest of the country shuts down to help contain the virus' spread.[27] However, labor shortages at any stage of the supply chain can create temporary food product shortages in affected markets. If labor shortages become severe, they could lead to wider multi- state, and possibly national, food shortages of affected products.

Many fruit and vegetable production activities are labor-intensive and require an adequate work force at key points in the products cycle—particularly at harvest—to successfully bring the crop to market. In addition, the U.S. agricultural sector relies on a large workforce to operate the production lines in food processing plants, including meat packing plants, as well as fruit and vegetable wholesale and distribution networks. This labor force performs supply chain activities including production, transportation, processing, warehousing, packaging, and retailing. Many of these supply chain activities cannot be automated or done remotely, but rely on workers being on site. For example, USDA reports that more than 1.5 million people worked in food processing in the United States in 2016.[28] Meat processing, which tends to be more labor intensive than other parts of the food sector, accounted for 500,000 of those employees.[29] Workers that are still planting and harvesting crops, or standing on an assembly line in a meat packing plant, during the coronavirus outbreak have a high risk of being infected with the disease given that they live, work, and travel in crowded conditions, and most do not have health care or paid sick leave.[30]

Another labor-intensive component of the food supply chain is federal safety inspection, which is undertaken by about 8,000 USDA safety

[27] Liz Crampton and Helena Bottemiller Evich, "White House points to CDC guidelines as more meatpacking workers infected," *Politico,* April 16, 2020.

[28] USDA, ERS, "Manufacturing: Food and Beverage Manufacturing," Processing & Marketing, accessed on April 23, 2020, https://www.ers.usda.gov/topics/food-markets-prices/processing-marketing/manufacturing/.

[29] Ibid.

[30] Liz Crampton, "Coronavirus impact on farm labor broadens push for federal action," *Politico*, April 3, 2020.

inspectors stationed at every agricultural manufacturing facility throughout the country, as well as about 3,800 safety inspectors from the Food and Drug Administration (FDA).[31]

As the food distribution network shifts more food products away from institutional outlets to grocery stores, labor at the retail level has come under greater stress.[32] Many grocery stores have begun implementing preventative measures, like reducing hours to give staff time to rest, clean, and stock shelves, while limiting exposure to customers.[33] All of these COVID-19 related measures tend to slow the food supply chain's throughput rate and thus have prolonged the period of empty or partially filled grocery store shelves.

Many Meat and Food Processing Plants Slow Operations or Close

Starting in mid-April, a surge in infections among workers in meat packing plants and other food processing plants led to multiple plant closures and contributed to unexpected surpluses of ready-for-market hogs, cattle, and poultry at the farm level.[34] As of May 1, news sources reported that at least 20 meatpacking and 5 food processing plants had been closed.[35] Due to a high degree of consolidation in the meat processing industry, a shutdown of four or five big plants could impact retail supplies.[36] On May 1, the Centers for Disease Control and Prevention (CDC) reported that 115 meat and poultry processing plants (with over 130,000 workers) had a combined 4,913 workers with confirmed cases of COVID-19, including 20 that had died from COVID-19.[37]

---

[31] Liz Crampton, "USDA Struggles to Counteract COVID-19," *Politico*, April 23, 2020.

[32] Lisa Held, "Food Distribution 101: What Happens When the Food Supply is Disrupted by a Pandemic," *Civil Eats*, April 15, 2020, https://civileats.com/2020/04/15/food-distribution-101-what-happens-when-the-food-supply-is-disrupted-by-a-pandemic/.

[33] Ibid.

[34] CRS Insight IN11366, *COVID-19 Disrupts U.S. Meat Supply; Producer Prices Tumble*.

[35] Leah Douglas, "Mapping Covid-19 in meat and food processing plants," *Food and Environment Reporting Network*, May 1, 2020, https://thefern.org/2020/04/mapping-covid-19-in-meat-and-food-processing-plants/.

[36] Josh Funk, "Virus Closes Some Meat Plants, Raising Fears of Shortages," Associated Press, April 13, 2020.

[37] Jonathan W. Dyal et al. "COVID-19 Among Workers in Meat and Poultry Processing Facilities—19 States, April 2020," Morbidity and Mortality Weekly Report, Early Release / Vol. 69, May 1, 2020, https://www.cdc.gov/mmwr/ volumes/69/wr/pdfs/mm6918e3-H.pdf.

Meat processing plant closures have two opposing effects: on the one hand, demand for livestock in the surrounding region is reduced and this tends to depress cash and futures prices, lowering prices that producers receive and that packers pay for market-ready livestock; on the other hand, the supply of consumer-ready product is reduced, which tends to raise wholesale and retail prices for the affected products. As evidence of this, USDA has reported a widening price gap between farm and wholesale prices for beef.[38]

On April 22, news sources reported that, with the closure of Tyson's pork processing plant in Waterloo, IA, about 15% of total U.S. pork processing capacity was off line.[39] According to an official with the Commodity Futures Trading Commission's (CFTC's) Livestock Marketing Task Force, most plants that were still open during this same period were operating at only 50% to 75% of normal production due to employee absenteeism.[40] Furthermore, the CFTC official noted that U.S. pork processing plants normally handled about 2.5 million hogs per week, but that slower operating line speeds had reduced that number to 2.1 million hogs, implying that an additional 400,000 market-ready hogs had to stay on the farm each week. By May 1, weekly hog slaughter had fallen to 1.5 million, implying that nearly a million hogs per week were backing up to the farm.[41]

On April 28, President Trump signed an executive order using authority under the Defense Production Act (DPA) of 1950, as amended (50 U.S.C. 4501 et seq.), and Section 301 of title 3, United States Code, to order the Secretary of Agriculture to take all appropriate action to ensure that meat and poultry processors continue operations consistent with the guidance for their operations jointly issued by the CDC and Occupational

---

[38] William Hahn, "Record increase in March farm-to-wholesale beef price spread driven by sharp wholesale price changes in mid-March 2020," *Chart of the Day*, USDA, ERS, April 23, 2020. See also "Cash Prices, Meat Cutout Diverge," *ProFarmer*, April 25, 2020.

[39] Liz Crampton, "USDA Struggles to Counteract COVID-19," *Politico*, April 23, 2020.

[40] Statement by Dave Amato of the CFTC, as cited in Jerry Hagstrom, "CFTC official: Meat Plants Operating at 60%; 400,000 Hogs per Week Backing Up," *The Hagstrom Report,* Volume 10, Number 111, April 23, 2020.

[41] USDA, Agricultural Market Service, "Estimated Weekly Meat Production Under Federal Inspection," *USDA Market News*, May 1, 2019.

Safety and Health Administration (OSHA).[42] Two issues associated with the reopening of closed plants are the availability of personal protective equipment (PPE) for plant workers, and the liability associated with hospitalization costs and/or deaths of infected plant workers.

On April 30, Secretary Perdue stated that USDA will ask meat processors to submit written plans to safely operate packing plants and review them in consultation with local officials.[43] Perdue said that USDA will work collaboratively with companies and state and local officials to set safety standards based on guidelines for workers released by the CDC and OSHA, and that USDA was working to insure the availability of the necessary PPE for plants to operate safely. The Secretary also said that President Donald Trump's executive order to open the plants will not remove legal liability, but that the CDC protection guidelines will give meat plants "a defensible answer" should they be sued, as long as they follow the guidance.[44]

Commodity Prices Plummet

Since mid-February, prices for many major agricultural commodities have plummeted (Figure 1). Commodity market price declines have been led, in part, by a precipitous fall in the price of crude oil (down 67% between January 2 and April 15), which feeds back into the U.S. market for ethanol and corn, and subsequently through the expanded market for livestock feedstuffs.

Prices for the livestock sector—cattle, hogs, poultry, and dairy—were hit particularly hard by the sudden economic shutdown and the associated dilemma of what to do with market-ready livestock, and because of the difficulty in diverting product from restaurants to retail outlets. During the

---

[42] White House, "Executive Order on Delegating Authority Under the DPA with Respect to Food Supply Chain Resources During the National Emergency Caused by the Outbreak of COVID-19," April 28, 2020, https://www.whitehouse.gov/presidential-actions/executive-order-delegating-authority-dpa-respect-food-supply-chain- resources-national-emergency-caused-outbreak-covid-19/.

[43] Mike Dorning, "Meat Plants to Reopen in 'Days Not Weeks,' USDA Boss Says," *Bloomberg*, April 30, 2020.

[44] Ibid.

January 2 to April 15 period, prices for lean hogs were down by 53.2%, live cattle 25.1%, and milk 20.6%.

The cotton and textile industries were negatively impacted by the widespread shutdown of retail businesses. Demand for clothing and apparel dropped precipitously with the economic closure.[45] This fed back into a collapse of demand that affects manufacturers, which affects cotton mills, and finally to contracts for cotton being canceled. U.S. cotton prices dropped by over 25% between January 2 and April 15 (Figure 1). USDA's Foreign Agricultural Service (FAS) reduced its forecast for global cotton consumption for 2020 by 6.4%.[46] According to FAS, spending on clothing is highly correlated to changes in GDP, and most economic forecasters are expecting strong declines in global GDP in the first half of 2020.

Reduced travel, slowing economic activity, and petroleum-product demand suppression related to the COVID-19 outbreak, combined with announced plans to increase crude oil supplies by Saudi Arabia and Russia in mid-April, contributed to the severe decline in crude oil prices.[47] Low oil prices contribute to lower agricultural prices via a strong biofuels link between the two sectors. Corn is both the world's foremost livestock feed grain and the principal feedstock in the production of the biofuel ethanol. Ethanol is blended with gasoline in the United States (at about a 10% share) for use in automobiles and light trucks. Thus, declining fuel demand contributes directly to falling prices for gasoline, ethanol, and corn.

Ethanol prices fell by nearly 33% from January 2 to April 15 (Figure 1). As of April 21, news sources report that nearly 30% of the nation's 204 biofuel plants have been idled since March 1, while many others have reduced their production volumes.[48]

---

[45] Bill Tomson, "COVID-19 Hammers Textiles as Cotton Farmers Hit the Fields," April 15, 2020.

[46] USDA, FAS, "COVID-19 Spurs Record Downward Adjustments to Global Demand," *Cotton: World Markets and Trade*, April 2020.

[47] CRS Insight IN11286 IN11286, *Low Oil Prices: Prospects for Global Oil Market Balance*.

[48] Mike Hughlett, "Nearly 30% of Ethanol Plants Idled, Including Four in Minnesota," *Star Tribune*, April 21, 2020.

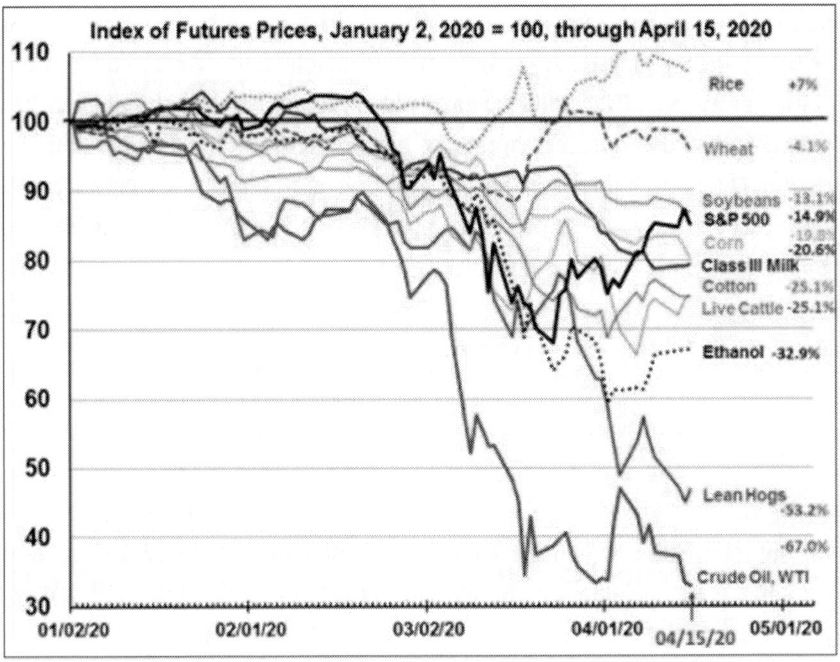

Source: Data are as of April 20, 2020. Indexes have been compiled by CRS from daily closing prices for nearby futures contracts on the Chicago Board of Trade (CBOT) under the CME Group as reported by RMBGroup.com/quotes for all commodities, except cotton which is traded on the ICE Futures U.S. exchange. Prices include live cattle and Class III Milk futures contracts for April 2020; corn, soybeans, cotton, wheat, rough rice, lean hogs, ethanol, and West Texas Intermediate (WTI) crude oil, contracts for May 2020; and S&P 500 E- Mini contracts for June 2020.

Notes: Class III milk is used primarily for cheese. Lean hogs are priced in carcass-weight equivalents. WTI crude oil is West Texas Intermediate (WTI) grade crude oil. It is unlikely that the entire downward price movement for all commodities is uniquely the result of the COVID-19 outbreak; there are other market factors that have contributed to the decrease in prices over this period as explained in the text.

Figure 1. Major Commodity Price Changes Since January 2020.

By April 25, the Renewable Fuels Association reported that about 46% of ethanol production capacity was idled.[49] Furthermore, the prices of

---

[49] Renewable Fuels Association (RFA) estimate from *Pro Farmer*, Weekly Report, April 25, 2020, p. 4. RFA reports that U.S. ethanol production was 15.8 billion gallons in 2019, http://www.ethanolrfa.org.

nearly all feed grains and oilseeds produced in the United States move in tandem with corn prices since they all compete for the same feed markets in consumption and farmland in production. Thus, this combination of the COVID-induced sudden disruption of normal agricultural demand and use, slowing U.S. and global economic activity, and sharply lower oil prices have placed strong downward pressure on commodity prices in international markets since the start of 2020.

International Market Export Restrictions

On top of these domestic disruptive factors, international markets for some major food items, such as rice and wheat, have experienced trade disruptions due to threats or actual imposition of protectionist policies on exports of major food products in certain important producer countries—including Russia (the world's leading wheat exporter), Kazakhstan, and Vietnam.[50] According to the World Trade Organization (WTO), 80 countries and customs territories so far have introduced export prohibitions or restrictions as a result of the COVID-19 pandemic.[51] In response, the United States, China, the European Union, and other members of the WTO representing over 60% of world agricultural exports pledged to refrain from imposing restrictions on the free flow of food out of their countries.[52]

By threatening to limit supplies to international markets, these protective policies have actually been supportive of rice and wheat prices in international markets (Figure 1). Since January, the nearby futures contract price for rough rice on the Chicago Board of Trade has actually risen by 7%, while the contract for wheat has fallen by 4%, unlike corn and soybean prices that have fallen 20% and 13%, respectively.

---

[50] Cullen S. Hendrix, "Wrong tools, wrong time: Food export bans in the time of COVID-19," Peterson Institute of International Economics, March 30, 2020, https://www.piie.com/blogs/realtime-economic-issues-watch/wrong-tools-wrong-time-food-export-bans-time-covid-19. On May 1, Vietnam lifted its export restrictions. Vietnam, Ministry of Industry and Trade, "PM Allows Rice Export Resumption From May 1," May 5, 2020.

[51] WTO, "WTO report finds growing number of export restrictions in response to COVID-19 crisis," April 23, 2020, https://www.wto.org/english/news_e/news20_e/rese_23apr20_e.htm.

[52] Doug Palmer, "WTO coalition urges countries to refrain from food export restrictions," *Politico*, April 22, 2020.

Current Assessments of Economic Damage

U.S. policymakers and business interests are concerned that the COVID-19 pandemic will inflict widespread economic harm on the U.S. and global economies.[53] It is still too early to make any definitive statements about what the eventual economic impacts will be on the U.S. economy or the agricultural sector, since it is unknown how long the disease will persist and what shape the economic recovery might take. For example, will the overall impact be V-shaped with a quick outbreak followed by a quick recovery, or will it be L-shaped with an elongated tail representing a slow recovery and a gradual reopening of businesses and retail outlets? Or be W-shaped if the virus recycles and re-emerges later in the summer or fall in a more virulent form—as did the H1N1 pandemic in 2009, or the 1918 flu pandemic—thus, causing a new round of shutdowns and economic closures?[54]

U.S. and Global GDP Outlook Revised Downward

As commerce slows, economic output is expected to follow with projections of a significant contraction in U.S. gross domestic product (GDP).[55] On April 29, the Bureau of Economic Analysis (BEA) reported the U.S. GDP (adjusted for inflation) had decreased by 4.8% during the first quarter of 2020.[56]

Many major financial institutions have also issued preliminary assessments of the economic impact of the COVID-19 pandemic with dire predictions. For example, in March, JP Morgan predicted a 25% decline in

---

[53] CRS Insight IN11235, *COVID-19: Potential Economic Effects*.

[54] Centers for Disease Control and Prevention, "2009 H1N1 Pandemic (H1N1pdm09 virus)," https://www.cdc.gov/flu/ pandemic-resources/2009-h1n1-pandemic.html; and "1918 Pandemic Influenza: Three Waves," https://www.cdc.gov/ flu/pandemic-resources/1918-commemoration/three-waves.htm.

[55] CRS In Focus IF10408, *Introduction to U.S. Economy: GDP and Economic Growth*.

[56] U.S. Bureau of Economic Analysis, U.S. Department of Commerce, Gross Domestic Product, 1st Quarter 2020 (Advance Estimate), April 29, 2020, https://www.bea.gov/news/2020/ gross-domestic-product-1st-quarter-2020-advance-estimate.

2nd quarter U.S. GDP.[57] The International Monetary Fund (IMF), in early April projected that the U.S. GDP would decline by 5% during 2020 (down 7.9 points from its January 2020 pre-COVID forecast of 2.9% growth).[58] The IMF also forecast that global GDP would decline by 3% (-6.3 points from January), and major economies would also see strong declines in GDP including the Euro-zone (-7.5%, -8.8 point from January), and Japan (-5.2%, -4.8 points from January).

On April 24, the Congressional Budget Office (CBO) released an update to its long-run baseline projection for the U.S. economic outlook that included a preliminary assessment of the economic impact of the COVID-19 pandemic.[59] CBO projected that the U.S. economy will experience a sharp contraction in the 2nd quarter of 2020—inflation-adjusted GDP is expected to decline by about 12% during the 2nd quarter, equivalent to a decline at an annual rate of -40% for that quarter. CBO also projected a 2.8% increase in real GDP in 2021. In addition, the U.S. unemployment rate is expected to average 11.4% for 2020 and 10.1% for 2021, consistent with CBO's current projection for a slow economic recovery.

Between March 14 and May 7, over 33 million American workers have filed first-time claims for unemployment benefits, according to the seasonally adjusted numbers of the U.S. Department of Labor.[60] According to news sources, numbers at that level indicate that over 20% of the U.S. labor force is suffering from layoffs, furloughs, or reduced hours during the coronavirus pandemic. In addition to reflecting the strong likelihood for high unemployment and impactful declines in consumer incomes through 2020, these forecasts also have important implications for the

---

[57] Ben Wick, "JPMorgan slashes its GDP forecast for next quarter, now sees a 25% contraction and an imminent recession despite 'Herculean' stimulus," *Business Insider*, March 26, 2020.

[58] IMF, "World Economic Outlook, April 2020—Chapter 1: The Great Lockdown," April 6, 2020.

[59] Phill Swagel, CBO Director, "CBO's Current Projections of Output, Employment, and Interest Rates and a Preliminary Look at Federal Deficits for 2020 and 2021," April 24, 2020.

[60] POLITICO Pro Breaking News: Labor Department reports 3.2M weekly jobless claims May 7, 2020.

rural, off-farm economy that so many farm households depend on (as described below).

Major Agricultural Commodity Sectors Suffer Substantial Harm

Several industry groups from the U.S. agricultural sector have released estimates of the economic damage experienced by producers and ranchers. Most of these early assessments are limited to evaluating the effect of the price decline on any unsold production of crops or livestock remaining under farmers' control, and the unexpected marketing costs of unsold products due to the shutdown of most institutional buying of agricultural products. Several of these damage assessments are summarized in Table A-1.

The livestock sector appears to be the hardest hit, as hog and cattle prices have dropped by 53% and 25% from January into mid-April, thus generating large sectoral losses estimated at $13.6 billion to $14.6 billion for the cattle/beef sector, and $5 billion for the hog sector. Early estimates of dairy sector losses exceed $8 billion, while the fresh produce sector losses are estimated at $5 billion. Other major commodity sector losses include corn ($4.7 billion to $6 billion), soybeans ($2 billion), cotton ($610 million to $3.5 billion), and the sheep and wool sector ($300 million).

The ethanol sector has not reported an overall dollar loss estimate, but reports that nearly 46% of its production capacity is offline (the United States produced 15.8 billion gallons of ethanol in 2019) and the price of ethanol has fallen by almost 33% (Figure 1).

Prices for several, but not all, of the affected commodities have turned upward slightly since mid-April when USDA's assistance program was announced. Corn prices are the most notable exception as they have continued their decline through the end of April. The corn price decline is partially driven by the catastrophic near collapse of the U.S. ethanol sector. Corn is the primary feedstock for U.S. ethanol production—nearly 38% of annual U.S. corn production is consumed by the ethanol sector.

U.S. Farm Income Projections Revised Downward

USDA's most recent U.S. farm economic outlook for 2020 (released on February 5, 2020) did not include the market effects of the COVID-19 pandemic.[61] USDA is not expected to release its next U.S. farm income projections until September 2, 2020. However, the Food and Agricultural Policy Research Institute (FAPRI), at the University of Missouri, released a preliminary assessment of the impact of COVID-19 on the U.S. farm income outlook in April.[62] FAPRI's preliminary projections assume a V-shaped recession where the market recovers quickly; market outcomes are driven largely by GDP and commodity price declines, and the supply chain disruptions described earlier in this chapter are not included in the analysis.[63] Macro factors include an inflation-adjusted 5% decline in consumer expenditures from FAPRI's January pre-COVID baseline projections, and a 10% decline in gasoline use.

Under these assumptions, FAPRI projects substantial price declines for all major grain, oilseed, and livestock commodities in 2019/20 and 2020/21, which result in large declines in farm revenue—including -\$11.9 billion in crop and -\$20.2 billion in livestock cash receipts (Table A- 1). The revenue declines are partially offset by declines in production expenses (-\$11.3 billion) and an increase in farm program payments (+\$2.3 billion) under the Agricultural Risk Coverage (ARC) and Price Loss Coverage (PLC) programs. However, the sum of the projected changes is for a large decline in U.S. net farm income of -\$20 billion for 2020 (down from FAPRI's January forecast of \$99 billion in 2020 net farm income, and compared to \$95.3 billion in 2019).[64] FAPRI does not include any payments under USDA's proposed CFAP. Although preliminary, FAPRI's

[61] CRS Report R46249, *U.S. Farm Income Outlook: February 2020 Forecast*.

[62] Pat Westhoff, Seth Meyers, Julian Binfield, and Scott Gerlt, "Early Estimates of the Impacts of COVID-19 on U.S. Agricultural Commodity Markets, Farm Income and Government Outlays," FAPRI-MU Report #02-20, April 13, 2020.

[63] This might be considered an optimistic or "best case" scenario, given the disease's rapid global spread through April, the economic havoc already reported by major financial institutions and U.S. government agencies, and the increasing likelihood that the negative economic effects will be prolonged at least into the summer of 2020.

[64] Pat Westhoff et al., "2020 U.S. Agricultural Market Outlook," FAPRI-MU Report #01-20, based on market conditions as of January 2020, but released March 2020.

early projections are indicative of the potentially large impact that COVID-19 may have on the U.S. agricultural sector.

Potential Impact on Farm Household Cash Flow

The COVID-19's economic impact is expected to vary across commodity sectors and regions based on the extent of price declines across commodities, the seasonality of production cycles, and off-farm work opportunities, as well as each household's level of near-term debt, tenure status, asset valuations, and other economic factors. Thus, each farm household's situation may be unique. This section briefly describes where COVID-19's economic impact may be felt most immediately. The potential impact on the farm household credit situation is not discussed here.[65] If the economic impact of COVID-19 persists into 2021, a broader range of economic factors are likely to be impacted, such as asset valuations and bankruptcies.

On-Farm Revenue, Expenses, and Federal Payments All Impacted

The principal market effects to date of COVID-19 have been the commodity price declines experienced in early 2020 (Figure 1), as well as lost sales and the costs associated with unexpected surplus animals, grains, and oilseeds held by farmers. The price declines, in particular, can have several potential effects on farm household income, including[66]

- the selling prices for 2019 crops still held in on-farm inventory;
- farm program payments for the 2019 crops under the marketing assistance loan (MAL), ARC, and PLC programs;
- prices and crop insurance payments for the 2020 crop;

---

[65] For background on this issue, see CRS Report RS21977, *Agricultural Credit: Institutions and Issues*.

[66] Schnitkey, G., K. Swanson, N. Paulson, C. Zulauf, and J. Coppess. "What We Know About Income Outlook for Crop Farms Given COVID-19," farmdoc daily (10):54, Department of Agricultural and Consumer Economics, University of Illinois at Urbana-Champaign, March 24, 2020.

- farm program payments for the 2020 crops under MAL, ARC, and PLC; and
- changes in input prices.

The sharp drop in commodity prices is expected to result in reduced farm household revenue. Revenue losses are expected to be partially offset by both increases in government payments under traditional farm revenue support programs (which are available for about two dozen designated program crops) and crop insurance, and by reductions in input expenses. However, the net effect is expected to be an overall decline in farm revenue compared with 2019. For farm operations carrying above average levels of debt, the restricted cash flow can cause severe financial stress. The emergency-response payment and purchase program announced by USDA (described below) is intended to help address, at least partially, the revenue decline and tightening cash flow situation for a wider array of farm households than usually receive government payments.

Off-Farm Income Impacted by Rural Economic Situation

The drop in prices for major farm commodities in early 2020 (Figure 1) suggests that farm revenues are likely to decline. However, of perhaps greater consequence to farm households has been the blow to off-farm income from the widespread economic shutdown and increases in unemployment. On average, 82% of farm household income comes from off-farm revenue sources.[67] As a result, the cash flows of farm households have been diminished from both the on- farm and off-farm effects of the COVID-19 pandemic.

Another blow to the rural economy is the strong decline in tax revenues, fees, and other sources of income, which hampers the ability of state and local governments to respond to the developing crisis through local programs and initiatives.[68]

---

[67] USDA, ERS, "2020 Farm Income Forecast," February 5, 2020. For details, see Figure 20 and related discussion in CRS Report R46249, *U.S. Farm Income Outlook: February 2020 Forecast*.

[68] CRS Insight IN11258, *State and Local Fiscal Conditions and Economic Shocks*.

FEDERAL RESPONSE TO COVID-19 FOR AGRICULTURE

This section reviews federal supplementary funding appropriated for assistance to the U.S. agricultural sector in response to the COVID emergency, and the USDA programs that were initiated with that supplementary funding. In addition to the supplementary funding, USDA announced that it was increasing certain flexibilities and extensions in several of USDA's farm programs—many authorized by the 2018 farm bill (P.L. 115-334)—as part of its effort to support the food supply chain.[69]

Also, USDA has established a "Coronavirus Disease (COVID-19)" web page that assembles information from a broad range of agriculture-related topics and issues, including the expanded program flexibilities.[70] The website includes "Frequently Asked Questions (FAQs)" for several prominent issues and agencies, links to additional resources, and a USDA COVID-19 Federal Rural Resource Guide.[71]

Supplemental Agriculture Appropriations

In March and April 2020, Congress passed and the President signed four supplemental appropriations acts in response to the COVID-19 pandemic:

- Coronavirus Preparedness and Response Supplemental Appropriations Act, 2020 (P.L. 116-123);
- Families First Coronavirus Response Act (FFCRA, P.L. 116-127);
- Coronavirus Aid, Relief, and Economic Security Act (CARES Act, P.L. 116-136); and

---

[69] USDA, "USDA Announces Coronavirus Food Assistance Program," Release No. 0222.20, April 17, 2020.

[70] USDA, "Coronavirus Disease (COVID-19)," online resource page, https://www.usda.gov/coronavirus.

[71] USDA, "COVID-19 Federal Rural Resource Guide," updated April 8, 2020, https://www.rd.usda.gov/sites/default/files/USDA_COVID-19_Fed_Rural_Resource_Guide.pdf.

- Paycheck Protection Program and Health Care Enhancement Act (P.L. 116-139).

These acts provide over $36 billion of new appropriations and policy changes in the jurisdiction of the Agriculture appropriations subcommittees,[72] including nearly $10 billion for agricultural assistance, about $26 billion for nutrition assistance programs, and $163 million for the Food and Drug Administration.[73]

Funds Directly Targeting Agriculture

The CARES Act provides $9.5 billion for USDA to "support agricultural producers impacted by coronavirus, including producers of specialty crops, producers that supply local food systems, including farmers' markets, restaurants, and schools, and livestock producers, including dairy producers."[74] This approach provides funding to the Secretary of Agriculture with general authority to respond to a crisis, and therefore is similar to emergency appropriations for wildfires and hurricanes in 2018 and 2019 in which USDA was tasked to develop a payment program from a general appropriation.[75]

The CARES Act also replenishes up to $14 billion of funding availability for the Commodity Credit Corporation (CCC). The CCC operates with a $30 billion line of credit with the U.S. Treasury.[76] In April 2020, USDA said that the CCC's borrowing authority was becoming low and that about $6 billion remained after having paid regular farm bill obligations and the final tranche of the 2019 Market Facilitation Program.[77] The $14 billion in the CARES Act is not new spending; rather, it would reimburse the CCC for past spending. After the funds are transferred,

---

[72] For context about regular annual appropriations, see CRS Report R45974, *Agriculture and Related Agencies: FY2020 Appropriations*.

[73] CRS In Focus IF11491, *Supplemental Appropriations for Agriculture and Related Agencies Due to COVID-19*.

[74] USDA, "USDA Announces Coronavirus Food Assistance Program," Release No. 0222.20, April 17, 2020.

[75] For example, see CRS In Focus IF11245, *FY2019 Supplemental Appropriations for Agriculture*.

[76] CRS Insight IN10941, *Commodity Credit Corporation: Q&A*.

[77] Successful Farming, "Smaller pot of money for coronavirus relief, says Perdue," April 9, 2020.

which requires waiting for a June 2020 financial statement, the CCC would have renewed access to more funding for future obligations.

For flexibility in the regular farm commodity support programs, the CARES Act allows Marketing Assistance Loans in FY2020 to be repaid over 12 months (rather than the usual nine months) to provide producers with flexibility in responding to disruptions.[78]

Funds Indirectly Targeting Agriculture and the Rural Economy

Access to food has been a concern during the pandemic, particularly in light of school closures. Rising unemployment may increase participation in the Supplemental Nutrition Assistance Program (SNAP) and other food assistance programs such as The Emergency Food Assistance Program (TEFAP), which provides aid to food banks and other emergency feeding organizations. Financial assistance through SNAP benefits or in-kind assistance through TEFAP may increase demand for agricultural products.

- The FFCRA and the CARES Act provide a total of $26 billion for various nutrition assistance programs and give USDA certain temporary flexibilities to increase program access and accommodate social distancing.[79] This includes $850 million for TEFAP, which is to increase USDA commodity purchases for food distribution.
- The FFCRA has language that, "During fiscal year 2020, the Secretary of Agriculture may purchase commodities for emergency distribution in any area of the United States during a public health emergency designation" (§1101(g)). CBO did not score this provision. USDA is using this authority for a $3 billion commodity purchase and distribution ("Farmers to Families Food Box") program (described later in this chapter).

---

[78] CRS In Focus IF11162, *2018 Farm Bill Primer: Marketing Assistance Loan Program.*

[79] CRS Insight IN11250, *USDA Domestic Food Assistance Programs' Response to COVID-19: P.L. 116-127, P.L. 116- 136, and Related Efforts*. See also the table in CRS In Focus IF11491, *Supplemental Appropriations for Agriculture and Related Agencies Due to COVID-19.*

- CBO estimated that SNAP policies in the FFCRA will increase mandatory spending by more than $21 billion over FY2020-FY2021, which is subject to appropriation and for which the CARES Act provides some funding.[80]

For rural development, the CARES Act provides $146 million, including $100 million for rural broadband grants, $25 million for rural telemedicine and distance learning, and $20.5 million to support rural business loans.[81]

For USDA agency operations, the CARES Act provides $141 million to six USDA agencies or offices. This includes $55 million for the Animal and Plant Health Inspection Service (APHIS), $45 million for the Agricultural Marketing Service (AMS), $33 million Food Safety and Inspection Service (FSIS), $4 million for the Foreign Agricultural Service (FAS), $3 million for the Farm Service Agency (FSA), and $750,000 for the Office of the Inspector General (OIG). The APHIS and AMS amounts are to replace user fee revenues that are expected to decline due to reduced air passenger traffic (APHIS) and because of reduced grading, inspections and audit services (AMS). The amounts for FSIS and FSA are to support temporary employees and adjustments to respond to COVID-19 workload demands. FAS received funding to repatriate staff from foreign postings during the pandemic.

Other provisions in the CARES Act outside the jurisdiction of agriculture appropriations may have provided loans and grants to certain agriculture-related businesses, such as through the Small Business Administration (SBA),[82] or to individuals through stimulus checks.[83] Prior to enactment of P.L. 116-139, the SBA reported that agricultural, forestry, fishing, and hunting businesses received $4.37 billion of Paycheck Protection Loans.[84]

---

[80] CBO, "Score of H.R. 6201, Families First Coronavirus Response Act," April 2, 2020.

[81] CRS In Focus IF11262, *The ReConnect Broadband Pilot Program*.

[82] CRS Insight IN11357, *COVID-19-Related Loan Assistance for Agricultural Enterprises*.

[83] CRS Insight IN11261, *COVID-19 and Stimulus Payments to Individuals: Potential Impacts of Direct Payments on Family Incomes*.

[84] Small Business Administration, "Paycheck Protection Program (PPP) Report." April 16, 2020.

USDA's Coronavirus Food Assistance Program (CFAP)

On April 17, 2020, Secretary of Agriculture Perdue announced the Coronavirus Food Assistance Program (CFAP), valued at $19 billion, to provide immediate financial relief to farmers, ranchers, and consumers in response to the COVID-19 national emergency.[85] According to Secretary Perdue, CFAP will include $16 billion in direct payments to producers that have been impacted by the decline in commodity prices and the disruption in food supply chains related to COVID- 19, and $3 billion in commodity purchases for distribution through food banks, faith-based organizations, and other nonprofit organizations. Direct payments are intended to partially offset the loss of market revenue from the price decline, and the unexpected carry-cost of unsold commodities for producers and ranchers of products that have been negatively affected by COVID-19. Commodity purchase and distribution programs serve the dual roles of supporting commodity market prices by temporarily increasing demand, and of expanding the availability of food to consumers who have lost their jobs or have limited resources.

CFAP funding is from three primary sources: the FFCRA, the CARES Act, and existing USDA authorities under the Commodity Credit Corporation (CCC) Charter Act.[86] In particular, Senator Hoeven, Chairman of the Senate Agriculture Appropriations Committee, stated that the $16 billion designated for the direct payments program derives from the $9.5 billion emergency program spending authorized by the CARES Act and an existing $6.5 billion balance in the CCC.[87] The $3 billion for the commodity purchase portion of CFAP derives from the FFCRA (§1101(g))

---

[85] USDA, "USDA Announces Coronavirus Food Assistance Program," Release No. 0222.20, April 17, 2020.

[86] For a discussion of USDA's authority under the CCC Charter Act, see CRS Report R44606, *The Commodity Credit Corporation: In Brief*.

[87] Senator John Hoeven (North Dakota), "USDA to Provide $16 Billion in Direct Assistance to Farmers and Ranchers, $3 Billion in Agriculture Purchases," News Release, April 7, 2020; hereafter as Senator Hoeven, News Release, April 7, 2020.

that authorizes USDA to purchase commodities for emergency distribution in the United States.[88]

CFAP Direct Payment Program

USDA has released limited information to date on the specifics of how CFAP will be implemented; however, information released by Senator Hoeven's office, coupled with news sources that have gleaned pieces of program information from USDA sources, could provide some context for understanding how CFAP's direct payments program may unfold.

Senator Hoeven's press release says that the $16 billion is to be allocated across four different commodity groupings as follows: livestock ($9.6 billion or 60%); row crops ($3.9 billion, %); specialty crops ($2.1 billion, 13.1%); and "other" crops ($500 million, 3.1%) (Figure 2). Direct payment spending under the livestock category would be further delineated as $5.1 billion for cattle, $2.9 billion for dairy, and $1.6 billion for hogs.

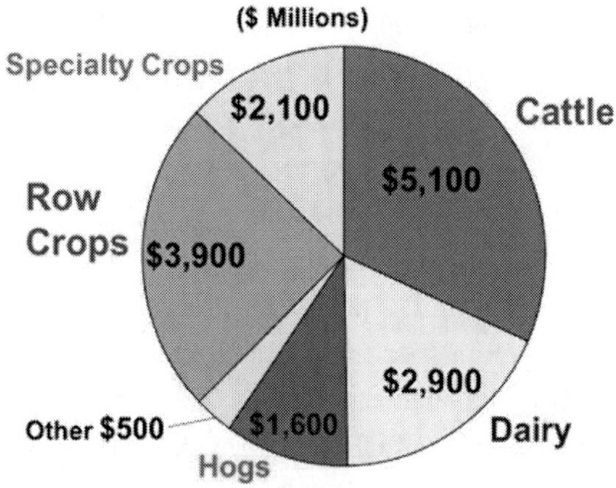

Source: Senator John Hoeven (North Dakota), "USDA to Provide $16 Billion in Direct Assistance to Farmers and Ranchers, $3 Billion in Agriculture Purchases," News Release, April 7, 2020.

Figure 2. CFAP Direct Payments by Commodity (According to Reports).

---

[88] Agricultural Marketing Service, "Notice of Funds Availability (NOFA); Purchase of Fruit, Vegetable, Dairy, and Meat Products Due to COVID-19 National Emergency-USDA Food Box Distribution Program," 85 *Federal Register* 23325, April 27, 2020.

To be eligible for CFAP direct payments, farmers and ranchers must produce commodities that have experienced at least a 5% price decrease between January and April of 2020. The specific prices and dates to be used for this calculation have not yet been announced, but using nearby futures contract prices as a guide would suggest that many commodities—including hogs, cattle, cotton, milk, corn, and soybeans—would relatively easily meet the 5%-decline threshold (Figure 1). Also, the "other" category was not specifically defined, but is expected to include commodities such as horticulture, hemp, sheep and goats, or any other commodity where a producer can show a revenue loss associated with at least a 5% price drop since January.

USDA has said the direct payments to producers will be based on estimated losses as measured by both: (a) market price declines, and (b) additional marketing costs associated with the unexpected oversupply of unsold production in calendar 2020. Under this structure, CFAP direct payments would be directly proportional, or "coupled," to actual production.[89]

Determination of the payment amounts and delivery mechanisms may develop similarly to the three tranches associated with payments under the Market Facilitation Programs (MFPs) in 2018 and 2019.[90] The payments are to represent the summation of the two different loss measures described above:

- First, payments, according to Senator Hoeven's news release, are to cover the market price declines of greater than 5% that occurred between January 1, 2020, and April 15, 2020.[91] The payment is to equal 85% of the loss—some policy analysts think that the loss could be calculated using some measure of the price decline times

---

[89] The nature of the linkage—coupled versus non-coupled—between program payments and producer production choices is relevant to how such payments may count against U.S. domestic support commitments at the WTO; for details, see CRS Report R45305, *Agriculture in the WTO: Rules and Limits on U.S. Domestic Support*.

[90] CRS Report R45310, *Farm Policy: USDA's 2018 Trade Aid Package*, and CRS Report R45865, *Farm Policy: USDA's 2019 Trade Aid Package*.

[91] Senator Hoeven, News Release, April 7, 2020.

the normal volume of the commodity marketed during that period.[92]

- Second, payments to cover future marketing losses from unsold product. This additional cost is valued as 30% of the expected losses from April 15 through the next two quarters (i.e., six months) or until mid-October. Again, the expected loss would most likely be calculated using some measure of the price decline times the normal volume of the commodity marketed during that period.

For its part, USDA has not provided details on: which prices will be used in the formula (such as local elevator prices, regional wholesale prices, or nearby futures contract prices); how the price decline will be measured (that is, will two specific dates be used, or will representative averages of prices in January and April be used); or how the share of production, characterized as "routine marketing for the relevant period" and eligibility for payments, will be measured.

Many of the program specifics are expected to be delineated in rulemaking. Since enforcement of social distancing remains in effect for the foreseeable future, producers may be asked to self-certify their losses.[93] If so, producers will need to save records and paperwork to demonstrate losses, especially producers that have destroyed their product (e.g., dumping of milk or plowing under specialty crops). It is expected that producers will be compensated for "dumped" milk, but whether compensation may be provided for depopulated (i.e., euthanized) livestock and poultry is uncertain. Furthermore, CFAP direct payments are expected to be limited to cover producers who own the commodity or product, thus animals raised under contact are not expected to be covered.[94]

USDA is expediting the rulemaking process for the direct payment program and expects to begin sign-up for the new program in early May.

---

[92] Michael Nepveux and John Newton, "UPDATE: What's in USDA's New Coronavirus Food Assistance Program?" Market Intel, American Farm Bureau, April 23, 2020.
[93] Ibid.
[94] Ibid.

CFAP direct payments are expected to be issued to producers by the end of May or early June.[95]

According to the Senator Hoeven press release, payment limits are expected at $125,000 per commodity grouping (livestock, row crops, and specialty crops), with an overall limit of $250,000 per individual or entity. Neither the CARES Act nor the underlying CCC authority requires payment limits. Applying payment limits at this point would be at USDA's discretion, as it chose to do when establishing the MFP and Wildfire and Hurricane Indemnity Program (WHIP) programs that were undertaken at the Secretary's discretion. The American Farm Bureau Federation expects that benefits received under traditional farm support programs such as ARC and PLC will not be added to CFAP payments when evaluating payment limits.[96] In other words, payment limits for CFAP are expected to be independent of other farm program benefits received by a farm.

CFAP Purchase and Distribution Program

In the press release that announced the CFAP, USDA designated $3 billion for a commodity purchase and distribution program. In subsequent announcements, the program has been called the USDA Food Box Distribution Program, and the USDA Farmers to Families Food Box Program.[97] The intention is to capture some of the supply chain and market disruption caused by the closure of restaurants, hotels, and other food service entities. Under the program, agricultural products are to be purchased from farmers and processors to support agricultural markets and reduce food waste. Products are to be distributed to food banks and other nonprofit organizations that serve those in need. This program is operated differently than, and separately from, existing USDA commodity purchase programs such as Section 32, TEFAP, and other Food and Nutrition

---

[95] Senator John Hoeven (North Dakota), "USDA to Provide $16 Billion in Direct Assistance to Farmers and Ranchers, $3 Billion in Agriculture Purchases," News Release, April 7, 2020.

[96] Michael Nepveux and John Newton, "UPDATE: What's in USDA's New Coronavirus Food Assistance Program?," Market Intel, American Farm Bureau, April 23, 2020.

[97] USDA Agricultural Marketing Service, "USDA Farmers to Families Food Box," https://www.ams.usda.gov/selling- food-to-usda/farmers-to-families-food-box, accessed May 1, 2020.

Service (FNS) food distribution programs.[98] However, food banks, school food authorities, and other nonprofits that participate in other FNS programs are to be eligible to receive food boxes through the USDA Farmers to Families Food Box Program.

USDA plans to purchase about $100 million per month of fresh produce, $100 million per month of dairy products, and $100 million per month of meat products (chicken and pork).[99] Because of the potential for food waste (lack of marketing options for ripe produce, dumping expressed milk, and euthanizing market-ready livestock) and high demand for food bank distribution during this pandemic discussed earlier in this chapter, USDA is expediting this purchase program relative to its usual commodity procurement and distribution timeline. Usually, the procurement-to-distribution timeline is two to five months, starting with product selection and identification and/or thorough analysis of market conditions for individual commodities, a solicitation period, review of applications, and manufacturing and delivery from producers to processors (vendors) and intermediaries that may reassemble or briefly store products for distribution and recipients.

The current food box program shortens the procurement-to-delivery to as little as one month, with an expected one week interval for each of the bidding, approval, and delivery stages. Solicitation for bids began on April 24, with bids due to USDA on May 1. Contracts are expected to be awarded on May 8, and initial program delivery may be as early as May 15.[100] The program is expected to operate through the end of 2020.

Another significant change is the product format. Under normal circumstances, products are often delivered from vendors (who sell to USDA) to recipient organizations in bulk formats that may require re-packaging before distribution to households. The CFAP purchase program intends for vendors to deliver household-ready boxes of the previously

---

[98] For comparison, see CRS Report RL34081, *Farm and Food Support Under USDA's Section 32 Program*, and CRS Report R45408, *The Emergency Food Assistance Program (TEFAP): Background and Funding*.

[99] Agricultural Marketing Service, "Notice of Funds Availability (NOFA); Purchase of Fruit, Vegetable, Dairy, and Meat Products Due to COVID-19 National Emergency-USDA Food Box Distribution Program," 85 *Federal Register* 23325, April 27, 2020.

[100] Ibid.

mentioned produce, meat and dairy products, or combinations thereof—ready for more convenient and immediate distribution ("off the truck and into the trunk") in order to support social distancing.[101]

In addition to the $3 billion CFAP purchase and distribution program, USDA announced on May 4, 2020, a plan to purchase $470 million of commodities with Section 32 authority that is at the discretion of the department.[102] USDA's initial press release for the CFAP had mentioned availability of $873 million in the Section 32 account.[103] The plan for this initial tranche is to solicit bids from vendors in June 2020 and begin deliveries as soon as July. In this purchase, USDA intends to buy $170 million of produce, $120 million of dairy products, $80 million of poultry, $70 million of fish, and $30 million of pork.

Next Steps

In addition to the $9.5 billion in funding to assist U.S. agricultural producers and ranchers with COVID-related losses, the CARES Act also provided $14 billion to replenish the CCC—this additional CCC spending authority is expected to be available in July after CCC prepares its June financial statement. The CCC borrows from the U.S. Treasury to finance its programs consistent with its permanent, indefinite authority to borrow up to $30 billion. Congress usually replenishes the CCC borrowing authority by annually appropriating funding to cover the CCC's net realized losses.[104] The supplemental reimbursement in the CARES Act could increase opportunities for USDA to use its executive authority in CCC to provide further direct support payments, as it did with trade aid payments in 2018 and 2019.[105]

---

[101] Agricultural Marketing Service, "Webinar discussing Coronavirus Food Assistance Program (CFAP) USDA Purchase & Distribution Program." April 21, 2020.

[102] USDA, "USDA Announces Additional Food Purchase Plans," press release, May 4, 2020.

[103] CRS Report RL34081, *Farm and Food Support Under USDA's Section 32 Program.*

[104] For more detailed information on the CCC, see CRS Report R44606, *The Commodity Credit Corporation: In Brief.*

[105] CRS Report R45865, *Farm Policy: USDA's 2019 Trade Aid Package.*

The CCC's annual borrowing authority has been fixed at $30 billion since 1987. On April 21, 2020, the American Farm Bureau released a report showing that, if adjusted for inflation, the CCC's borrowing authority would be $67.5 billion in 2020. In Congress, Senator John Hoeven has called for increasing the CCC's borrowing authority to $50 billion,[106] while H.R. 6728 would raise CCC's borrowing authority to $68 billion.[107] Increases in CCC's borrowing authority could be permanent or temporary (for certain fiscal years, or the duration of the public health emergency). While farm bills designate CCC to make various types of congressionally directed payments,[108] it is USDA's use of its discretionary authority in recent years that has put pressure on the CCC borrowing limit.[109]

More USDA Assistance

At present, USDA appears to have used nearly all of its available borrowing authority in the CCC in composing the CFAP payments. By July, USDA may be expected to use some of the $14 billion in supplemental CCC funding in the CARES Act, and its general CCC authority, to provide additional support to the agriculture sector. However, the nature and timing of any further support has not yet been announced by USDA.

After it completes the $470 million Section 32 purchase described above, USDA would be expected to have about $400 million remaining in the Section 32 account for the rest of the fiscal year, based on the $873 million that USDA mentioned in its initial CFAP announcement.

USDA may also be directed by Congress to provide certain future support to agriculture. Representative Austin Scott has introduced a bill

---

106 Senator Hoeven, "Phase III COVID-19 Response [could] Include $50 Billion in Funding Authority to Help Ag Producers, Ranchers: Provides $30 Billion CCC Replenishment, Increases Cap by $20 Billion for USDA COVID-19 Response," press release, March 22, 2020. See also, Jerry Hagstrom, "Hoeven: USDA rule on COVID farm payments sent to OMB," *The Hagstrom Report*, April 27, 2020.

107 See also, Jerry Hagstrom, "Peterson arranges hog depopulations, plans CCC increase with strings," *The Hagstrom Report*, April 28, 2020.

108 CRS Report R45425, *Budget Issues That Shaped the 2018 Farm Bill.*

109 CRS Insight IN11168, *The CCC Anomaly in an FY2020 Continuing Resolution.*

(H.R. 6611) that would provide an additional $50 billion of funding (separate from the CCC, and similar to the $9.5 billion in the CARES Act) for COVID-related agricultural assistance programs.

Rural COVID-19 Task Force Proposed

Many Members of Congress have expressed concerns that rural America is not well prepared to handle the COVID-19 emergency, and that rural problems feed over into agriculture.[110] For example, in April, Senate Democrats released a report warning that more isolated rural areas of the United States "face disproportional challenges that put them at high risk," and that laid out their own rural policy ideas.[111] Several U.S. Representatives and Senators have asked USDA to establish a Rural COVID-19 Task Force to help identify rural needs and tailor the allocation of resources to address them.[112] They suggest that the Task Force could consist of a diverse group of experts and representatives from all sectors of rural areas, including agriculture, health care including mental health, and the private sector. In particular, the Rural COVID-19 Task Force would help to identify rural challenges, develop strategies and policy recommendations, assemble a guide of available federal programs and resources, consult with the USDA and congressional committees, and provide oversight on the distribution of funding.

In their letter, the Members noted that people living in rural America are more likely to be uninsured, advanced in age, and have pre-existing medical conditions. Rural hospitals and health systems often have fewer ICU beds and resources available to handle an increased demand on rural health care infrastructure during a pandemic. According to the letter, one in five Americans living in rural areas are people of color, who have been disproportionately affected by the current crisis. Many rural areas are without reliable internet access, which limits their ability to work or attend school remotely.

---

[110] "Rural Relief Ideas for the Next Stimulus Plan," *Politico*, April 28, 2020.

[111] Senators Debbie Stabenow and Joe Manchin, "Rural America and COVID-19: Small Towns and Rural Communities are at High Risk," April 23, 2020.

[112] Congressional letter to Secretary of Agriculture Perdue regarding the establishment of a "Rural COVID-19 Task Force," April 16, 2020.

U.S. Agricultural Stakeholders

In March and early April, prior to USDA's announcement of CFAP, many industry groups from the U.S. agricultural sector had written public letters to Secretary Perdue and USDA detailing their concerns and the need for federal assistance in response to the economic damage that has hit their different industries as a result of the COVID-19-related emergency. Many of these letters included industry estimates of their sectoral economic losses due to COVID-19 (several of these estimates have been compiled by CRS and are presented in Table A-1).

Following the announcement of CFAP, most industry groups that are targeted for CFAP assistance have expressed appreciation for the aid that USDA has announced. At the same time, several agriculture-related industries do not appear to be eligible for support under the CFAP—including ethanol, poultry, sheep and lamb, specialty livestock such as mink, or horticultural products.[113] Also, agriculture industry groups almost universally note that CFAP can be only the first step in the federal response, as the amounts provided fall well short of industry loss estimates for most sectors.

Another industry concern involves the announced payment limits associated with CFAP direct payments. Industry groups have stated that a $125,000 payment limit would severely restrict needed aid to individual producers. Some Senators and Representatives have followed up on industry concerns about payment limits by sending a letter to the Administration asking that the limits be removed for livestock, dairy, and specialty crop producers.[114] The issue of payment limits may be addressed in the next round of USDA COVID-19 assistance or related bills (e.g., H.R. 6611).

---

[113] Michael Nepveux and John Newton, "UPDATE: What's in USDA's New Coronavirus Food Assistance Program?," Market Intel, American Farm Bureau, April 23, 2020.

[114] Jerry Hagstrom, "Senators, House members oppose farm payment limits," April 24, 2020.

ISSUES FOR CONGRESS

Immediate congressional concerns involve how to channel assistance to those industries affected by COVID-19. This involves identifying affected industries, as well as measuring the extent of their losses, and providing some measure of compensation to help foster survival and recovery. Another immediate concern may be monitoring and oversight of the large sums of taxpayer money that will be flowing out through the CCC. Producer self-certification of losses may create a moral hazard and an incentive to over-report losses.

Congress may consider using its investigative authority to better understand the nature and institutional rigidities inherent in the current food supply chain, and to ascertain whether there is a role for the federal government to help facilitate food supply chain management. Such considerations may include an evaluation of whether there is a need to facilitate less reliance by the private sector on low-cost, just-in-time supply chains. This model has proven to be inflexible in responding to the COVID-19 emergency, which requires rapid product transition from bulk institutional buyers to consumer-ready retail outlets, and greater regional storage capacity to hold temporary surges in unsold product. The potential costs to taxpayers of supporting a more flexible supply chain could include expanded regional warehousing, cold storage, food bank storage, etc. Expediting the food assistance supply chain is also being proposed through "farm-to-food bank" programs (S. 3605). These actions could be coupled with new federal programs designed to increase demand by temporarily expanding SNAP benefits and/or federal food purchases when certain market or economic conditions are triggered.

Congress may also want to consider whether the current farm safety net programs authorized by the farm bill's Title I—targeting of program crops—are sufficiently flexible and responsive when confronted with sudden, widespread price declines and an abrupt cessation of institutional purchases. For example, payments under the current ARC and PLC programs are delayed nearly 13 months after the program crop's harvest and reflect a 12-month average price that may not fully capture the potential short-term price drop related to the COVID outbreak. Program modifications—such as the inclusion of an early partial payment for ARC and PLC—could offer greater flexibility in responding to short-term cash flow problems for farm households.[115]

As one alternative, ARC and PLC could be supplemented or replaced by a new payment program that would be: short-term in nature; would better reflect local market conditions to capture disparities in regional economic harm; and would rely on market conditions both to trigger payments and to determine the size of those payments. The payment triggers could be set at catastrophic levels such that they would only be triggered under unique circumstances such as a major price plunge sparked by an event of the magnitude of COVID-19. The payment formula could be designed such that USDA could make payments under such a program before or shortly after harvest.

In addition, Congress may also consider the long-term effects that might result from the COVID- 19 emergency, particularly if the economic recovery is slow and lengthy. Such long-term effects may include heightened indebtedness and potential bankruptcies by farm households across the agricultural sector, as well as accelerated industry consolidation and altered consumption patterns. Without robust agricultural production to serve as the engine of growth, rural America might experience a slower recovery than the rest of the country. Prolonged depressed market conditions due to widespread layoffs, limited employment opportunities, and sustained reductions to rural wages and incomes would provide a weak background to foster agriculture's eventual recovery.

Congress may also consider that rural banks tend to be smaller and less well-capitalized than banks catering to urban and suburban markets.

---

[115] Such partial payments were part of the former target-price deficiency payment program. USDA based such early partial payments on USDA price forecasts made in November, only three months into the marketing year. The final payment would use the full 12-month market year payment. Producers could be subject to repayment obligations if the initial November forecast over-estimated the final program payment.

Farm loan debt forgiveness (e.g., S. 3602) and loan repayment flexibilities[116] are also in discussion. A slow rural economic recovery might reduce business and consumer confidence, leading to a reduction in spending and investment, and a tightening of financial conditions that could further slow the return to economic normalcy in rural areas.

APPENDIX. ASSESSING THE ECONOMIC IMPACT OF COVID-19 ON U.S. AGRICULTURE

Studies Project Deep Losses across Major Commodity Sectors

Several universities, think tanks, and commodity groups have released early assessments of COVID-19's potential impact on selected commodity sectors (Table A-1). Most of these early assessments are limited to evaluating the effect of the price decline on any unsold production remaining under farmers' control, and the unexpected surplus of unsold products due to the shutdown of most institutional buying of agricultural products.

Table A-1. Estimates of COVID-19 economic impact, various commodity sectors (Changes from pre-COVID scenario; $M = $ millions)

Category	Sector Loss ($M)	Subsector Loss ($M)	Loss Period Covered	Source #
U.S. Cattlemen's Association (4/17/20)				1
Cattle / Beef	-$14,627		2020	
Cow/calf		-$4,425	2020	
Stocker/Background		-$5,551	2020	
Feedlot		-$4,652	2020	
National Cattlemen's Beef Association (4/14/20)				2
Cattle / Beef	-$13,617		2020/2021	

116 USDA, "Farm Loan Flexibilities" at https://www.farmers.gov/coronavirus, accessed May 7, 2020.

Table A-1. (Continued)

Category	**Sector Loss ($M)**	**Subsector Loss ($M)**	**Loss Period Covered**	**Source #**
Cow/calf		-$8,127	2020/2021	
Stocker/Background		-$2,476	2020	
Feedlot		-$3,014	2020	
National Pork Producers Council (4/14/20)				3
Hog Sector	-$5,000		2020	
National Milk Producers Federation (4/10/20)				4
Dairy Sector	-$8,627		2020	
American Sheep Industry Association (3/27/20)				5
Sheep & Wool Sector	-$125 to -$350		March to May	
National Cotton Council (4/16/20)				6
Cotton Sector	-$3,000 to -$3,500		2020	
Ninety Fresh Produce Organizations (4/6/20)				7
Fresh Produce Sector	-$5,000 (-$1,000/week)		Late March-Early May	
Colorado State University (3/31/20)				8
U.S. Local & Regional Food Markets	-$689		March to May	
Renewable Fuels Association (4/25/20)				9
Category	Sector Loss ($M)	Subsector Loss ($M)	Loss Period Covered	Source #
Ethanol: Lost sales	~46% of plants offline		As of April 24	
Ethanol: Price decline	~33% decline		Jan. 2–Apr.15	
FarmDoc Webinar Forecast (4/25420)				10
Corn Sector	-$6,000		2020	
Cotton Sector	-$750		2020	
Soybeans Sector	-$2,000		2020	
Food and Policy Research Institute: (FAPRI) Farm Income Projections with COVID-19 (4/13/20)				11
Change from FAPRI January Baseline			2020	
U.S. Crop Receipts	-$11,850		2020	
Corn		-$4,715	2020	
Soybeans		-$2,050	2020	
Wheat		-$400	2020	
Cotton		-$610	2020	
Other crops		-$4,075	2020	
U.S. Livestock Receipts	-$20,240		2020	
Cattle		-$9,570	2020	

Category	Sector Loss ($M)	Subsector Loss ($M)	Loss Period Covered	Source #
Hogs		-$2,244	2020	
Poultry		-$4,054	2020	
Milk		-$3,972	2020	
Other livestock		-$400	2020	
U.S. Production Expenses	-$11,260		2020	
Feed		-$2,820	2020	
Purchased Livestock		-$3,830	2020	
Fuel & Fertilizer		-$2,200	2020	
Other expenses		-$2,410	2020	
Government Payments	+$2,300		2020	
Other Net Farm Income	-$1,540		2020	
Total U.S. Net Farm Income	-$20,080		2020	
Center for Agriculture and Rural Development (CARD), (4/20/20)				12
State of Iowa	-$6,640		2020	
Ethanol (lost sales)		-$2,500	2020	
Hogs		-$2,100	2020	
Corn		-$788	2020	
Fed cattle		-$658	2020	
Ethanol (price decline)		-$347	2020	
Soybeans		-$213	2020	
Calves & feeder cattle		-$34	2020	
Florida Department of Agriculture (4/15/20)				13
State of Florida	-$523		March to May	
Lettuce		-$5 to -$7	March to May	
Green beans		-$40 to -$50	March to May	
Cabbage		-$24	March to May	
Zucchini & squash		-$17 each	March to May	
Peppers		-$10	March to May	
Cucumbers		-$33 to -$8	March to May	
Sweet corn		-$48	March to May	
Watermelon		-$95	March to May	
Tomatoes		-$165	March to May	

Table A-1. (Continued)

Category	Sector Loss ($M)	Subsector Loss ($M)	Loss Period Covered	Source #
Blueberries		-$45 to -$50	March to May	
Radishes		-$1.5	March to May	

Source: Compiled by CRS from various sources, see Notes.

Notes: This table is not a comprehensive list of COVID-19-related studies, but represents an assemblage of such studies of several major agricultural sectors of the United States, readily available to the public, as listed by "Source #" in the table:

1. "Estimating COVID-19 Effects on the Cattle Markets," Custom Ag Solutions as commissioned by the United States Cattlemen's Association, released April 17, 2020.
2. "Economic Damages to the U.S. Beef Cattle Industry Due to COVID-19," Oklahoma Cooperative Extension Service and National Cattlemen's Beef Association, April 14, 2020.
3. National Pork Producers Council, April 14, 2020; https://nppc.org/hog-farmers-face-covid-19-financial-crisis/.
4. The estimate is based on the price decline of -$3.90 per cwt. for the all milk price reported by USDA in the March and April WASDE reports (World Agricultural Supply and Demand Estimates) times the national milk marketing (skim-milk basis) of 212.2 billion pounds. This estimate was suggested to CRS in a phone conversation with an official of the National Milk Producers Federation on May 5, 2020.
5. Estimates include $125 million in immediate damages for sheep producers and a $350 million spread over the entire 2020 lamb crop, Red River Farm Network, "Sheep Markets Feel the Repercussions of COVID-19," May 4, 2020, https://www.rrfn.com/2020/05/04/sheep-markets-feel-the-repercussions-of-covid-19/.
6. National Cotton Council as cited in "NCC: Cotton lost $3 billion, needs aid from emergency fund," The Hagstrom Report, April 17, 2020.
7. Eighty-five industry organizations representing a variety of produce growers in a letter to USDA, "Produce Market Stabilization Program," April 6, 2020.
8. Dawn Thilmany, Becca Jablonski, Deb Tropp, Blake Angelo, and Sarah Low, "Mitigating Immediate Harmful Impacts of COVID-19 on Farms and Ranches selling through Local and Regional Food Markets," Policy Briefs, Food Systems, Colorado State Univ., March 31, 2020.
9. Renewable Fuels Association (RFA) estimate from Pro Farmer, Weekly Report, April 25, 2020, p. 4.
10. Jonathon Coppess and Nick Paulson, "Speculation: Potential Row Crop Payments," FarmDoc Webinar: An Update on Policy and Political Outlook for Coronavirus Pandemic Relief, Department of Agricultural and Consumer Economics, University of Illinois at Urbana-Champaign, April 24, 2020.
11. Pat Westhoff, Seth Meyers, Julian Binfield, and Scott Gerlt, "Early Estimates of the Impacts of COVID-19 on U.S. Agricultural Commodity Markets, Farm Income and Government Outlays," FAPRI-MU Report #02-20, University of Missouri, April 13, 2020.
12. Chad E. Hart, Dermot J. Hayes, Keri L. Jacobs, Lee L. Schulz, and John M. Crespi, "The Impact of COVID-19 on Iowa's Corn, Soybean, Ethanol, Pork, and Beef Sectors," CARD Policy Brief, 20-PB 28, April 2020.
13. "Florida Seasonal Crop COVID-19 Impact Assessment," Florida Department of Agriculture and Consumer Services, April 15, 2020.

In: COVID 19: Agricultural and Food Issues ISBN: 978-1-53618-561-4
Editor: Kristian Goldschmidt

Chapter 8

USDA's Coronavirus Food Assistance Program (CFAP) Direct Payments*

Randy Schnepf

Summary

On April 17, 2020, Secretary of Agriculture Sonny Perdue announced that the U.S. Department of Agriculture (USDA) would initiate the Coronavirus Food Assistance Program (CFAP) to provide immediate financial relief to farmers, ranchers, and consumers in response to the COVID-19 national emergency. CFAP consists of two distinct initiatives: a $16 billion direct payment program for agricultural producers that have been impacted by the decline in commodity prices and the disruption in food supply chains related to COVID-19, and a $3 billion food purchase and distribution program. This chapter focuses on the larger of these two CFAP initiatives—the $16 billion direct payment program. The CFAP direct payment program derives its funding from two sources: $9.5 billion

* This is an edited, reformatted and augmented version of a Congressional Research Service publication R46395, prepared for Members and Committees of Congress dated June 16, 2020.

emergency program spending authorized by the Coronavirus Aid, Relief, and Economic Security Act (CARES Act; P.L. 116-136) and an existing $6.5 billion balance from the Commodity Credit Corporation (CCC) Charter Act of 1948 (P.L. 80-806; 15 U.S.C. 714 *et seq.*).

On May 19, 2020, USDA released the final rule that details CFAP's $16 billion direct payment program, including the list of eligible commodities, eligibility requirements for producers, payment calculations, and application procedures. In its announcement, USDA stated that it would begin accepting applications on May 26, through the Farm Service Agency (FSA), from agricultural producers who have suffered losses. To be eligible, an application must be submitted to any FSA county office by the close of business on August 28, 2020. Payments to eligible producers are expected to be made soon after each application is processed. USDA's FSA will be the principal agency charged with implementing CFAP's direct payment program. FSA is to be assisted by the Agricultural Marketing Service (AMS) with respect to matters dealing with producers of specialty crops.

To be eligible for a payment, a commodity must have suffered a price loss of at least 5% during the mid-January to mid-April period, or was subject to additional significant marketing costs for unexpected supply chain disruptions, including unsold inventories. Specialty crops may also qualify for a payment if product was shipped off the farm, but spoiled prior to receipt of payment due to COVID-19-caused disruption of the food supply chain. For an individual or legal entity to be eligible, they must complete an application to determine the quantities affected and to meet certain other criteria, including conservation compliance, share in the risk of profit and loss from the farm's operation, and an average Adjusted Gross Income (AGI) for 2016, 2017, and 2018 that is less than $900,000, unless at least 75% of AGI is from farming and ranching activities.

USDA is to track payments made from the CARES Act and the CCC Charter Act funds separately, to ensure that the payments are consistent with each respective authority. As a result, every eligible commodity may receive a payment based on two different types of losses or costs: (1) income losses due to mid-January-to-mid-April price declines, which is to be partially offset by payments from CARES Act funds; and (2) unexpected marketing costs due to food supply chain disruptions, which is to be partially offset by payments from CCC funds.

USDA is to pay an initial tranche equal to 80% of the aggregate payment to each qualifying producer. If funds remain available after the initial payment to all eligible applicants, and depending on whether USDA adds other commodities to the list, USDA is to disburse the remaining funds such that the initial and final payments do not exceed a total of $9.5 billion of CARES Act funds and $6.5 billion of CCC funds. Funds may be prorated if necessary to stay within these limits. There is a payment limitation of $250,000 per person or entity for all commodities combined. Applicants who are corporations, limited liability companies, or limited partnerships may qualify for payments of up to $750,000 if at least three members each actively provide at least 400 hours of personal labor or personal management to the farming operation.

Congress may be interested in evaluating whether CFAP payments mitigate the damage caused by the COVID-19 pandemic for individual farming operations. Will additional assistance be necessary later this year? Is additional assistance needed to support certain affected industries that are not eligible under USDA's initial rollout? Another immediate concern may be oversight of the large sums of taxpayer money that are to flow out through the USDA and the CCC. Producer self-certification of losses may create an incentive to over-report losses.

Introduction

On April 17, 2020, Secretary of Agriculture Sonny Perdue announced that the U.S. Department of Agriculture (USDA) would initiate the Coronavirus Food Assistance Program (CFAP), valued at $19 billion, to provide immediate financial relief to farmers, ranchers, and consumers in response to the COVID-19 national emergency.[1]

---

[1] USDA, "USDA Announces Coronavirus Food Assistance Program," Release No. 0222.20, April 17, 2020.

CFAP Includes Two Programs

According to Secretary Perdue, CFAP consists of two distinct initiatives: a $16 billion direct payment program for agricultural producers that have been impacted by the decline in commodity prices and the disruption in food supply chains related to COVID-19, and a $3 billion food purchase and distribution program for affected consumers referred to as the Farmers to Families Food Box Program.

These two programs derive their specific funding from different sources, and are being implemented by different agencies within USDA. The $16 billion designated for the direct payments program is funded from $9.5 billion of emergency spending authorized by the Coronavirus Aid, Relief, and Economic Security Act (CARES Act; P.L. 116-136) and an existing $6.5 billion balance from the Commodity Credit Corporation (CCC) Charter Act of 1948 (P.L. 80-806; 15 U.S.C. 714 *et seq.*).[2] The $3 billion for the commodity purchase portion of CFAP is funded from the Families First Coronavirus Response Act (FFCRA; P.L. 116-127, §1101(g)) that authorizes USDA to purchase commodities for emergency distribution in the United States.[3]

USDA, through its Agricultural Marketing Service (AMS), began food purchase and distribution activities within a week of announcing the CFAP. In contrast, the direct payment component of CFAP is being implemented by USDA's Farm Service Agency (FSA). FSA went through a longer (albeit expedited) rulemaking process to determine how payments would be made.

---

[2] USDA, FSA, "Coronavirus Food Assistance Program," 85 *Federal Register* 30825, May 21, 2020.

[3] Agricultural Marketing Service, "Notice of Funds Availability (NOFA); Purchase of Fruit, Vegetable, Dairy, and Meat Products Due to COVID-19 National Emergency-USDA Food Box Distribution Program," 85 *Federal Register* 23325, April 27, 2020.

Report Overview

This chapter describes the details of the larger of these two CFAP initiatives—the direct payment program. It describes how this program is funded and administered, including specific details on the calculation and implementation of payments. The report begins with a brief overview of the CFAP direct payment program, including its policy objectives, funding sources, and implementing agencies. The next section describes the details of the CFAP direct payment program, including eligibility requirements for both producers and commodities, application procedures, payment calculations, and the program's timeline. The report ends with a brief discussion on issues of potential interest to Congress. An Appendix at the end of the report includes supplementary tables that provide details on USDA's derivation of the payment rates for affected commodities under both CARES Act and CCC Charter Act criteria, as well as additional USDA forms that must be completed by producers for payment eligibility.

An earlier report by CRS provides information on how the COVID-19 pandemic affected the U.S. food supply chain—including its deleterious impact on agricultural producers—and precipitated the congressional and USDA actions described here.[4] This earlier report also provides details on the CFAP food purchase and distribution initiative and its Farmers to Families Food Box program. Those events and programs are not reviewed here.

CFAP Direct Payment Program Overview

USDA Authorized to Design and Implement CFAP

Under the CARES Act, Congress delegated broad authority to USDA to develop and administer a support program in response to COVID-19.[5] In

---

[4] CRS Report R46347, *COVID-19, U.S. Agriculture, and USDA's Coronavirus Food Assistance Program (CFAP).*

[5] The CARES Act (P.L. 116-136), Division B—Emergency Appropriations For Coronavirus Health Response And Agency Operations, Title I, Agricultural Programs, Office of the Secretary.

particular, the CARES Act provides $9.5 billion for USDA to "prevent, prepare for, and respond to coronavirus by providing support to agricultural producers impacted by coronavirus, including producers of specialty crops, producers that supply local food systems, including farmers' markets, restaurants, and schools, and livestock producers, including dairy producers."[6] This approach provides funding to the Secretary of Agriculture with general authority to respond to a crisis, and therefore is similar to emergency appropriations for wildfires and hurricanes in 2018 and 2019 in which USDA was tasked to develop a payment program from a general appropriation.[7]

Similarly, USDA interprets the CCC Charter Act as providing broad authority to use CCC funds in support of U.S. agriculture.[8] Based on these broad authorities, USDA developed the CFAP direct payment program.

On May 19, 2020, USDA released a final rule to implement CFAP's $16 billion direct payment program including eligibility requirements for commodities and producers, application procedures, and payment calculations.[9] In its announcement, USDA stated that it would begin accepting applications on May 26, through the Farm Service Agency (FSA), from agricultural producers who have suffered losses.[10] All applications must be received by August 28, 2020.

FSA is to be the principal agency charged with implementing CFAP's direct payment program. AMS is to assist FSA with respect to specialty crops issues, such as determinations of sales prices and marketing costs.

---

[6] USDA, "USDA Announces Coronavirus Food Assistance Program," Release No. 0222.20, April 17, 2020. Published in the Federal Register on May 21, 2020 (85 *Federal Register* 30825).

[7] For example, see CRS In Focus IF11245, *FY2019 Supplemental Appropriations for Agriculture*.

[8] For a discussion of USDA's authority under the CCC Charter Act, see CRS Report R44606, *The Commodity Credit Corporation: In Brief*.

[9] USDA, "USDA Announces Details of Direct Assistance to Farmers through the Coronavirus Food Assistance Program," News Release, May 19, 2020.

[10] For up-to-date program information, including application forms and payment processing information, visit the USDA FSA web portal for CFAP at https://www.farmers.gov/cfap.

CFAP direct payments are to provide financial assistance to producers and ranchers of agricultural commodities that suffered from price declines of at least 5% between mid-January and mid-April, or who face additional significant marketing costs associated with unexpected carrying-costs of unsold commodities or spoilage due to market supply chain disruptions. The CFAP direct payment program stands in contrast to congressionally authorized farm bill support programs, supplemental assistance, and other USDA-initiated payment programs for both its scale ($16 billion) and its complicated payment formulation.[11] The CFAP separates COVID- related injury into two components (income losses due to price declines, and unexpected marketing costs due to food supply chain disruptions), each with its own distinct funding source—the CARES Act and the CCC, respectively. The payments calculated for each of these two funding sources for each eligible commodity are combined into a single payment per producer, with an initial tranche equal to 80% of the combined payment for which each qualifying producer is potentially eligible. This unusual payment formulation is explained in this chapter.

USDA could provide additional financial assistance for the U.S. agricultural sector beyond CFAP later in the summer when a replenishment payment of $14 billion for the CCC becomes available. Congress is also considering providing additional support, including support for additional agricultural commodities and sectors that were not initially eligible under CFAP.[12] These potential additional efforts are not discussed in this chapter.

---

[11] The proposed CFAP direct payments program would exceed previous USDA ad hoc direct payment programs of large scale, including the market loss assistance payments of 1999 ($3 billion), 2000 ($11 billion), and 2001 ($5.5 billion), source: USDA, Economic Research Service (ERS), Farm Income and Wealth Statistics, https://www.ers.usda.gov/data-products/farm-income-and-wealth-statistics/; and the market facilitation payment (MFP) programs of 2018 ($8.6 billion) and 2019 ($14.5 billion), source: USDA, FSA, Market Facilitation Program, https://www.farmers.gov/manage/mfp.

[12] On May 15, 2020, the House passed the HEROES Act (H.R. 6800), which would authorize $16.5 billion in additional direct payments by the Secretary of Agriculture. Among its provisions, the bill would compensate producers who have had to dispose of livestock and poultry that could not be sold due to processing disruptions, and would provide special assistance targeted to dairy producers. It further includes aid targeted to biofuels plants, fruit and vegetable growers, and local agriculture.

Funding for CFAP Direct Payments

Funding for CFAP's direct payment program is from two sources: $9.5 billion from the CARES Act, and $6.5 billion obtained with existing USDA authorities under the CCC Charter Act.[13] USDA states that each of these funding sources has its own specific authority and, as a result, USDA is to use the funds from each source in a manner that USDA interprets as being consistent with that authority.[14]

- USDA is to use the $9.5 billion of CARES Act funding to compensate producers for commodity and livestock losses due to price declines of at least 5% that occurred between mid-January 2020 and mid-April 2020. In the case of specialty crops, the funding is to compensate for losses associated with products that were shipped from the farm by April 15, but spoiled due to market disruptions, resulting in no payment.
- USDA is to use $6.5 billion of CCC funding to compensate producers for unexpected costs due to on-going market disruptions and to assist with the transition to a more orderly marketing system as the pandemic wanes.

Thus, each of these funding sources are to be used for different types of losses or costs—CARES Act payments for price declines, and CCC payments for unexpected marketing costs. Such marketing costs may include the removal or disposition of surplus agricultural commodities, the purchase of materials and facilities required to produce and market agricultural commodities, and aid in the development of new markets, marketing facilities, and uses for such commodities.

---

[13] For a discussion of USDA's authority under the CCC Charter Act, see CRS Report R44606, *The Commodity Credit Corporation: In Brief.*

[14] USDA, FSA, "Coronavirus Food Assistance Program," 85 *Federal Register* 30825, May 21, 2020.

USDA has stated that it will track payments made from the CARES Act and the CCC Charter Act funds separately, to ensure that the payments are consistent with each respective authority. Nonetheless, USDA is to combine payments made under both CARES Act and CCC funds into a single payment for eligible producers. Thus, a producer will need to complete only one CFAP payment application.

Eligibility Criteria for CFAP Direct Payments

USDA has developed criteria to determine which commodities and which producers are eligible for CFAP support.[15] Eligibility criteria for producers builds off of general USDA program eligibility criteria, whereas criteria identifying eligible commodities is largely unique to the CFAP program and varies by commodity groups.

Eligible Commodities

According to USDA, only commodities produced in the United States are eligible for CFAP direct payments. In addition, USDA has specified certain eligibility criteria for each of the major agricultural commodity groups that includes a price decline of at least 5% from January 15, 2020, to April 15, 2020, and/or unexpected marketing costs (including spoilage) as listed in Table 1. Based on an analysis of these criteria, USDA has determined that the commodities listed in Table 1 are eligible for either a price-decline or marketing-cost payment.

[15] Ibid.

Table 1. Eligibility Criteria and Eligible Commodities by Category

Category	Eligibility Criteria and Eligible Commodities
Non-Specialty Crops[a]	**Eligibility Criteria:** Non-specialty crop producers who have suffered at least a 5% price decline between mid-January and mid-April (see **Table A-1**), and who face increased marketing costs for unpriced inventories as of January 15, 2020 (see **Table 2**). **Eligible Non-Specialty Crops:** Malting barley, canola, corn, upland cotton, millet, oats, soybeans, sorghum, sunflowers, durum wheat, and hard red spring wheat.
Livestock Commodities	**Eligibility Criteria:** Livestock producers who have suffered at least a 5% price decline between mid-January and mid-April (see **Table A-1**), and who face increased marketing costs for unpriced inventories as of January 15, 2020 (see **Table 2**). **Eligible Livestock Commodities:**[b] Cattle, hogs and pigs, and sheep (lambs and yearlings).
Wool	**Eligibility Criteria:** Wool producers who have suffered at least a 5% price decline between mid-January and mid-April (see **Table A-1**), and who face increased marketing costs for unpriced inventories as of January 15, 2020 (see **Table 2**). **Eligible Wool:** Graded (clean basis) and ungraded (paid on a greasy basis).
Dairy	**Eligibility Criteria:** All dairy operations with milk production in January, February, and/or March 2020; as well as any dumped milk production during the months of January, February, and March 2020. **Eligible Dairy:** All milk produced during the first quarter of 2020.
Specialty Crops[c]	**Eligibility Criteria:** Producers of specialty crops are eligible for CFAP payments in any of the following three categories: (1) Had crops that suffered at least a 5% price decline between January 15, 2020, and April 15, 2020;[d] (2) Had produce that shipped by April 15, 2020, but subsequently spoiled due to loss of marketing channel;[e] and (3) Had shipments that did not leave the farm by April 15, 2020 (for example, were harvested but are sitting in crates on the farm), or mature crops that remained unharvested by that date (for example, were plowed under) due to lack of buyers, and which have not been and will not be sold[f]. **Eligible Specialty Crops:** **Fruit:** apples, avocados, blueberries, cantaloupe, grapefruit, kiwifruit, lemons, oranges, papaya, peaches, pears, raspberries, strawberries, tangerines, tomatoes, watermelons. **Vegetables:** artichokes, asparagus, broccoli, cabbage, carrots, cauliflower, celery, sweet corn, cucumbers, eggplant, garlic, iceberg lettuce, romaine lettuce, dry onions, green onions, peppers, potatoes, rhubarb, spinach, squash, sweet potatoes, taro. **Nuts:** almonds, pecans, walnuts. **Other:** beans, mushrooms.
Other	Certain other commodities (including those with insufficient information) will be evaluated through a Notice of Funds Availability (NOFA) process.

Source: USDA, FSA, "Coronavirus Food Assistance Program," Final Rule, 85 *Federal Register* 30825, May 21, 2020.

Notes: Non-specialty crops, livestock, and wool that meet the 5% price decline criteria are assumed to have met the marketing cost criteria.

a. According to USDA's CFAP rule (85 *Federal Register* 30825, May 21, 2020), non-specialty crops are those crops listed in this table, and any crops announced in a NOFA. The term excludes crops intended for grazing.

b. See the notes to **Table A-1** for a details on eligible animal species.

c. According to USDA's CFAP rule (85 *Federal Register* 30825, May 21, 2020), specialty crop are those crops listed in this table, and any crops for which funds are made available and announced in a NOFA. The term excludes crops intended for grazing.

d. To verify a qualifying price decline for a specialty crop, producers must maintain records, such as a bill of sale, documenting the price received for the crop.

e. To verify market spoilage, producers must obtain documentation such as a letter from the buyer explaining non-payment or other record validating non-payment (this applies to producers who have met contractual obligations in delivering the crop to the buyer, but have not been paid).

f. To verify unsold inventories, if requested, producers must provide supporting documentation, which AMS will use to substantiate claims on a case-by-case basis.

Request for Additional Commodities

USDA is to consider additional crops for eligibility for CFAP payments by collecting information on potentially eligible crops.[16] Producers of commodities not included on the original CFAP list who assert they are eligible may submit information to USDA for consideration in decisions about including additional commodities. Approximately $637 million or about 4% of the CFAP's $16 billion in funding is available to provide assistance to producers of other commodities that are identified through the Notice of Funds Availability (NOFA) process.[17] USDA has stated that it is particularly interested in obtaining information about nursery products, aquaculture products, and cut flowers.

Ineligible Commodities

Non-specialty crops and livestock that did not suffer a price decline of at least 5% from mid- January 2020 to mid-April 2020 are not eligible for CFAP. Specifically, this includes sheep more than two years old, eggs/layers, soft red winter wheat, hard red winter wheat, white wheat, rice, flax, rye, peanuts, feed barley, Extra Long Staple (ELS) cotton, alfalfa, forage crops, hemp, and tobacco.[18] USDA may reconsider the excluded commodities if credible evidence supports a 5% price decline or COVID-related marketing costs. Any crop intended for grazing is excluded from eligibility for CFAP.

Eligible Producers

According to USDA rules, to qualify for a CFAP payment, an eligible individual or legal entity must meet the following criteria:

---

[16] For more details, see the information posted at USDA's CFAP web portal at https://www.farmers.gov/cfap, including a USDA Fact Sheet, "Consideration of Additional Commodities," available under "Resources."

[17] USDA, "Coronavirus Food Assistance Program: Cost Benefit Analysis," May 14, 2020.

[18] For more information, see additional USDA CFAP program information at https://www.farmers.gov/cfap.

- Each producer or legal entity of qualifying commodities must complete a CFAP application form and provide any required documentation (as specified below).[19]
- A producer must share in the risk of producing a qualifying crop or livestock, and is entitled to share in the crop or livestock available for marketing between January 15, 2020, and April 15, 2020, or April 16, 2020, through May 14, 2020. A contract grower who does not own the livestock is to be considered a producer if the contract allows the grower to have risk in the livestock.
- The producer must be in compliance with conservation provisions—including highly erodible land conservation (Sodbuster) and wetland conservation (Swampbuster) provisions.[20]
- A producer must be either a U.S. citizen or a resident alien; however, a foreign person may qualify if they provide sufficient land, capital, and active personal labor to the farming operation. USDA has established that 400 hours of active personal labor or active personal management meets this standard.
- There is no requirement to have crop insurance coverage or coverage under the Noninsured Crop Disaster Assistance Program.
- A person or legal entity, other than a joint venture or general partnership,[21] is ineligible for payments if the person's or legal entity's average adjusted gross income (AGI), using the average of the AGIs for the 2016, 2017, and 2018 tax years, is more than $900,000, unless at least 75% of that person's or legal entity's average AGI is derived from farming, ranching, or forestry-related activities.
- A producer must not have a controlled substance violation.

---

[19] USDA, FSA, "Coronavirus Food Assistance Program," Final Rule, 85 *Federal Register* 30825, May 21, 2020.

[20] See CRS Report R42459, *Conservation Compliance and U.S. Farm Policy*.

[21] With respect to joint ventures and general partnerships, this AGI provision will be applied to each member of the joint venture and general partnership.

Table 2. CFAP Payment Formula by Commodity Category

	Quantity Measure	Factor	Payment Rate (PR)[a]
Non-Specialty Crops—payment equals sum of:			
1	Unpriced inventory[b] as of 1/15/20	x 50% x	Cares Act PR (**Table A-1**)
2	Unpriced inventory as of 1/15/20	x 50% x	CCC PR (**Table A-1**)
Specialty Crops—payment equals sum of:			
1	Quantity sold during 1/15/20 to 4/15/20	x 100% x	Cares Act PR-1 (**Table A-2**)
2	Quantity shipped, but spoiled:[c] 1/15/20 to 4/15/20	x 100% x	Cares Act PR-2 (**Table A-2**)
3	Unpriced quantity on farm:[d] 1/15/20 to 4/15/20	x 100% x	CCC PR-3 (**Table A-2**)
Cattle—payment equals sum of:			
1	Unpriced head sold during 1/15/20 to 4/15/20	x 100% x	Cares Act PR (**Table A-1**)
2	Unpriced inventory, max during 4/16/20 to 5/14/20[e]	x 100% x	CCC PR (**Table A-1**)
Hogs and Pigs—payment equals sum of:			
1	Unpriced head sold during 1/15/20 to 4/15/20	x 100% x	Cares Act PR (**Table A-1**)
2	Unpriced inventory, max during 4/16/20 to 5/14/20	x 100% x	CCC PR (**Table A-1**)
Dairy—payment equals sum of:			
1	1st quarter milk production during 1/15/20 to 4/15/20[f]	x 100.0% x	Cares Act PR (**Table A-1**)
2	1st quarter milk production during 1/15/20 to 4/15/20[g]	x 101.4% x	CCC PR (**Table A-1**)
Lambs and Yearlings—payment equals sum of:			
1	Unpriced head sold during 1/15/20 to 4/15/20	x 100% x	Cares Act PR (**Table A-1**)
2	Unpriced inventory, max during 4/16/20 to 5/14/20	x 100% x	CCC PR (**Table A-1**)
Wool—payment equals sum of:			
1	Unpriced inventory as of 1/15/20	x 50% x	Cares Act PR (**Table A-1**)
2	Unpriced inventory as of 1/15/20	x 50% x	CCC PR (**Table A-1**)

Source: USDA, FSA, "Coronavirus Food Assistance Program," Final Rule, 85 *Federal Register* 30825, May 21, 2020.

Notes: Farmers and ranchers will self-certify their claims regarding inventories, sales, and shipments. *Unpriced* means any production or inventory that is not subject to an agreed-upon price in the future through a forward contract, agreement, or similar binding document.

a. Payment rates (PR) for each commodity are listed in **Table A-1** and **Table A-2** at the end of this chapter.

b. The quantity harvested, but as-yet-unpriced inventory, for which the producer has vested ownership. The inventory quantity may not exceed 50% of the producer's 2019 total production of the commodity.

c. Quantity of the harvested crop that was shipped but was unpaid or spoiled prior to receiving a payment due to a disruption of the marketing channel.

d. Quantity of the harvested crop that did not leave the farm (was unsold) or was donated and the quantity of mature crops that remained unharvested.

e. The highest inventory number between April 16 and May 14, 2020.

f. Dairy operations that dissolved on or before March 31, 2020, are eligible for the CARES Act payment rate.

g. Second quarter milk production is approximated by applying an adjustment factor of 1.014 to the 1st quarter production. Dairy operations that dissolved after March 31, 2020, are eligible for a prorated payment for the number of days that the operation commercially marketed milk in the second quarter.

CFAP Direct Payment Calculations

A producer's total CFAP payment is determined by first, multiplying the relevant quantity of each affected commodity by its respective payment rate. The payments from the two rates are then combined into a single payment per eligible commodity for each eligible producer (Table 2). Second, the total CFAP payment is obtained by summing payments across all of a producer's commodities. These calculations require both payment rates and applicable quantities as described in the box "Deriving the CFAP Payment Rates (PRs) and Payment Quantities."

Examples of how this payment formula may be applied to certain commodities are available in the footnoted references.[22] Under this structure, CFAP payments are directly proportional, or "coupled," to actual production.[23]

Projected CFAP Direct Payment Outlays

USDA estimates that, once adjustments have been made to account for payment limits, the $16 billion in CFAP funding is to be allocated across four different commodity groupings as follows: livestock ($9.4 billion or 58.9%); row crops ($3.5 billion, 21.9%); specialty crops ($2.4 billion, 15.0%); and "other" commodities ($670 million, 4.2%) (Figure 1).[24] Direct payment spending under the livestock category can be further delineated as $5.1 billion for cattle, $2.8 billion for dairy, and $1.6 billion for hogs.

---

[22] Paulson, N., G. Schnitkey, J. Coppess, C. Zulauf, and K. Swanson, "Coronavirus Food Assistance Program (CFAP) Rules Announced," farmdoc daily (10): 95, Dept. of Agricultural and Consumer Economics, University of Illinois at Urbana-Champaign, May 22, 2020; see also Shelby Myers, "Reviewing the CFAP Payment Formula for Non-Specialty Crops," Market Intel, American Farm Bureau Federation, May 28, 2020.

[23] The nature of the linkage—coupled versus non-coupled—between program payments and producer production choices is relevant to how such payments may count against U.S. domestic support commitments at the World Trade Organization (WTO); for details, see CRS Report R45305, *Agriculture in the WTO: Rules and Limits on U.S. Domestic Support*.

[24] USDA, "Coronavirus Food Assistance Program: Cost Benefit Analysis," Table 11. Estimated Net CFAP Payments by Commodity Group, after payment limitations, May 14, 2020.

Deriving the CFAP Payment Rates (PRs) and Payment Quantities

The CFAP payment for each commodity is based on separate payment rates for each of the two funding sources—CARES Act and CCC. The different payment rates for each commodity and their underlying derivation are described in Table A-1 and Table A-2.

CARES Act PR: Based on the Price Decline between Mid-January to Mid-April

The CARES Act PR reflects the income loss associated with the price decline for each commodity between mid- January and mid-April. USDA estimates the price-loss payment rate for each commodity as a portion of the price decline between the week of January 13-17, 2020, and the week of April 6-9, 2020, using futures contract prices or AMS-reported cash prices depending on availability.

CCC PR: Based on Unexpected Marketing Costs

The market-cost PR (CCC PR) for each commodity is based on USDA's projection of costs likely to be incurred for marketing the surplus inventory due to disrupted markets.

Payment Quantities Are Approximated by Inventories or 1st Quarter Sales

To approximate losses, the CARES Act and CCC PRs are multiplied by an applicable measure of the affected quantity of an eligible commodity (depending on the commodity as shown in Table 2).

For non-specialty crops and wool, the payment quantity for both CARES Act PR and CCC PR is approximated by using the lower of either:

(1) the unsold quantity of unpriced inventory[25] as of January 15, 2020, for which the producer has vested ownership, or (2) 50% of the producer's 2019 total production of the commodity. For livestock (including cattle, hogs and pigs, and lambs and yearlings),

- the payment quantity for the CARES Act PR is approximated by the number of animals sold during the period of January 15, 2020, to April 15, 2020; and
- the payment quantity for the CCC PR is approximated by the maximum unpriced inventory of animals during the period from April 16, 2020, to May 14, 2020.

Specialty crops may be eligible for any of three types of PRs (**Table A-2**):[26]

- CARES Act PR-1 reflects the mid-January to mid-April price decline, and is applied to the producer-certified sales (in pounds) during the period of January 15, 2020, to April 15, 2020; and
- CARES Act PR-2 is applied to producer-certified product that has been shipped from the farm by April 15, 2020, but that spoiled due to COVID-related market disruptions prior to receiving payment.[27]
- CCC PR-3 reflects COVID-related marketing costs and is applied to unsold inventory—estimated by the producer-certified maximum unpriced inventory from April 16, 2020, to May 14, 2020.

[25] *Unpriced* means any production or inventory that is not subject to an agreed-upon price in the future through a forward contract, agreement, or similar binding document.

[26] If a specialty crop's price did not decline by at least 5% during this period, it is not eligible for CARES Act PR-1, but may be eligible for CARES Act PR-2 or CCC PR-3. See **Table A-2** for examples of commodities that did not experience a sufficiently large price decline, but were still eligible for spoilage loss or marketing loss payments.

[27] The two different types of CARES Act payments are labeled "CARES Act PR-1" for price losses, and "CARES Act PR-2" for spoilage losses in **Table A-2**.

For dairy:

- the payment quantity for the CARES Act PR is approximated by using the quantity of milk sold during the 1st quarter of 2020; and
- the payment quantity for the CCC PR is theoretically applied to the quantity of milk sold during the 2nd quarter—which is approximated by applying a growth rate of 1.014% to 1st quarter sales.

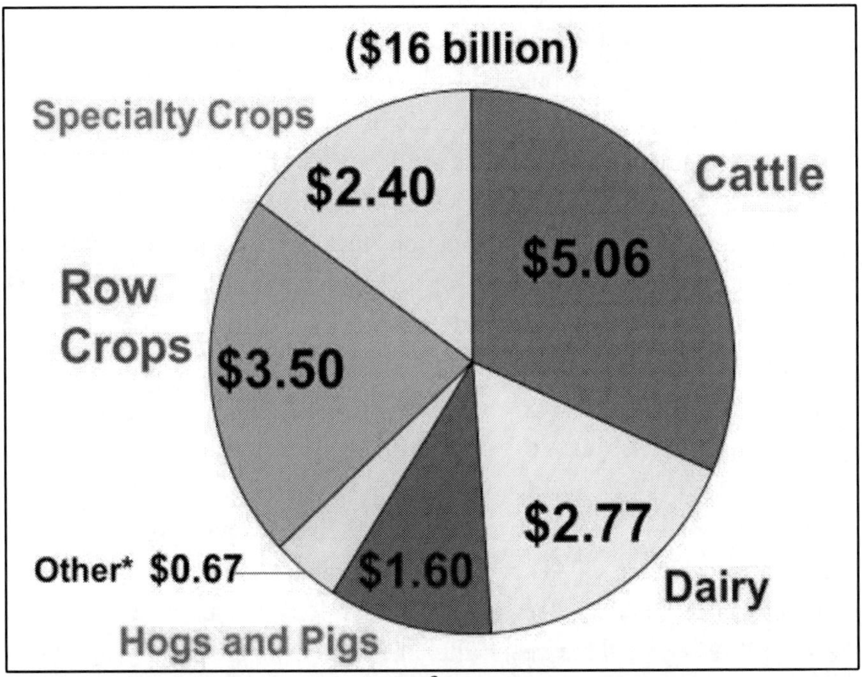

f

Source: USDA, "Coronavirus Food Assistance Program: Cost Benefit Analysis," Table 11. Estimated Net CFAP Payments by Commodity Group, after payment limitations, May 14, 2020, p. 20.

Note: *Other includes commodities that may be identified through the NOFA process.

Figure 1. Projected CFAP Net Payments by Commodity After Payment Limitations.

CFAP Application Process

USDA accepted CFAP applications starting on May 26, 2020. Producers should apply through their local FSA Service Center; however, an application may be submitted to any FSA county office.[28] To be eligible, an application must be submitted by the close of business on August 28, 2020. Payments to an eligible producer are expected to be made soon after each application is processed, according to USDA.

In addition to the application form, several other standard USDA forms must be on file for all persons and entities requesting CFAP benefits (Table A-3). However, for producers that are existing FSA program participants, most of these supplementary forms are likely already on file at their local FSA Service Center. In addition to these forms, if requested by USDA, an applicant must provide documentation that verifies the quantities of production and/or inventory used in the application, and that establishes the applicant's ownership share and value at risk for the commodity.

Producer Data Subject to Verification

Producers are expected to self-certify their sales and inventories that are used to calculate CFAP payments.[29] Producers who are approved for participation in CFAP are required to retain documentation in support of their application for three years after the date of approval. This includes records and paperwork to demonstrate losses, especially producers who have destroyed their product (e.g., dumping of milk or plowing under specialty crops).

Participants receiving CFAP payments or any other person who furnishes such information to USDA must permit authorized representatives of USDA or the Government Accountability Office (GAO),

---

[28] CFAP application forms, as well as a "CFAP Payment Calculator" to assist producers with the application process, are available from USDA at https://www.farmers.gov/cfap.

[29] USDA, FSA, "Coronavirus Food Assistance Program," Final Rule, 85 *Federal Register* 30825, May 21, 2020.

during regular business hours, to enter the agricultural operation and to inspect, examine, and make copies of books, records, or other items for the purpose of confirming the accuracy of the information provided by the participant.

Initial Payment Tranche of 80%

USDA is to pay out an initial tranche of 80% of an eligible participant's total potential CFAP payments. By issuing initial payments, FSA's goal is to provide assistance to those eligible participants who immediately apply for assistance, while trying to ensure that CFAP payments do not exceed the $16 billion funding limit. If funds remain available after the initial payment to all eligible applicants, USDA is to disburse the remainder of available funding such that the initial and final payments do not exceed a total of $9.5 billion for CARES Act funds and $6.5 billion for CCC funds. Funds may be prorated if necessary to stay within these limits.

Payment Limits

CFAP payments are, in general, subject to a per-person and per-legal-entity payment limitation of $250,000,[30] which is expected to result in some eligible commodity producers not receiving the entire calculated CFAP payment. For example, USDA projects that, based on its CFAP payment formula without payment limits, over $19 billion in CFAP payments would be made.[31] However, after applying the CFAP payment limit criteria, USDA projects that a net of $16 billion in CFAP payments will be made.

---

[30] This compares with a payment limit of $125,000 per individual under traditional farm programs from Title I of the 2018 farm bill (P.L. 115-334). See CRS Report R46248, *U.S. Farm Programs: Eligibility and Payment Limits*.

[31] USDA, "Coronavirus Food Assistance Program: Cost Benefit Analysis," May 14, 2020, p. 8.

The CFAP payment limitation applies to the total amount of CFAP payments made with respect to all eligible commodities of an individual or entity. However, unlike payment limits for other FSA programs where corporate entities are limited to a single payment limit,[32] USDA has elected to apply special payment limitation rules to CFAP participants that are corporations, limited liability companies, and limited partnerships. Under the special payment limitation rules, these corporate entities may receive CFAP payments up to $750,000 if three or more different individual owners or shareholders of the legal entity each contributed at least 400 hours of active personal labor or active personal management (or combination thereof) with respect to the production of 2019 commodities.[33] Under normal USDA rules governing recordkeeping requirements in regards to meeting the "actively engaged in farming" criteria, all persons that provide any management to the farming operation and seek to qualify as a farm manager—eligible for payments up to the individual payment limit—are required to maintain contemporaneous records or activity logs of their management activities.[34]

For a corporate entity,

- in general, the payment limit for the entity is $250,000;
- the payment limit for the entity is $500,000 if two different individual owners or shareholders of the legal entity each contributed at least 400 hours of active personal labor or active personal management (or combination thereof) with respect to the production of 2019 commodities; and
- the limit is $750,000 if three or more different individual owners or shareholders of the legal entity each contributed at least 400 hours of active personal labor or active personal management (or combination thereof) with respect to the production of 2019 commodities.

---

[32] CRS Report R46248, *U.S. Farm Programs: Eligibility and Payment Limits.*

[33] USDA, FSA, "Coronavirus Food Assistance Program," Final Rule, 85 *Federal Register* 30825, May 21, 2020.

[34] See CRS Report R44656, *USDA's Actively Engaged in Farming (AEF) Requirement*.

Neither the CARES Act nor the underlying CCC authority requires payment limits. Applying payment limits was done at USDA's discretion, as it also chose to do when establishing the Market Facilitation Payment and Wildfire and Hurricane Indemnity Program programs that were undertaken at the Secretary's discretion.[35] Benefits received under traditional farm support programs such as the Acreage Risk Coverage and Price Loss Coverage programs are not to be added to CFAP payments when evaluating payment limits.[36] In other words, payment limits for CFAP are independent of other farm program benefits received by a farm.

ISSUES FOR CONGRESS

Four issues may be of potential interest to Congress:

1. An immediate congressional concern could involve monitoring the implementation of the CFAP to ensure that program funding is fairly allocated across all affected agricultural sectors and all injured producers within those sectors, to the maximum extent possible.
2. Congress may also be interested in evaluating whether CFAP payments help mitigate the damage caused by the COVID-19 pandemic. Will additional assistance be necessary later this year? Is additional assistance needed to support certain affected industries that are not eligible under USDA's initial CFAP rollout?
3. Another immediate concern may be monitoring and oversight of the large sums of taxpayer money that will be flowing out through the USDA and the CCC. Producer self-certification of the farm data needed to calculate losses may create an incentive to over-report losses.

---

[35] See CRS Report R45310, *Farm Policy: USDA's 2018 Trade Aid Package*, CRS Report R45865, *Farm Policy: USDA's 2019 Trade Aid Package*, and CRS In Focus IF11539, *Wildfires and Hurricanes Indemnity Program (WHIP)*.

[36] USDA, "Coronavirus Food Assistance Program: Cost Benefit Analysis," May 14, 2020, p. 8.

4. Congress may evaluate whether CFAP will serve as a useful template for future disaster response programs.

Appendix. Supplementary Tables

Table A-1. CFAP Payment Rates (PRs): Non-Specialty Crops, Dairy, & Livestock

Commodity	Unit	Price Change ($ per unit)				Payment Rates	
		Average: Jan. 13-17	Average: Apr. 6-9	Decline (PD)[a]	% change	CARES Act	CCC
Non-Specialty Crops[b]						**50% PD**	**55% PD**
Barley (malting)	Bu.	$4.27	$3.60	$0.67	15.7%	$0.34	$0.37
Canola	Lb.	$0.17	$0.15	$0.02	11.8%	$0.01	$0.01
Corn	Bu.	$3.93	$3.30	$0.63	16.0%	$0.32	$0.35
Durum wheat	Bu.	$5.83	$5.46	$0.37	6.3%	$0.19	$0.20
Hard red spring wheat	Bu.	$5.64	$5.28	$0.36	6.4%	$0.18	$0.20
Millet	Bu.	$6.19	$5.57	$0.62	10.0%	$0.31	$0.34
Oats	Bu.	$3.04	$2.74	$0.30	9.9%	$0.15	$0.17
Sorghum	Bu.	$3.73	$3.14	$0.59	15.8%	$0.30	$0.32
Soybeans	Bu.	$9.47	$8.57	$0.90	9.5%	$0.45	$0.50
Sunflowers	Lb.	$0.18	$0.15	$0.03	16.7%	$0.02	$0.02
Upland cotton	Lb.	$0.72	$0.54	$0.18	25.0%	$0.09	$0.10
Dairy[c]						**80% PD**	**25% PD**
Dairy	Cwt.	$17.61	$11.72	$5.89	33.4%	$4.71	$1.47
Livestock[d]						**80% PD**	**Costs**[e]
Slaughter cattle: fed[f]	Head	$1,736	$1,469	$267	15.4%	$214	$33
Slaughter cattle: mature[g]	Head	$744	$630	$114	15.3%	$92	$33
Feeder cattle: < 600 lbs.[h]	Head	$812	$685	$127	15.6%	$102	$33
Feeder cattle: >600 lbs.[i]	Head	$1,107	$934	$173	15.6%	$139	$33
All other cattle[j]	Head	$812	$685	$127	15.6%	$102	$33
Pigs[k]	Head	nal	na	$35	58%	$28	$17
Hogs[m]	Head	na	na	$23	21%	$18	$17
Lambs and yearlings[n]	Head	na	na	$41	26%	$33	$7
Wool[o]						**50% PD**	**55% PD**
(graded, clean)	Lb.	$5.04	$3.62	$1.42	28.2%	$0.71	$0.78
(non-graded, greasy)[p]	Lb.	—	—	—	—	$0.36	$0.39

Sources: USDA, FSA, "Coronavirus Food Assistance Program," Final Rule, 85 *Federal Register* 30825, May 21, 2020; and USDA, FSA, "Coronavirus Food Assistance Program Cost-Benefit Analysis, May 14, 2020. According to USDA, commodity prices are drawn from several sources.

For non-specialty crops with futures market data, quotes for May futures contracts from various exchanges are used: Chicago Mercantile Exchange (CME) for corn, soybeans, and oats; Minneapolis Grain Exchange for hard red spring (HRS) wheat; and the Intercontinental Exchange for cotton and canola. Sorghum is calculated as 95% of the corn futures price, durum wheat is 103.4% of the HRS wheat futures price, sunflowers is the soybean oil

CME futures price divided by two, plus one cent. AMS data is used for other non-specialty crops where futures contracts are not traded.

For specialty crops, price data was provided by AMS and represents an average of all units shipped of domestic production, whether conventional or organic. The raw data source for the prices is the AMS Market News Portal, https://www.ams.usda.gov/market-news/fruits-vegetables. The prices are for the shipping point if available, or terminal market if not. For any particular crop, shipping point and terminal market prices are not mixed.

For livestock prices, price data was provided by AMS for cattle, hogs, milk (Class III and Class IV), and lamb and yearlings. The weighting of milk class prices to represent all-milk prices is described in the notes below. Adjustments to per-pound prices to account for live weight for cattle, hogs, and pigs are described in the notes below. For wool, instead of a futures or cash price, the difference in the Eastern Market Indicator, as reported by AMS in the *National Wool Review*, for the same two week periods is used to calculate the price decline.

Notes: PD = price decline; na = not available.

a. The price decline (PD) is the difference between the weekly average of the futures contract prices (or weekly average of the cash prices if the futures prices are unavailable) for the weeks of January 13-17, 2020, and April 6-9, 2020. Non-specialty crops and livestock commodities whose price decline is at least 5% are listed in this table and are eligible for a CFAP payment as described in **Table 2**.

b. For all non-specialty crops, the CARES Act payment rate is 50% of the PD; the CCC payment rate is 55% of the PD.

c. Dairy prices use an average of Class III (60% weight) and Class IV (40% weight) prices; the CARES Act payment rate is 80% of the calculated price decline; the CCC payment rate is 25% of the decline.

d. For livestock, the CARES Act payment rate is 80% of the price decline.

e. The CCC payment rate for livestock is based on projected costs likely to be incurred for marketing the 2020 inventory after April 15, 2020, due to disrupted markets.

f. Fed cattle are cattle that are market-ready (with a weight of 1,200 lbs. or more) and intended for slaughter. The price loss per-head of fed cattle is approximated by first calculating the difference between the weekly average prices for January 13-17, 2020, and April 6-9, 2020. These values are multiplied by 14 (assuming an average weight of 1,400 lbs. or 14 cwt) to approximate the equivalent per-head values for fed cattle.

g. Mature cattle are either cull cattle from a dairy herd, or breeding livestock that have been removed from inventory and are intended for slaughter. Similar to the procedure used for fed cattle, per-head prices for mature cattle are approximated by multiplying the per-cwt calculated average prices and their difference by 50% and 12 because culled cattle typically weight about 1,200 lbs. and are worth approximately half of fed cattle.

h. Feeder cattle are young cattle that are taken off pasture and brought to a feedlot for finishing weight gain to achieve market-ready status prior to slaughter. For feeder cattle under 600 lbs., the calculated average prices and their difference are multiplied by 5.5 to represent the average weight of 550 lbs.

i. Similarly, for feeder cattle over 600 lbs., the calculated average prices and their difference are multiplied by 7.5 to represent 750 lbs., the average weight when feeder cattle are placed on feed.

j. Commercially raised or maintained bovine animals not meeting definitions of other cattle categories, excluding beefalo, bison, and animals used for dairy production.

k. Pigs are any swine that weighs less than 120 lbs.

l. USDA did not provide the price averages used to calculate the price decline for pigs, hogs, and lambs and yearlings.

m. Hogs are any swine that weights more than 120 lbs. For hogs, the negotiated purchase prices as reported by AMS (https://mpr.datamart.ams.usda.gov/) were used to calculate the price change between the two periods.

n. Lambs and yearlings are all sheep less than two years of age.

o. Wool means the fiber sheared from a live sheep and includes, unless noted otherwise, graded and nongraded wool. Graded wool is paid on a clean basis, and ungraded wool is paid on a greasy basis. For all wool, the CARES Act payment rate is 50% of the PD; the CCC payment rate is 55% of the PD.

p. USDA uses a 50% conversion rate from greasy wool to clean wool, thus the payment rates for greasy wool are set at half the value of clean wool.

Table A-2. CFAP Payment Rates (PRs): Specialty Crops

Commodity	Price Change[a] ($ per lb.)				Payment Rates (PR)		
	Average Price: Jan. 13-17	Average Price: Apr. 6-10	Price Decline (PD)[b]	% change	CARES Act PR-1: for Sales Losses[c] 80% of PD	CARES Act PR-2: for product spoilage due to market disruption[d]	CCC PR-3: for unsold inventory[e]
	($/lb)	($/lb)	($/lb)	%	PR-1 ($/lb)	PR-2 ($/lb)	PR-3 ($/lb)
Almonds	$1.90	$1.58	$0.32	16.8%	$0.26	$0.57	$0.11
Apples	—	—	—	—	—	$0.18	$0.03
Artichokes	$1.64	$0.81	$0.83	50.6%	$0.66	$0.49	$0.10
Asparagus	—	—	—	—	—	$0.38	$0.07
Avocados	—	—	—	—	—	$0.14	$0.03
Beans	$0.55	$0.33	$0.22	40.0%	$0.17	$0.16	$0.03
Blueberries	—	—	—	—	—	$0.62	$0.12
Broccoli	$1.62	$0.84	$0.78	48.1%	$0.62	$0.49	$0.10
Cabbage	$0.22	$0.16	$0.05	27.3%	$0.04	$0.07	$0.01
Cantaloupe	—	—	—	—	—	$0.10	$0.02
Carrots	$0.38	$0.35	$0.03	7.9%	$0.02	$0.11	$0.02
Cauliflower	$1.02	$0.89	$0.14	12.7%	$0.11	$0.31	$0.06
Celery	—	—	—	—	—	$0.07	$0.01
Corn, sweet	$0.43	$0.31	$0.11	27.9%	$0.09	$0.13	$0.03
Cucumbers	$0.50	$0.34	$0.16	32.0%	$0.13	$0.15	$0.03
Eggplant	$0.50	$0.40	$0.09	20.0%	$0.07	$0.15	$0.03
Garlic	—	—	—	—	—	$0.85	$0.17
Grapefruit	—	—	—	—	—	$0.11	$0.02
Kiwifruit	—	—	—	—	—	$0.32	$0.06
Lemons	$0.70	$0.61	$0.09	12.9%	$0.08	$0.21	$0.04
Lettuce, iceberg	$0.50	$0.25	$0.25	50.0%	$0.20	$0.15	$0.03
Lettuce, romaine	$0.40	$0.31	$0.09	22.5%	$0.07	$0.12	$0.02
Mushrooms	—	—	—	—	—	$0.59	$0.11
Onions, dry	$0.18	$0.16	$0.02	11.1%	$0.01	$0.05	$0.01
Onions, green	—	—	—	—	—	$0.30	$0.06

Commodity	Price Change[a] ($ per lb.)				Payment Rates (PR)		
	Average Price: Jan. 13-17	Average Price: Apr. 6-10	Price Decline (PD)[b]	% change	CARES Act PR-1: for Sales Losses[c] 80% of PD	CARES Act PR-2: for product spoilage due to market disruption[d]	CCC PR-3: for unsold inventory[e]
	($/lb)	($/lb)	($/lb)	%	PR-1 ($/lb)	PR-2 ($/lb)	PR-3 ($/lb)
Oranges	—	—	—	—	—	$0.14	$0.03
Papaya	—	—	—	—	—	$0.32	$0.06
Peaches	$1.05	$0.96	$0.09	8.6%	$0.08	$0.32	$0.06
Pears	$0.58	$0.49	$0.09	15.5%	$0.08	$0.18	$0.03
Pecans	$3.10	$2.76	$0.34	11.0%	$0.28	$0.93	$0.18
Peppers, bell type	$0.73	$0.56	$0.17	23.3%	$0.14	$0.22	$0.04
Peppers, other	$0.73	$0.55	$0.18	24.7%	$0.15	$0.22	$0.04
Potatoes	—	—	—	—	—	$0.04	$0.01
Raspberries	—	—	—	—	—	$1.45	$0.28
Rhubarb	$3.42	$3.23	$0.19	5.6%	$0.15	$1.03	$0.20
Spinach	$1.22	$0.77	$0.45	36.9%	$0.37	$0.37	$0.39
Squash	$1.30	$0.40	$0.90	69.2%	$0.72	$0.39	$0.08
Strawberries	$2.40	$1.35	$1.05	43.8%	$0.84	$0.72	$0.14
Sweet potatoes	—	—	—	—	—	$0.18	$0.04
Tangerines	—	—	—	—	—	$0.22	$0.04
Taro	—	—	—	—	—	$0.23	$0.05
Tomatoes	$1.26	$0.46	$0.80	63.5%	$0.64	$0.38	$0.07
Walnuts	—	—	—	—	—	$0.45	$0.09
Watermelons	—	—	—	—	—	$0.02	—

Sources: USDA, FSA, "Coronavirus Food Assistance Program," Final Rule, 85 *Federal Register* 30825, May 21, 2020; and USDA, FSA, "Coronavirus Food Assistance Program Cost-Benefit Analysis, May 14, 2020.

Notes: Specialty crops include, but are not limited to, those commodities listed in this table. An "—" signifies that either a price increase occurred between the two periods used to calculate the price change, or the price decrease was less than 5%, hence no payment will be made under CARES Act PR-1. However, the commodity may be eligible for payments under CARES Act PR-2 and CCC PR-3.

a. Weekly AMS cash prices are used to calculate the average prices for the two periods.

b. The price decline is the difference in the weekly average of cash prices between the weeks of January 13-17, 2020, and April 6-10, 2020. Commodities whose price decline is equal to or greater than 5% are eligible for a CFAP payment based on the CARES Act PR-1 payment rate as described in **Table 2**.

c. The CARES Act payment rate (PR-1) for price losses equals 80% of the calculated price decline.

d. The CARES Act payment rate (PR-2) for market spoilage equals 30% of the lost value of any shipment.

e. The CCC payment rate (PR-3) for unsold inventory equals 5.875% of the crop's value. The 5.875% is calculated as 25% of the average price decline (of 23.5%) across specialty crops for which data was available.

Table A-3. Additional USDA Forms for CFAP Eligibility

Form	Purpose
CCC-901	**Member Information** CCC-901 identifies members of a farm or ranch that is a legal entity. Member Information will be completed by legal entities and joint operations to collect the following: member names, addresses, and Tax Identification Numbers (TINs); and citizenship status. If any member of the legal entity is a foreign person, CCC-902 must be completed by the legal entity or joint operation.
CCC-902	**Farm Operating Plan** For CFAP purposes only the following fields on the CCC-902 form are required to be completed: names, addresses, and TINs; citizenship status; and contributions to the farming operation for foreign persons.
CCC-941	**Average Adjusted Gross Income (AGI) Certification and Consent to Disclosure of Tax Information** CCC-941 information is used to evaluate whether a producer is in compliance with AGI criteria.
CCC-942	**Certification of Income from Farming, Ranching and Forestry Operations** CCC-942 must be completed if a producer exceeds the AGI criteria of $900,000, but derives at least 75% of income from farming activities.
AD-1026	**Highly Erodible Land Conservation (HELC) and Wetland Conservation (WC) Certification** AD-1026 certifies compliance with required conservation provisions.
AD-2047	**Customer Data Worksheet Request for Business Partner Record Change** AD-2047 provides basic customer contact information.
SF-3881	**ACH Vendor/Miscellaneous Payment Enrollment Form (Direct Deposit)** SF-3881 collects a producer's banking information to allow USDA to make payments to via direct deposit.

Source: USDA, Coronavirus Food Assistance Program, web portal, https://www.farmers.gov/cfap.

Note: For producers that are existing FSA program participants, most of these forms are likely already on file at their local FSA Service Center. In addition to these forms, if requested by USDA, an applicant must provide documentation that verifies the quantities of production and/or inventory used in the application, and that establishes the applicant's ownership share and value at risk for the commodity.

In: COVID 19: Agricultural and Food Issues ISBN: 978-1-53618-561-4
Editor: Kristian Goldschmidt

Chapter 9

USDA DOMESTIC FOOD ASSISTANCE PROGRAMS' RESPONSE TO COVID-19: P.L. 116-127, P.L. 116-136, AND RELATED EFFORTS*

Randy Alison Aussenberg and Kara Clifford Billings

INTRODUCTION

U.S. Department of Agriculture (USDA) Food and Nutrition Service (FNS) programs are often part of emergency response efforts, providing program flexibilities, foods for distribution, and benefits for redemption. Emergencies generate different FNS responses, which can vary with states' requests. During the COVID-19 pandemic, access to food—particularly in light of school closures—has been a concern for many. Some also view the Supplemental Nutrition Assistance Program (SNAP) as a force for economic stimulus. This Insight discusses related provisions of the second

* This is an edited, reformatted and augmented version of Congressional Research Service Publication No. IN11250, Updated March 31, 2020.

and third COVID-19 legislative packages, which supplement FNS's prior COVID-19 response with new funds and authorities.

USDA-FNS Responses To-Date

FNS maintains a COVID-19 response website. The website includes a state-by-state list of waivers, some of which USDA has issued using new authorities under P.L. 116-127 (discussed below).

Some options already existed under current law. For example, during unanticipated school closures, school districts that normally operate the National School Lunch Program and School Breakfast Program may transition to serving meals through the summer meals programs.

In addition, USDA may authorize emergency feeding programs during a presidentially declared disaster or emergency. USDA has approved requests from certain states and Indian tribal organizations to operate a Disaster Household Distribution program.

Families First Coronavirus Response Act (P.L. 116-127)

Supplemental Appropriations for WIC and TEFAP (Division A)

P.L. 116-127 provides a $500 million supplemental appropriation for the Special Supplemental Nutrition Program for Women, Infants, and Children (WIC). It also provides a $400 million supplemental appropriation for The Emergency Food Assistance Program (TEFAP), of which $100 million can be used for food distribution costs.

Child Nutrition Programs (§§2102, 2202)

Under §12(l) of the Richard B. Russell National School Lunch Act (42 U.S.C. §1760(l)), states and service providers may request, and USDA may grant, waivers of child nutrition program requirements. P.L. 116-127 temporarily expands these authorities.

§2102 allows USDA to grant waivers that increase federal costs for the purpose of providing meals and snacks during a COVID-19-related school closure. In addition, §2202 allows USDA to issue national waivers that apply to all states opting in. §2202 allows USDA to waive nutritional requirements in child nutrition programs if there is a food "supply chain disruption" due to COVID-19. §2202 also specifies that USDA may grant waivers to allow non-congregate feeding in the Child and Adult Care Food Program. The waiver authority under §2202 expires on September 30, 2020.

SNAP

SNAP Benefits during School Closures (§1101)

P.L. 116-127 includes what has been called "P-SNAP" (Pandemic Supplemental Nutrition Assistance Program), an option allowing USDA approval of state plans to provide SNAP benefits, when a school is closed five or more days, to households with children who would normally receive free or reduced-price school meals (in an amount equal to at least five days' of free meal reimbursements).

This option was last available in FY2010 (during the H1N1 flu pandemic), enacted in an FY2010 appropriations law (P.L. 111-80, §746). However, no SNAP state agencies ever administered P-SNAP.

Work-Related Requirements (§2301)

SNAP has work-related eligibility requirements, the strictest being a time limit for nondisabled adults (ages 18 to 49) without dependents ("ABAWDs") who work less than 80 hours per month. §2301 partially suspends this time limit nationwide during the period of the Secretary of Health and Human Services' public health emergency declaration, allowing participants who would have lost eligibility due to the time limit to continue to receive benefits.

Separately, on March 13, 2020, a federal court blocked enforcement of the part of a USDA final rule that would make it more difficult for states to use labor statistics to waive the time limit, changes that were to go into effect April 1. The court "determined that aspects of the Final Rule are likely unlawful ... [and] USDA will be enjoined from implementing those aspects of the Final Rule nationwide pending final judicial review." Part of the court's rationale was the global pandemic.

Benefit Increases, Administrative Requirements (§2302)

§2302 provides for temporary SNAP benefit increases during the public health emergency. It requires USDA to grant SNAP state agencies' requests that are supported "with sufficient data (as determined by [USDA])." The increases are "to address temporary food needs not greater than the applicable maximum monthly allotment for the household size." Previously, a nationwide increase to SNAP benefits was included in the American Recovery and Reinvestment Act of 2009 (P.L. 111-5).

The bill also allows USDA to adjust (through guidance and based on states' requests) administrative requirements like issuance methods and reporting requirements.

WIC (§§2203, 2204)

In addition to the supplemental funding provided, P.L. 116-127 gives USDA further authority to grant waivers allowing WIC participants to get certified (or recertified) without being physically present at the WIC clinic (normally required). Waiver requests are to be made by state agencies to USDA.

Also, the bill authorizes USDA to grant waivers from program administrative requirements that a state determines "cannot be met due to COVID-19" and is "necessary to provide assistance" under WIC.

Grants for Nutrition Assistance Funding for Certain Territories (§1102)

Northern Mariana Islands, Puerto Rico, and American Samoa do not operate a SNAP program; instead, they operate programs funded by Nutrition Assistance Program block grants in lieu of SNAP. §1102 provides $100 million for grants to these territories for nutrition assistance "in response to a COVID-19 public health emergency," as presumably they would not have the authority to operate P-SNAP or other options in the bill that increase SNAP benefits.

CARES Act (P.L. 116-136)

The third supplemental package provided additional funding for FNS programs "to prevent, prepare, and respond to coronavirus" as follows:

- $8.8 billion for child nutrition programs.
- $15.8 billion for the SNAP account. This includes $15.5 billion for SNAP participation should it exceed earlier budget estimates; $100 million for the Food Distribution Program on Indian
- Reservations (FDPIR) ($50 million for facilities/equipment, $50 million for food); and $200 million for nutrition assistance grants to Northern Mariana Islands, Puerto Rico, and American Samoa.
- $450 million for TEFAP, up to $150 million of which may be used for food distribution costs.

In: COVID 19: Agricultural and Food Issues
ISBN: 978-1-53618-561-4
Editor: Kristian Goldschmidt

Chapter 10

COVID-19: SUPPLY CHAIN DISRUPTIONS IN THE U.S. FRUIT AND VEGETABLE INDUSTRY: IN BRIEF*

Renée Johnson

INTRODUCTION

The U.S. fruit and vegetable industry entered 2020 following a difficult 2019 characterized by both lower prices and reduced production in some sectors. By broad market category, the U.S. Department of Agriculture (USDA) reported that grower prices for both fruits and tree nuts[1] and vegetables and pulses[2] began 2020 at low levels, with expectations for continued price declines.

* This is an edited, reformatted and augmented version of Congressional Research Service Publication No. R46348, dated May 8, 2020.

[1] USDA, Economic Research Service (ERS), Fruit and Tree Nuts Outlook, FTS-370, March 2020, https://www.ers.usda.gov/publications/pub-details/?pubid=98169.

[2] USDA, ERS, Vegetables and Pulses Outlook, VGS-363, September 2019, https://www.ers.usda.gov/publications/pub-details/?pubid=95065.

In mid-March 2020, the foodservice industry, which accounts for a substantial share of the produce industry's sales, was largely shut down as most states closed all but essential businesses in response to the coronavirus (COVID-19) public health emergency. This situation resulted in major supply chain disruptions for the U.S. produce industry. Reportedly, up to one-half of the produce industry sells its products to foodservice distributors (e.g., Sysco Corporation, U.S. Foods), according to former California Agricultural Commissioner A.G. Kawamura,[3] which in turn supply schools, restaurants, and other institutional customers (*Figure 1*). The sudden decrease in the foodservice market left many produce growers with nowhere to sell their crops. Given the highly perishable (non-storable) nature of most fresh fruits and vegetables, many growers were unable to pivot quickly to other marketing channels. Some growers reportedly resorted to plowing under their fields to avoid incurring the added costs of harvesting, cooling, and packaging. Some sectors are reporting plans to reduce acreage due to loss of foodservice demand.[4]

Another up to one-half of the produce industry reportedly relies on sales to larger retail chain stores (e.g., Walmart or membership clubs such as Costco) with some produce growers selling nearly all of their production to some combination of the foodservice market and larger retail chain stores. Fruits and vegetables also are sold to wholesale distributors and produce markets that serve independent grocery stores and restaurants. Direct-to-consumer sales of fruits and vegetables comprise a small portion of total U.S. produce sales. Available data from the 2017 *Census of Agriculture* indicate that the value of all food sold directly to consumers accounted for an estimated 3% of the value of total U.S. agricultural

[3] Comments made at the Farm Foundation's *Virtual Forum: Challenges & Opportunities for Agriculture in a Post- Pandemic World*, April 28, 2020, https://www.farmfoundation.org/forums/virtual-forum-challenges-opportunities-for-agriculture-in-a-post-pandemic-world/. Precise quantitative estimates are not available given the sheer number of fruit and vegetable crops, and may likely vary by sector and growing region.

[4] See, for example, University of California Agricultural Issues Center, "How Coronavirus Broke the Agricultural Supply Chain," posted by J. Goettsch, April 24, 2020, https://aic.ucdavis.edu/2020/04/24/the-great-potato-giveaway/.

production.[5] *Figure 1* broadly depicts the U.S. value chain and the various produce marketing channels.

The U.S. produce sector is highly streamlined and specialized. Many perishable produce items are pre-sold and packaged in bulk to certain specifications, and producers cannot readily shift such items from one marketing channel to another. Most produce is planted, harvested, packed, and shipped to a precise schedule to replenish stores' inventories "just in time," according to a University of California report.[6] This limits the industry's flexibility in responding to market disruptions. Cash flow concerns pose an added challenge to produce growers, according to A.G. Kawamura, especially if a business's supplier or buyer defaults on a commitment, or is unable to deliver supplies or move outgoing products due to the current pandemic.[7]

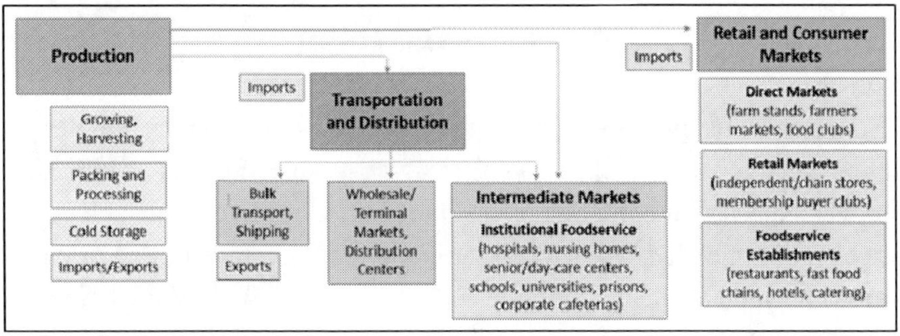

Source: CRS with information from USDA's U.S. Fresh Produce Markets: Marketing Channels, Trade Practices, and Retail Pricing Behavior (AER 825), September 2003, and National Policy & Legal Analysis Network to Prevent Childhood Obesity (NPLAN), Providing Fresh Produce in Small Food Stores, 2014.

Figure 1. U.S. Fruit and Vegetable Value Chain and Marketing Channels.

---

[5] USDA, *2017 Census of Agriculture*, 2019 (Table 2), https://www.nass.usda.gov/Publications/AgCensus/2017/Full_Report/Volume_1,_Chapter_1_US/.

[6] E.M. Bruno, R.J. Sexton, and D. A. Sumner, "The Coronavirus and the Food Supply Chain," *ARE Update*, Vol. 23, No. 4, Mar/Apr, 2020, https://giannini.ucop.edu/publications/are-update/issues/2020/23/4/the-coronavirus-and-the- food-supply-chain/.

[7] Comments made at the Farm Foundation's Virtual Forum: Challenges & Opportunities for Agriculture in a Post- Pandemic World, April 28, 2020.

Challenges for the U.S. produce industry are further compounded by ongoing agricultural labor concerns. Fruit and vegetable production is labor-intensive and few automation/mechanization options are available for handling perishable fruits and vegetables. On April 15, 2020, in an effort to lessen labor disruption impacts on the industry, the Department of Homeland Security and USDA announced a temporary final rule to change certain H-2A visa requirements.[8] Related concerns involve the potential effects if farmworkers and other produce industry workers become ill from the virus, as happened in some U.S. meatpacking plants. COVID-19 outbreaks have disrupted meat processing and caused some shortages of certain meat products in grocery stores.[9]

Market Overview Based on Limited Aggregated Data

Analysis of producer price changes for fruits and vegetables is complicated given the sheer number of crops and markets in the U.S. produce industry. Spot market prices for selected fruits and vegetables (i.e., price paid in the physical market, and not a futures/contract price or paper transaction) are available through USDA's Specialty Crop Market News.[10] These pricing data cover roughly 790 commodities across 1,450 terminal markets in 11 U.S. locations.[11] A compilation of these data from USDA, as posted by *The Packer* magazine,[12] suggests that prices for all fruits and

---

[8] USDA, DHS and USDA Move to Protect American Farmers and Ensure Continued Flow of America's Food Supply, Press Release No. 0219.20, April 15, 2020, https://www.usda.gov/media/press-releases/2020/04/15/dhs-and-usda-move-protect-american-farmers-and-ensure-continued.

[9] For example, see CRS Insight IN11366, *COVID-19 Disrupts U.S. Meat Supply; Producer Prices Tumble*.

[10] USDA, Agricultural Marketing Service (AMS), "Specialty Crops, Custom Reports," https://www.marketnews.usda.gov/mnp/fv-home.

[11] USDA, AMS, "Specialty Crops Market News," AMS PowerPoint presentation, September 19, 2019. Access to data to run a custom report for a specific crop in a specific market is available at https://marketnews.usda.gov/mnp/fv- report-config-step1?type=termPrice.

[12] Tom Karst, "How will we be changed by COVID-19?" *The Packer*, May 4, 2020, https://www.thepacker.com/ article/how-will-we-be-changed-covid-19.

vegetables in April 2020 averaged more than 10% lower compared to April 2019. These aggregate data mask differences in price performance for individual crops.

Other more readily accessible data are available from the Federal Reserve Bank of St. Louis based on aggregated Producer Price Indexes (PPI) for farm goods compiled by the U.S. Bureau of Labor Statistics.[13]

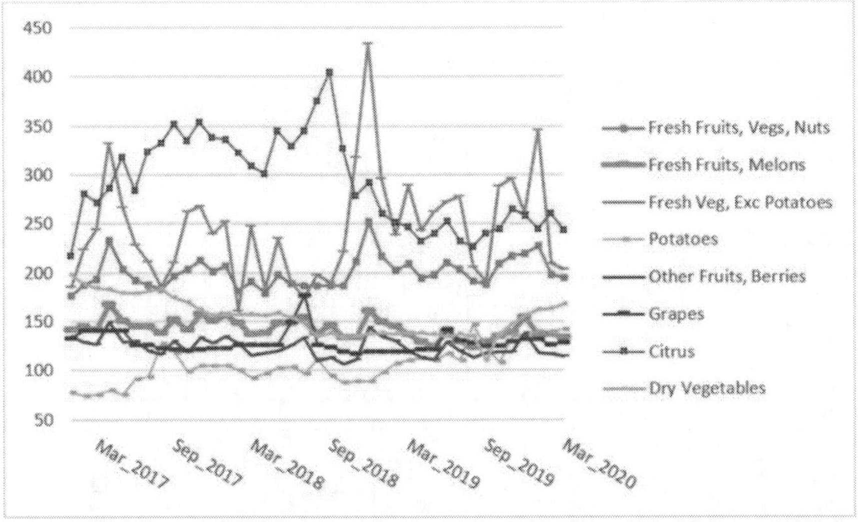

Source: CRS from Federal Reserve Economic Data, Producer Price Indexes (PPI) for Farm Products (https://fred.stlouisfed.org/release/tables?rid=46&eid=142105#snid=142107). Data are monthly, not seasonally adjusted. March 2020 are the most recent available data.

Figure 2. Producers Prices Indexes (PPI) for Selected Fruits and Vegetables.

Figure 2 shows monthly PPI for selected product groupings (March 2015 through March 2020). Data are not yet available for April. In aggregate, PPI for all fruits and all vegetables have roughly followed seasonal pricing cycles from previous years. As shown, price changes vary depending on the crop. Price variability may also be by growing location and other factors. A University of California report, for example, shows how recent price fluctuations differ among red potatoes (marked by

[13] See https://fred.stlouisfed.org/release/tables?rid=46&eid=142105#snid=142107.

increasing prices), green bell peppers (decreasing prices), and navel oranges (stable prices), which the researchers attributed to each crop's relative storability.[14]

Table 1. Sales of Fruit,Vegetables, and Tree Nuts, Selected States (2017) (Sales in millions of dollars)

	Fruits, Tree Nuts, Berries		Vegetables, Melons, Potatoes			
State	Number of Farms	Sales	Number of Farms	Sales	Total Sales	Share
California	35,087	19,708.7	5,048	8,167.8	27,876.5	58%
Washington	5,237	3,614.9	2,437	1,094.9	4,709.8	10%
Florida	6,650	1,298.7	1,835	1,284.1	2,582.8	5%
Idaho	532	25.1	1,355	1,147.1	1,172.2	2%
Oregon	4,923	612.1	2,112	539.2	1,151.4	2%
Arizonaa	818	(D)	1,115	1,009.1	1,009.1	2%
Georgia	3,865	421.8	1,899	566.4	988.2	2%
Michigan	2,876	440.7	3,089	535.1	975.8	2%
New York	3,083	399.8	3,588	378.7	778.5	2%
Wisconsin	1,942	177.4	3,184	543.0	720.4	1%
North Carolina	2,205	109.1	3,508	553.4	662.6	1%
Texas	5,714	213.3	2,221	352.4	565.7	1%
Minnesota	1,118	25.8	2,860	383.3	409.1	1%
New Jersey	949	141.3	1,390	222.5	363.8	1%
Pennsylvania	2,978	171.6	4,266	187.3	358.9	1%
All Other	32,017	1,221	35,413	2,620	3,840	8%
Total U.S.	109,994	28,581.4	75,320	19,584	48,165	100%

Source: CRS from USDA Quick Stats (https://quickstats.nass.usda.gov/). Reflects most recent available data from the 2017 Census of Agriculture. States ranked based on share of total fruit, vegetable, and tree nut sales. Individual state figures may not add to total U.S. figures due to rounding.

Notes: (D) = Withheld to avoid disclosing data for individual operations. Totals shown exclude withheld totals for some states to avoid disclosing individual operations. Vegetable category includes sweet potatoes. Tree nuts exclude peanuts. State total reflects vegetable, melon, and potato sales only, as sales data for fruits, tree nuts, and berries not disclosed.

Table 1 ranks the leading fruit, vegetable, and tree nut producing states in terms of the number of farms and farm-gate sales in 2017. At the farm

[14] Ellen M. Bruno, Richard J. Sexton, and Daniel A. Sumner, "The Coronavirus and the Food Supply Chain," *ARE Update*, Vol. 23, No. 4, Mar/Apr, 2020, https://giannini.ucop.edu/ publications/are-update/issues/2020/23/4/the- coronavirus-and-the-food-supply-chain/.

level, the value of U.S. market sales for all fruits, vegetables, and tree nuts totaled $48.2 billion in 2017.[15]

USDA Assistance

On April 17, 2020, USDA announced the Coronavirus Food Assistance Program (CFAP), which is to provide $19 billion in emergency aid to farmers and ranchers to address ongoing market disruptions.[16] CFAP includes $16 billion in direct payments to producers and $3 billion in purchases. Details have not been finalized by USDA, but Senator John Hoeven issued a statement,[17] according to which the produce sector may expect about $2.1 billion in direct payments and one-third of the $3 billion in purchases. In addition to the CFAP, USDA has announced its plans to buy $470 million of agricultural commodities, including $170 million in specialty crops, for delivery beginning in July.[18] Purchases include: asparagus ($5 million), orange juice ($25 million), pears ($5 million), potatoes ($50 million), prunes ($5 million), raisins ($15 million), strawberries ($35 million), sweet potatoes ($10 million) and tart cherries ($20 million). USDA's purchase plan is to establish "Farmers to Families Food Boxes" consisting of fresh fruits and vegetables, meat, and/or dairy products, or some combination of these products for delivery to food banks and other nonprofits,[19] which could help provide additional distribution and procurement options for U.S. agricultural producers. USDA stated it

---

[15] Based on data from the most recent USDA 2017 *Census of Agriculture.* Data are not readily available of the value of retail sales for fruit and vegetables.

[16] USDA, "*USDA Announces Coronavirus Food Assistance Program,* Press Release No. 0222.20, April 17, 2020, https://www.usda.gov/media/press-releases/2020/04/17/usda-announces-coronavirus-food-assistance-program.

[17] Senator John Hoeven, "Hoeven: USDA to Provide $16 Billion in Direct Assistance to Farmers and Ranchers, $3 Billion in Agriculture Purchases," News Release, April 17, 2020, https://www.hoeven.senate.gov/news/news-releases/ hoeven-usda-to-provide-16-billion-in-direct-assistance-to-farmers-and-ranchers-3-billion-in-agriculture-purchases.

[18] USDA, AMS "USDA Announces Additional Food Purchase Plans," Press Release, May 4, 2020, https://www.ams.usda.gov/press-release/usda-announces-additional-food-purchase-plans.

[19] For more information see https://www.ams.usda.gov/selling-food-to-usda/farmers-to-families-food-box.

would begin purchasing an estimated $100 million per month each for fresh produce, livestock (pork and chicken), and dairy products. USDA has said the purchase and distribution of these commodities will begin in May 2020.

For the direct payment portion of the CFAP, according to Senator Hoeven's statement, agricultural producers would receive a payment based on 85% of price losses that occurred between January 1 and April 15, 2020, and a second payment for 30% of losses occurring from April 15 through the following two quarters. Producers may be subject to a payment limit of $125,000 per commodity and an overall limit of $250,000 per individual, but USDA has not yet indicated this specifically. The statement notes that USDA is expediting its rulemaking to begin enrollment in May and to distribute initial payments by early June.

PRODUCER RESPONSE

Industry groups have expressed appreciation for the support that USDA announced, but claim the amounts fall short of expected industry losses. Some industry stakeholders assert that a $250,000 payment limit per entity (as described in Senator Hoeven's statement) is insufficient for the produce industry's high-value crops. On April 23, 2020, 28 Senators sent a letter to President Trump requesting these limits be removed for specialty crop, livestock, and dairy producers.[20] A group of 126 Members of the House sent a similar letter.[21]

Beyond the USDA food purchase plans, the U.S. fruit and vegetable industry continues to consider alternative produce distribution options to expand beyond the more traditional foodservice and chain store sectors, including various direct-to-consumer sales and streamlined government

[20] Letter to President Donald J. Trump from several members of the U.S. Senate, April 23, 2020, https://www.feinstein.senate.gov/public/index.cfm?a=files.serve&File_id=2A2717EC-8FD0-4CFF-86D7- 98406E2E5E49.

[21] Letter to President Donald J. Trump from several members of the U.S. House of Representatives, April 23, 2020, https://panetta.house.gov/sites/ panetta.house.gov/ files/documents/ 2020. April_23_CongressmanPanetta_Letter_CFAPPaymentLimits.pdf.

procurement options. The United Fresh Produce Association (UFPA), an industry trade association, maintains a COVID-19 resources page with information for the fresh produce industry with specific information on how to access these and other support programs.[22]

---

[22] See https://www.unitedfresh.org/coronavirus-fresh-produce-industry-resource-page/.

In: COVID 19: Agricultural and Food Issues ISBN: 978-1-53618-561-4
Editor: Kristian Goldschmidt

Chapter 11

COVID-19 DISRUPTS U.S. MEAT SUPPLY; PRODUCER PRICES TUMBLE*

Joel L. Greene

INTRODUCTION

U.S. livestock and poultry producers entered 2020 with an optimistic outlook for prices and income. Then in mid-March the food service sector, which accounts for a substantial share of meat consumption, was largely shut down as most states closed all but essential businesses. A temporary surge in retail meat purchases offset some of the reduction in food service demand. However, in early April, the situation worsened for producers as COVID-19 outbreaks began spreading in meatpacking plants around the United States, disrupting meat processing and leading to some shortages of meat products in grocery stores. The reduction in slaughter is weighing on livestock prices, reflecting shrinking meatpacker demand for slaughter-ready livestock and poultry.

---

* This is an edited, reformatted and augmented version of Congressional Research Service Publication No. IN11366, dated April 29, 2020.

On April 12, 2020, Smithfield Foods, the largest pork processor in the United States, indefinitely closed its Sioux Falls, SD, pork processing plant because of the spread of COVID-19 among plant workers. With an estimated daily slaughter capacity of 19,500 head, the Sioux Falls plant accounts for 4% of estimated daily U.S. hog slaughter. More pork slaughter capacity has been lost since then because of COVID-19 outbreaks in plants operated by other processors. On April 13, JBS USA announced it was closing its Greeley, CO, beef plant until April 24 because of an outbreak of COVID-19 among plant workers. *Cattle Buyers Weekly* estimates daily slaughter capacity of 6,000 head at the Greeley plant, which would be 4%- 5% of daily national cattle slaughter capacity. Other beef plants have had to close due to COVID-19 outbreaks among workers. Outbreaks of COVID-19 have hit beef, pork, and poultry plants across the country. As of April 27, at least 15 meatpacking plants were reported closed. Some plants have shut down temporarily to clean and reconfigure work stations; others have not set reopening dates.

Reportedly, some plants have implemented onsite temperature monitoring of employees, spread out workers within plants, set up plexiglass dividers between workers who normally operate in tight spaces, and spread the work of one shift over two shifts, thus slowing operating speed. Even so, some meat plant employees have asserted that packers have not implemented adequate protective measures in their plants. On April 26, 2020, the Occupational Safety and Health Administration updated guidelines to address workplace conditions for meat plant workers. On April 28, 2020, President Trump announced that he plans to sign an executive order that would use the Defense Production Act to classify meat processing as critical infrastructure to keep production plants open.

PRODUCTION SHORTFALL

The Commodity Futures Trading Commission reported in its April 22, 2020, Agricultural Advisory Committee meeting that U.S. meat plants were operating at about 60% of capacity. The estimate is based on plant closures combined with open plants that are running at 50%-75% of normal capacity as plants slow operations to protect workers. Steve Meyer of Kerns and Associates has estimated that nearly 30% of pork processing capacity will be offline as of April 24 because of COVID-19 outbreaks. The shutdown of some meat slaughter plants may have a cascading effect on consumer products, because slaughter plants often supply intermediate products to food plants for further processing, thus leading to a sequence of supply disruptions.

Estimated meat production (beef, pork, and chicken) for the week ending April 25 declined 13%, compared with the same week a year ago, while during the three weeks since April 4, meat production was 10% lower than the same period last year. As long as COVID-19 disruptions continue, there may be shortages of meat products in some grocery stores. However, there are ample numbers of cattle, hogs and broiler chickens to meet demand, especially given reduced demand from institutional markets such as schools and restaurants. As of April 9, the U.S. Department of Agriculture (USDA) forecasts record U.S. meat and poultry production of 108 billion pounds in 2020, or 3% more than in 2019. USDA also reported that stocks of beef, pork, and chicken in cold storage warehouses exceeded 2 billion pounds at the end of March. This quantity amounts to about one to two weeks of meat supply. The industry operates on a "just in time" delivery system, and this quantity is about 6% more than a year ago. If the shutdown of livestock processing facilities spreads to additional plants, or if shutdown periods are extended, then both meat supplies and livestock prices could be further reduced.

PRICE DECLINES

At the farm level, the disruptions at meat plants have triggered rapid declines in livestock and poultry prices to producers over the past few weeks, lowering returns that were originally projected to increase compared with 2019. In February 2020, USDA forecast a 5% increase to $136 billion in cash receipts for livestock and poultry producers in 2020 compared with 2019. With prices falling since February, USDA will likely significantly reduce the cash receipts forecast later in the year. Cash market prices for cattle, hogs, and broilers have fallen during April by between 10% and nearly 30% (*Table 1*).

Table 1. Cattle, hog, and broiler prices

	2019	**Jan**	**Feb**	**Mar**	**April 1-23**	**% change, March to April**
5-Area Steer, $/cwt	116.78	123.89	118.59	112.48	101.00	-10%
National Hog Carcass, $/cwt	47.95	58.69	54.81	58.86	48.00	-18%
Broilers, Wholebird, ¢/lb.	88.60	90.56	80.64	79.35	56.00	-29%

Source: USDA's benchmark prices for cattle, hogs, and broilers as reported monthly in the World Agriculture Supply and Demand Estimates. The price data are from USDA's Agricultural Marketing Service.

Notes: Reported annual average for 2019 and reported monthly prices for January-March 2020. CRS compiled the April estimates from average daily prices for steers and hogs and average weekly prices for broilers available from April 1-23. Cwt = 100 pounds.

Based on cattle and hog futures prices on the Chicago Mercantile Exchange (CME), the outlook for producer prices have trended down since January 2020, with the June cattle contract price falling more than 30% and the May hog contract falling by 38% (see *Figure 1* and *Figure 2*). The outlook for fall prices (as of April 24), based on CME October contracts, are 20% lower for cattle and nearly 30% lower for hogs than at the beginning of the year.

PRODUCER LOSSES

In April 2020, the Food and Agricultural Policy Research Institute, which provides analysis of agricultural markets and policies to Congress, estimated a decline of $20 billion in cash receipts for the livestock industry in 2020 due to COVID-19. Of this total, cattle comprised $9.5 billion, hogs

$2.2 billion, and poultry $4.1 billion. Trade groups representing the cattle and hog industries have released higher estimates of income losses this year, amounting to $13.6 billion and $5 billion, respectively. The sheep industry has estimated its losses at more than $300 million.

Source: RBM Group, CME, Contract LEM20, April 23, 2020.

Figure 1. Live Cattle Futures, June Contract.

USDA Assistance

On April 17, 2020, USDA announced the Coronavirus Food Assistance Program (CFAP), which is to provide $19 billion in emergency aid to farmers and ranchers to address ongoing market disruptions. CFAP includes $16 billion in direct payments to producers and $3 billion in purchases of fresh produce, meat, and dairy products for food banks and other feeding programs. In its announcement, USDA states that it will begin purchasing an estimated $100 million per month each for livestock (pork and chicken), dairy products, and fresh produce. USDA has indicated that the purchase and distribution of these commodities could begin by mid-May.

Source: RBM Group, CME, HEK20 contract, April 23, 2020.

Figure 2. Live Hog Futures, May Contract.

USDA has not provided operational details about the direct payments to producers. However, on April 17, Senator Hoeven issued a statement that USDA would provide a total of $9.6 billion to livestock and dairy producers, of which $5.1 billion is for cattle, $2.9 billion for dairy, and $1.6 billion for hogs. The remaining $6.4 billion would be distributed to other agricultural producers. According to Senator Hoeven's statement, agricultural producers would receive a payment based on 85% of price losses that occurred between January 1 and April 15, 2020, and a second payment for 30% of losses occurring from April 15 through the following two quarters. Producers would be subject to a payment limit of $125,000 per commodity and an overall limit of $250,000 per individual. The statement notes that USDA is expediting its rulemaking to begin signup in early May and distribute initial payments by the end of May and early June.

PRODUCER RESPONSE

Industry groups have expressed appreciation for the aid that USDA announced but note that the amounts fall short of industry loss estimates for cattle and hogs. Some industry stakeholders assert that the $125,000 payment limit will severely restrict needed aid to individual producers. On April 23, 28 Senators sent a letter to the President Trump requesting that these limits be removed for livestock, dairy, and specialty crop producers. A group of 126 Members of the House sent a similar letter.

In: COVID 19: Agricultural and Food Issues ISBN: 978-1-53618-561-4
Editor: Kristian Goldschmidt

Chapter 12

PANDEMIC WEAKENING MILK PRICES; INDUSTRY CALLS FOR POLICY ACTION*

Joel L. Greene

INTRODUCTION

Milk price prospects for U.S. dairy producers in 2020 have weakened as the U.S. economy deteriorates under the expanding Coronavirus 2019 (COVID-19) pandemic. The year began with a positive outlook for the dairy industry that included higher milk prices and increased income for dairy producers in 2020, according to the U.S. Department of Agriculture (USDA). The dairy industry has gone through several years of relatively low milk prices, leading some dairy farms to exit the industry and two large dairy processors to file for bankruptcy. In 2019, according to USDA, the number of licensed dairy herds declined from 37,468 to 34,187, a nearly 9% drop. Almost every state lost dairies last year.

* This is an edited, reformatted and augmented version of Congressional Research Service Publication No. IN11326, dated April 10, 2020.

As a safety net for dairy producers, Congress enacted Section 1401 of the 2018 farm bill (P.L. 115-334) establishing the Dairy Margin Coverage (DMC) program. The program guarantees dairy producers could receive payments when the margin or difference narrows between the USDA-reported all milk price and a calculated feed ration value based on the prices of corn, soybean meal, and alfalfa hay. The DMC replaced the 2014 farm bill's (P.L. 113-79's) Margin Protection Program (MPP). The MPP program made very few payments to dairy producers during the 2015-2017 period ($12 million), and many producers considered the program ineffective as a safety net during years of relatively low milk prices. Congress amended MPP in the Bipartisan Budget Act of 2018 (BBA; P.L. 115-123) which resulted in higher MPP payments in 2018 ($253 million).

Building on these BBA changes, the DMC program expanded the margin guarantee by allowing producers to select a margin, ranging from $4.00 to $9.50 per hundredweight (cwt; 100 pounds) of milk at a lower premium cost. Producer-paid premiums for higher levels of margin coverage were significantly reduced in DMC compared with MPP to provide a greater incentive for dairy producers to participate in the program. For each month when the actual margin falls below a producer's selected DMC margin, the producer receives a payment from USDA for the difference between the two for the dairy's monthly milk production history.

As a result of these program changes, 82% of dairy operations participated in the DMC in 2019, with most selecting the highest margin level of $9.50. Dairies received DMC payments of more than $312 million in 2019, compared to about $265 million during the 2015-2018 period under MPP. USDA made margin payments during the first six months of 2019, when the margin averaged $8.57. Higher all milk prices during the last half of the year resulted in an average margin of $10.46 per cwt, well above $9.50, which negated any payments. Against the backdrop of higher milk prices in the latter half of 2019, and an expectation that milk prices would rise further in 2020, 48% of dairy operations signed up for DMC for the 2020 calendar year program.

The economic impacts of COVID-19 on dairy markets have drastically altered the outlook for milk prices. Milk purchases at retail were strong in March 2020, but probably insufficient to offset the loss of milk and dairy product sales due to the closure of schools and reduced food service demand. The National Milk Producers Federation (NMPF) estimates the eroding outlook for milk prices since mid-February could result in losses of $6 billion in total gross farm income for dairy producers in 2020.

Prices for the CME April Class III milk futures contract (milk used for cheese and indicative of future milk prices) dropped 20% from a contract high of $18.00 per cwt on January 24, 2020, to $14.45 per cwt on April 1, 2020. *Figure 1* shows how expectations have changed for dairy margins for the remainder of 2020 based on USDA's DMC Decision Tool, which calculates daily forecasts of milk and feed prices.

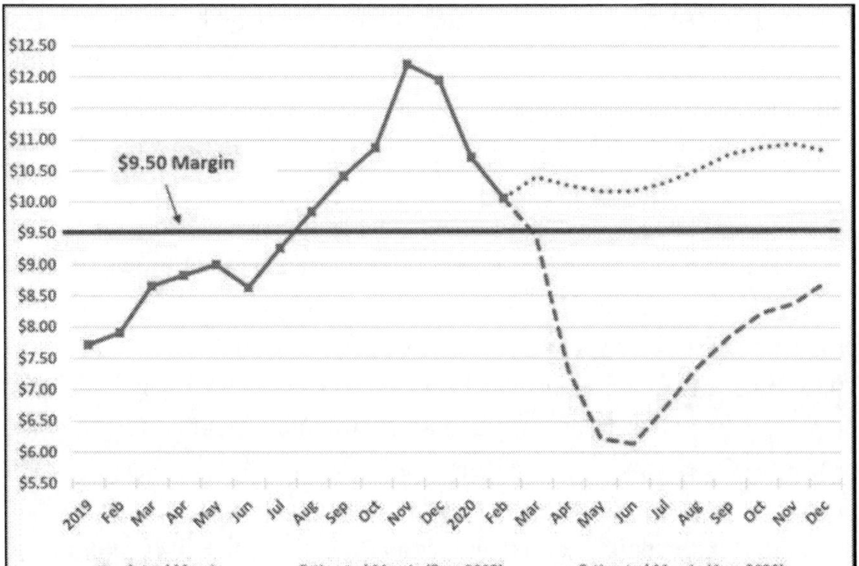

Source: USDA, Farm Service Agency, Dairy Margin Coverage Program.

Note: Estimates are from USDA's DMC Decision Tool model results on December 27, 2019 and April 1, 2020.

Figure 1. Dairy Margin Coverage Program. Monthly Margins for 2019 and 2020, $ per cwt.

Dairy producers may use the Decision Tool to determine how to participate in the DMC program. Heading into 2020, most dairy analysts expected the all milk-feed value margin to be well above the top margin level of $9.50 per cwt, with only slight possibilities of DMC payments during the year; thus contributing to the drop-off in participation in 2020.

The CARES Act (P.L. 116-136) provides USDA with $9.5 billion to "support agricultural producers impacted by coronavirus, including producers of specialty crops, producers that supply local food systems, including farmers' markets, restaurants, and schools, and livestock producers, including dairy producers." USDA is in the process of developing a program of emergency relief for dairy producers, and other producers economically harmed by the coronavirus pandemic, and is receiving policy suggestions from dairy industry stakeholders.

On April 6, 2020, the NMPF and the International Dairy Foods Association (IDFA), which represent milk producers and dairy processors, respectively, released a proposed plan of how USDA could support the dairy industry. These two groups say that milk supplies exceed demand by 10%, and expect this surplus to increase with the spring flush (a seasonal increase in milk production).

A key provision of the plan is that milk producers cut milk production by 10% from their March 2020 output. For dairies that cut production, the plan also proposes that USDA pay producers $3 per cwt on 90% of milk production. These payments would occur from April through September 2020, except for months in which the average of Class III and Class IV prices (prices representing cheese and butter/nonfat dry milk, respectively) exceed $16 per cwt. The plan also calls for USDA to indemnify dairy producers who dump or remove milk from commercial use because of disruptions in the dairy market, and reopen signup for the DMC program to give producers a second opportunity to enroll.

For dairy processors, the plan calls for USDA to provide recourse (repayable) loans to support processors holding excessive stocks during this time of market disruption, and to offer forgivable loans if processors follow certain conditions, such as continuing to buy milk from producers and maintaining staffing.

The plan also requests that USDA make substantial purchases of dairy products for food banks to help stabilize dairy markets and increase food security. Further, the industry asks that USDA grant state-based flexibility (e.g., restrictions on dairy product container size and fat content) for feeding programs such as Women, Infants, and Children (WIC) and Summer Food Service program for students. Also, the plan requests that USDA increase funding for the Supplemental Nutrition Assistance Program (SNAP), and for several other feeding programs, including international food aid.

In response to the plan, the American Dairy Coalition has called for regionalization of the production cuts, reasoning that seasonal production falls in the south during the summer, so northern tier dairy producers would bear the brunt of cuts to meet the 10% target compared with March production. Also, some analysts have questioned whether the DMC program enrollment should be reopened to milk producers (April 3, 2020, AgriTalk Live) because it would be a precedent for farm programs that generally operate with set deadlines for signup.

In: COVID 19: Agricultural and Food Issues ISBN: 978-1-53618-561-4
Editor: Kristian Goldschmidt

Chapter 13

EXECUTIVE ORDER ON THE FOOD SUPPLY CHAIN AND THE DEFENSE PRODUCTION ACT: FAQS*

Nina M. Hart

INTRODUCTION

On April 28, 2020, President Trump issued an executive order (EO) invoking the Defense Production Act (DPA) to address the food supply chain for meat and poultry products during the national emergency caused by the COVID-19 outbreak. The release of the EO was preceded by some state and congressional calls for such an order, and significant media coverage about the potential contents of such an order. This chapter addresses some frequently asked questions about the EO's contents, legal bases, and potential effects.

* This is an edited, reformatted and augmented version of Congressional Research Service Publication No. LSB10456, dated May 1, 2020.

WHAT DOES THE EO DO AND HOW DOES IT USE THE DPA?

The EO designates "meat and poultry in the food supply chain" as "critical and strategic materials" under section 101(b) of the DPA. This section of the DPA permits the President to control distribution of designated materials in the civilian market, if: (1) the material is determined to be essential to the national defense, and (2) that the requirements of the national defense cannot be met through other means "without creating a significant dislocation of the normal distribution" of the designated material "to such a degree as to create appreciable hardship." Having made such a finding, the EO authorizes the Secretary of Agriculture to "take all appropriate action under [section 101(b)] to ensure that meat and poultry processors continue operations consistent with the guidance for their operations jointly issued by the [Centers for Disease Control and Prevention (CDC)] and [the Occupational Safety and Health Administration (OSHA)]." Based on a plain reading of the text, the EO itself does not order any plants to remain open. Rather, it permits the Secretary to issue regulations, orders, or take other actions to address continued operation of processors, while also ensuring that processors adhere to the CDC/OSHA guidance.

The EO further authorizes the Secretary to "identify additional specific food supply chain resources" as "critical and strategic materials" under the DPA. Based on this authorization, the Secretary could conceivably extend the reach of the EO beyond meat and poultry to any other commodity or resource within a food supply chain, provided these commodities or resources satisfy the requirements of section 101(b).

Finally, the EO provides additional authority under section 101(a) of the DPA to allow the Secretary to prioritize contracts or "allocate materials, services, and facilities as deemed necessary or appropriate to promote the national defense." To implement this part of the EO, the President has also delegated authority to the Secretary to, among other things, issue regulations and orders, require information or records from any person, and seek to enforce DPA orders (discussed in more detail below). The U.S. Department of Agriculture (USDA) currently has regulations, known as the Agriculture Priorities and Allocations System (APAS), in place to set out how it will implement section 101 during emergencies and non-emergencies, as directed by President Obama's Executive Order 13603. President Trump's new EO seeks to ensure that USDA may create new or amended regulations if USDA believes it necessary to address the COVID-19 emergency (i.e., USDA may rely on existing regulations, but could also issue new or amended regulations).

How does an Agency Implement a DPA EO?

There is not necessarily a standard way of implementing an EO implicating the DPA, as much depends on the contents of any EO and the fact that agencies have discretion to issue regulations and orders within the scope of their authority, as they believe appropriate. That said, USDA appears, at least initially, to be attempting to implement the EO via negotiation as opposed to more coercive measures (e.g., orders directed to specific processors or regions). On April 29, 2020, USDA issued a statement that it intends to first "work with meat processing [plants] to affirm they will operate in accordance with the CDC and OSHA Guidance, and then work with state and local officials to ensure that these plants are allowed to operate."

Are There Penalties for Processors That Fail to Comply with DPA Orders?

The DPA empowers the President and agencies to enforce DPA orders in several ways. First, they may seek injunctions or restraining orders in federal court to require an entity to do or refrain from doing something in violation of a DPA order. Second, they may promulgate regulations imposing liability on non-compliant parties. These regulations may incorporate the statutory penalties set out in section 103 of the DPA, which states that "any person who willfully" violates the statutory or regulatory provisions involving Title I of the DPA (i.e., allocation or prioritization of contracts and orders against hoarding supplies) may be fined up to $10,000 or imprisoned for not more than one year, or both.

USDA currently has regulations addressing compliance under its APAS program. If USDA chooses to rely on these regulations to implement the prioritization and allocation of orders under the current EO, it could also seek to rely on the penalty components of these provisions if processors refuse to comply with DPA orders.

DOES THE EO OR DPA LIMIT THE LEGAL LIABILITY OF PROCESSORS?

The EO contains no reference to legal liability or immunity from lawsuits for affected processors and, on its own, does not grant immunity to processors. However, the DPA contains a provision that provides parties subject to DPA orders with a defense against liability, which applies regardless of whether an EO expressly references it. Section 707 of the DPA states as follows:

> No person shall be held liable for damages or penalties for any act or failure to act resulting directly or indirectly from compliance with a rule, regulation, or order issued pursuant to this chapter, notwithstanding that any such rule, regulation, or order shall thereafter be declared by judicial or other competent authority to be invalid.

As an initial matter, this provision is not a blanket grant of immunity from liability for all actions that an entity may take under a DPA order. Rather, the act (or failure to act) in question must "result[] directly or indirectly from compliance" with an EO or other rule, regulation, or order issued by an agency. Thus, while an entity may raise a DPA order as a defense if sued for actions taken while complying with that order, whether that entity will ultimately be granted immunity by a court is a fact-specific inquiry. This is evident from a case in which a federal court of appeals rejected a company's argument that it was immune from liability for toxic waste cleanup costs because it was producing Agent Orange under a DPA order for the federal government.

Moreover, whether section 707 extends beyond breach of contract situations to tort law (e.g., negligence or product defect claims) remains debated and under-litigated. On its face, section 707 does not differentiate between contract and tort law, which may indicate it could apply to both types of disputes provided the act in question resulted "directly or indirectly" from compliance with a DPA order. As one federal district court has noted, Congress originally designed section 707 to apply only to contract issues, but Congress later removed the section's express reference to contractual liability. Given this legislative history, it may be that Congress intended that entities subject to DPA orders could raise section 707 as a defense in both contract and tort litigation. But, as that same district court pointed out, it may be "unlikely that Congress would work such a major change in tort law without being explicit about it."

In sum, the precise scope and applicability of section 707 is unclear for at least two reasons: (1) it is undecided whether and how far it extends beyond contract disputes to tort litigation, and (2) whether an action in dispute results "directly or indirectly" from compliance with a DPA order is fact-specific and must be determined on a case-by-case basis.

DOES OSHA'S ENFORCEMENT POLICY LIMIT LEGAL LIABILITY FOR PROCESSORS?

OSHA's interim enforcement response plan, as clarified by a statement of enforcement policy, does not affect processors' legal obligations. The response plan and enforcement policy are mechanisms used to set out broadly OSHA's enforcement priorities and indicate how it will exercise its enforcement discretion.

Thus, although OSHA indicates that it "does not anticipate citing employers that adhere to" the CDC/OSHA guidance for operations, this is not a guarantee of non-enforcement or immunity.

Potentially to address processor concerns about legal liability resulting from continuing operations, OSHA has indicated a willingness to entertain requests from processors that are sued for alleged workplace exposure to COVID-19 to participate in the litigation, provided the processor being sued has "demonstrated good faith attempts to comply" with the CDC/OSHA guidance. However, this is not a promise to participate in litigation. Further, OSHA has extended a similar offer to employees, indicating it may opt to participate on behalf of the employees "if their employer has not taken steps in good faith" to comply with the CDC/OSHA guidance.

In: COVID 19: Agricultural and Food Issues ISBN: 978-1-53618-561-4
Editor: Kristian Goldschmidt

Chapter 14

COVID-19 AND USDA FARM LOAN FLEXIBILITIES*

Jim Monke

INTRODUCTION

On May 21, 2020, the U.S. Department of Agriculture (USDA) temporarily expanded the Disaster Set- Aside (DSA) provision to allow flexibility for farm loan repayment due to the economic effects of the Coronavirus Disease 2019 (COVID-19) pandemic. The set-aside provision allows a borrower to move a loan payment owed to USDA's Farm Service Agency (FSA) to the end of the loan or, in the case of an annual operating loan, to extend the payment by a year. Interest continues to accrue on the deferred principal; neither the interest nor the principal is forgiven. The set-aside (deferment) is meant to provide financial relief and cash flow flexibility during a crisis.

* This is an edited, reformatted and augmented version of Congressional Research Service Publication No. IN11415, dated June 3, 2020.

Earlier during the pandemic, on March 26, 2020, USDA announced certain flexibilities in its loan-making and servicing procedures. These included relaxing deadlines, accommodating social distancing, and temporarily suspending loan accelerations and new foreclosures.

DISASTER SET-ASIDE PROVISION

Disaster set-aside has been available to FSA farm loan borrowers since 1994 (7 C.F.R. §766.51-61). The underlying authority (Section 331A of the Consolidated Farm and Rural Development Act, 7 U.S.C. §1981a) gives the Secretary of Agriculture discretion to defer principal and interest to forgo foreclosure if a borrower is temporarily unable to make payment due to circumstances beyond the borrower's control. USDA implemented the regulations to apply to natural disasters but has made temporary exceptions for losses due to low prices in 1998 and 1999 (64 *Federal Register* 392, January 5, 1999; and 65 *Federal Register* 31248, May 17, 2000) and now for the COVID-19 pandemic in 2020.

The DSA affects loans made and serviced by FSA, referred to as *direct loans*. The DSA covers direct farm ownership (real estate) loans and direct farm operating loans but may also apply to conservation and emergency loans that are also direct. The DSA does not apply to *guaranteed loans* that are made and serviced by an originating lender such as the Farm Credit System (FCS) or commercial banks. DSA is intended to allow borrowers to recover from losses without incurring additional debt or liquidating assets. The cost to the government is less than forgiving debt. Normally, one set-aside installment is allowed for each loan, but the temporarily expanded provision allows for a second set-aside if one already exists.

Farm Loan Servicing

Loan servicing refers to the process followed under law when a borrower is unable to make payments on a loan, the loan becomes delinquent, or the lender initiates foreclosure. Federal law prescribes a series of borrower rights for USDA farm loans, including notification, restructuring, write-down, and homestead protection.

For direct loans, USDA stated in a press release on March 26, 2020, that it would "temporarily suspend loan accelerations, non-judicial foreclosures, and referring [new] foreclosures to the Department of Justice" during the public health emergency.

For guaranteed loans, USDA is being more flexible working with originating lenders (e.g., commercial banks or the FCS), when the lenders request, to consider future lines of credit, temporary payment deferrals, or forbearances on foreclosures.

Availability of FSA Farm Loans

As of May 28, 2020—two-thirds through FY2020—FSA reported that about half of its farm loan funding authority remained unused. This incorporates both FY2020 appropriations and carryover balances. For operating loans, about 60% or more of direct and guaranteed loan authorities remain. For farm ownership loans, about one-quarter of the direct and guaranteed loan authorities remain, and less than 5% remains for borrowers who are not beginning farmers or members of socially disadvantaged groups. For emergency loans, nearly all available loan authority remains, since program use depends on natural disasters that have been few through the fall and winter and because a public health emergency is not among the criteria for making them available (*Table 1*).

For many farms at this point in the crop year, annual operating lines of credit were secured before the public health emergency was declared. The funding remaining, especially for operating loans, indicates that sufficient authority has likely been available for this crop year. The COVID-19 pandemic is undercutting farm income rather than immediately raising loan demand for many farms. Nonetheless, lower farm incomes may lead to future loan repayment difficulties, as indicated by the DSA and loan servicing flexibilities discussed above. Payments on many farm loans may not be due until later in the calendar year after harvest.

Table 1. USDA Farm Loans: FY2020 Authority and Availability Remaining (dollars in millions)

	FY2020 Loan Authority	Availability as of May 28, 2020	Percentage Remaining
Operating loans			
Direct loans	2,463	1,462	59%
Guaranteed loans	3,764	2,847	76%
Farm ownership loans			
Direct loans	1,875	523	28%
Guaranteed loans	2,750	802	29%
Emergency loans (direct)	122	121	99%

Source: FSA, farm loan programs funding (accessed June 1, 2020).

Note: FY2020 loan authority includes annual appropriations and carryover balances.

FSA Farm Loan Portfolio

On April 30, 2020, FSA released updated data about its outstanding loan portfolio and loan delinquencies (*Table 2*). FSA holds $12.8 billion of direct loans (including farm operating and farm ownership), which is about 3% of the $415 billion sector-wide farm debt total in 2019. The FSA delinquency rate on repayment remains steady at about 5%, especially on a seasonal basis compared to the previous year. This is generally a high rate relative to bank loans but is not abnormal for FSA borrowers, who are by definition unable to obtain credit elsewhere.

FSA guarantees $16.6 billion of loans made by other lenders, which is about 5% of an estimated $340 billion of farm loans made by commercial banks and the FCS. The delinquency rate for these loans is steady at 1.6%, consistent with the underlying loan performance at those lenders. Growth in the FSA portfolio since last year is faster than the growth in the sector-wide debt balance, indicating that more farms are experiencing financial stress as the farm economy has slowed even before the pandemic.

Table 2. USDA Farm Loan Portfolio and Delinquency Rates (dollars in millions)

Direct Loans			**Guaranteed Loans**	
	Portfolio	% Delinquent	Portfolio	% Delinquent
Sept. 30, 2018	11,437	4.7%	15,741	1.4%
Apr. 30, 2019	11,485	5.5%	15,761	1.6%
Sept. 30, 2019	12,250	4.5%	16,435	1.6%
Apr. 30, 2020	12,752	5.0%	16,583	1.6%

Source: FSA, farm loan programs loan servicing data as of 9/30/2019, and farm loan programs direct and guaranteed data as of 4/30/2020.

Notes: FSA direct loans are about 3% of the $415 billion sector-wide farm debt total in 2019. FSA guarantees are about 5% of an estimated $340 billion of the farm loans that are issued by the Farm Credit System and commercial banks.

In: COVID 19: Agricultural and Food Issues ISBN: 978-1-53618-561-4
Editor: Kristian Goldschmidt

Chapter 15

COVID-19-RELATED LOAN ASSISTANCE FOR AGRICULTURAL ENTERPRISES*

Robert Jay Dilger and Sean Lowry

INTRODUCTION

The Coronavirus Aid, Relief, and Economic Security Act (CARES Act; P.L. 116-136) created the Small Business Administration's (SBA's) Paycheck Protection Program (PPP) and Emergency Economic Injury Disaster Loan (EIDL) grants to provide short-term, economic relief to certain small businesses and nonprofits.

Important note: The SBA started accepting PPP and Emergency EIDL grant applications on April 3, 2020. The SBA stopped accepting new PPP and Emergency EIDL grant applications on April 16, because funding provided for these programs by the CARES Act was either exhausted (PPP) or nearing exhaustion (Emergency EIDL grants). The SBA resumed accepting PPP applications on April 27, following enactment of the

* This is an edited, reformatted and augmented version of Congressional Research Service Publication No. IN11357, updated May 5, 2020.

Paycheck Protection Program and Health Care Enhancement Act (P.L. 116- 139). The act provided an additional $321.335 billion for the PPP and $10 billion for Emergency EIDL grants.

The SBA continued to process Emergency EIDL grant applications submitted prior to April 16. However, because of the large number of Emergency EIDL grant applications submitted, the SBA did not accept new Emergency EIDL grant applications until May 4, 2020, and then only on a limited basis to accommodate agricultural enterprises. P.L. 116-139 included a provision (Division A, §101) to temporarily permit agricultural enterprises, as defined by Section 18(b) of the Small Business Act (15 U.S.C. §647(b)), with not more than 500 employees to receive Emergency EIDL grants and EIDLs through December 31, 2020.

This chapter will be updated to reflect any legislative changes affecting these programs.

PPP Loan Terms and Eligibility

PPP loans feature a two-year term at 1% interest, the waiver of the SBA's up-front loan guarantee and annual servicing fees, relaxed underwriting requirements, deferred payments for six months (interest does accrue), and loan forgiveness of up to 100% of the loan's principal amount under specified conditions related to the borrower's retention of employees and wages and the use of the funds for specified purposes, such as payroll and employee benefits.

PPP loans can be used for payroll costs; costs related to the continuation of group health care benefits during periods of paid sick, medical, or family leave, and insurance premiums; employee salaries, commissions, or similar compensations; payments of interest on any mortgage obligation (excluding any prepayment of or payment of principal on a mortgage obligation); rent (including rent under a lease agreement); utilities; interest on any other debt obligations that were incurred before February 15, 2020; and refinancing an SBA EIDL loan made between January 31, 2020, and April 3, 2020.

Agricultural enterprises, defined in Section 18(b) of the Small Business Act as "small business concerns engaged in the production of food and fiber, ranching, and raising of livestock, aquaculture, and all other farming and agricultural-related industries," may apply for a PPP loan if the applicant meets the PPP's eligibility requirements:

- any SBA 7(a) loan guarantee program-eligible small business;
- any business, 501(c)(3) nonprofit organization, 501(c)(19) veteran's organization, or tribal business not currently eligible that has not more than 500 employees; or, if applicable,
- the SBA's size standard in number of employees for the industry in which they operate.

Sole proprietors, independent contractors, and eligible self-employed individuals are also eligible to receive a covered loan.

SBA PPP guidance notes that, in addition to having 500 or fewer employees, agricultural producers, farmers, and ranchers are eligible for the PPP if the business fits within its revenue-based industry size standard, which, for most agricultural enterprises, is average annual receipts of no more than $1 million. Additionally, they can qualify for PPP loans if their business meets SBA's "alternative size standard," which is currently (1) maximum net worth of not more than $15 million, and (2) the business's average net income after federal income taxes (excluding any carry-over losses) for the two full fiscal years before the date of the application is not more than $5 million.

As of April 16, 2020, the SBA reports that 46,334 PPP loans, totaling $4.37 billion (1.28% of the total amount approved), went to agricultural, forestry, fishing, and hunting businesses.

EIDL Terms and Eligibility and Emergency EIDL Grants

EIDLs, if related to economic damages caused by the COVID-19 pandemic, feature a loan term of up to 30 years with 3.75% interest for small businesses and 2.75% interest for private nonprofit organizations. The loans may be used for fixed debts (rent, etc.), payroll, accounts payable, and some bills that could have been paid had the disaster (in this case, the COVID-19 pandemic) not occurred.

The CARES Act authorizes the SBA to provide an advance EIDL payment of up to $10,000 to borrowers adversely affected by the COVID-19 pandemic. The advance payment, referred to as an emergency EIDL grant in the CARES Act, does not have to be repaid, even if the applicant is subsequently denied an EIDL. Due to high demand, the SBA limited EIDL advance payments to $1,000 per employee, capped at $10,000.

Small businesses, most private nonprofit organizations, small agricultural cooperatives, small aquaculture businesses, and nurseries deriving less than 50% of their annual receipts from the production of nursery or other agricultural products are EIDL eligible. The CARES Act temporarily expands, through December 31, 2020, EIDL eligibility to include startups, cooperatives, and eligible ESOPs (employee stock ownership plans) with fewer than 500 employees, sole proprietors, and independent contractors.

Historically, agricultural enterprises, other than small agricultural cooperatives, small aquaculture enterprises, and eligible small nurseries, have not been eligible for EIDL because language was added to Section 18 of the Small Business Act prohibiting the SBA from duplicating "the work or activity of any other department or agency of the Federal Government," to exclude agricultural enterprises. At one time, the Small Business Act specifically mentioned the Farmers Home Administration as a source of program duplication, but that reference has been removed. Section 18 also specifies that if "loan applications are being refused or loans denied by such other department or agency responsible for such work or activities...

then, for purposes of this section, no duplication shall be deemed to have occurred."

As mentioned, P.L. 116-139 temporarily permits agricultural enterprises with not more than 500 employees to receive EIDL and Emergency EIDL grants through December 31, 2020.

The SBA stopped accepting new EIDL and Emergency EIDL grant applications on April 16, 2020, because funding provided by the CARES Act for these programs was nearly exhausted. The SBA continued to process EIDL and Emergency EIDL grant applications submitted prior to April 16 on a first- in, first-out basis. Because of the large number of EIDL and Emergency EIDL grant applications submitted, the SBA did not accept new EIDL and Emergency EIDL grant applications until May 4, and then only on a limited basis to accommodate agricultural businesses provided eligibility by the Paycheck Protection Program and Health Care Enhancement Act.

As mentioned, the SBA is also processing EIDL and Emergency EIDL grant applications from agricultural businesses that had submitted prior to the legislative change. These agricultural businesses do not need to reapply. All other EIDL and Emergency EIDL grant applications that were submitted before the SBA stopped accepting new applications on April 15 are being processed on a first-in, first-out basis.

As of April 24, 2020, the SBA reports that it had approved 38,984 COVID-19-related EIDL loans, totaling $7.9 billion, and nearly 1.2 million Emergency EIDL grants, totaling $4.8 billion.

In: COVID 19: Agricultural and Food Issues ISBN: 978-1-53618-561-4
Editor: Kristian Goldschmidt

Chapter 16

USDA RURAL DEVELOPMENT AND COVID-19: SUPPLEMENTAL FUNDING AND AGENCY ACTIONS*

Alyssa R. Casey

INTRODUCTION

The U.S. Department of Agriculture (USDA) Rural Development agency (RD) administers loan, grant, and technical assistance programs that support infrastructure, housing, and business development in rural areas. In response to the COVID-19 pandemic, Congress provided supplemental funding to certain RD programs. In addition, RD has taken a number of actions to provide relief for current program participants and assist potential applicants in applying for program funding.

---

* This is an edited, reformatted and augmented version of Congressional Research Service Publication No. IN11391, dated June 1, 2020.

CARES ACT PROVISIONS RELATED TO USDA RD PROGRAMS

Congress included supplemental funding for, and provisions related to, RD programs in the Coronavirus Aid, Relief, and Economic Security Act (CARES Act, P.L. 116-136). Division B of this act includes $25 million for the Distance Learning and Telemedicine (DLT) Program and $100 million for grants under the ReConnect broadband pilot program. It also provides a $20.5 million loan subsidy for the Business and Industry (B&I) Loan Guarantee Program, which will support approximately $1 billion in loan guarantees to businesses in eligible rural areas. The *loan subsidy* is the estimated cost of providing a certain level of loans, including administrative costs and the potential costs of borrower defaults.

In January 2020, prior to enactment of the CARES Act, RD began accepting applications for the second round of ReConnect funding. In March 2020, after President Trump signed the CARES Act into law, RD announced that it would use the supplemental CARES Act funding to award additional grants under the second round of ReConnect. The application window for the second round of ReConnect was open January 31-April 15, 2020. In April 2020, RD announced a second application window for FY2020 DLT funding and stated that it will use the supplemental CARES Act funding in addition to unused funds from the first FY2020 DLT application window. The application window for DLT is open April 14-July 13, 2020. In May 2020, RD began accepting applications for CARES Act B&I loan guarantees. RD changed certain terms from the traditional B&I Loan Guarantee Program, including increasing the maximum guarantee from 80% to 90% of the loan amount and increasing the maximum term for working capital loans from seven years to 10 years.

The CARES Act's foreclosure moratorium, single-family and multifamily mortgage forbearance provisions, and eviction moratorium apply to RD rural housing programs, including single-family and multifamily direct and guaranteed loans and rural rental assistance. Prior to enactment of the CARES Act, RD had implemented its own foreclosure moratorium and forbearance options as well as more limited eviction protections for its multifamily and single-family programs.

USDA RD Actions

In addition to the CARES Act, RD has taken a number of actions to respond to COVID-19. RD announced that lenders administering certain guaranteed loan programs may offer 180-day loan payment deferrals without prior agency approval. Borrowers of certain RD direct loans may request payment deferrals or other payment assistance. RD also extended application deadlines for certain loan and grant programs to allow applicants additional time to submit applications during the pandemic. On April 13, 2020, USDA published the COVID-19 Federal Rural Resource Guide, a compilation of federal funding and resources potentially available to rural entities for COVID-19 response. For the most up-to-date information on RD actions to respond to COVID-19, see their COVID-19 response webpage.

Policy Considerations

As Congress continues to respond to the impacts of COVID-19 in rural areas, it may consider a number of policy options related to RD programs. The CARES Act provided supplemental funding to Department of Housing and Urban Development rental assistance programs but not to RD rural rental assistance (Section 521). Some Members of Congress have called for additional funding for rural rental assistance in anticipation of increased funding needs as tenant incomes decline during the current economic downturn.

In response to economic hardship due to COVID-19 that some RD loan borrowers may be experiencing, RD has offered some payment flexibilities for current borrowers, as described above. Some stakeholders have proposed that RD also temporarily defer principal and interest payments on all RD direct loans.

Some RD borrowers provide utilities—such as electricity, water, or broadband—to rural customers. Many utility providers in rural areas were already facing low profit margins prior to COVID-19 due to few paying customers in some sparsely populated areas. Deferrals on RD loans could help utility providers that lack extensive financial reserves provide payment flexibility to customers and potentially avoid disconnecting utilities for customers struggling to pay their bills due to job losses or reduced incomes related to COVID-19.

School and workplace closures during COVID-19 have highlighted the so-called digital divide, or the gap between those with internet access and those without, and its implication for activities such as teleworking, distance learning, and buying groceries online. Congress may consider additional funding for RD programs that finance telemedicine facilities and broadband deployment. However, broadband deployment projects often take years to complete and would not bring immediate relief to rural residents lacking broadband during COVID-19. (The Federal Communications Commission, which also administers broadband deployment and adoption programs, has expanded certain broadband and telemedicine programs to respond to COVID-19.) The current economic downturn is also negatively impacting state and local fiscal conditions, causing a decline in state and local revenues and a simultaneous increase in demand for public services, including unemployment insurance and medical care. Congress could consider providing supplemental appropriations to RD programs that finance local public infrastructure, such as public utilities or health and safety facilities, to reduce the financial burden on state and local governments.

Entities eligible for RD programs—such as local governments, nonprofit organizations, and small businesses in rural areas—often lack the personnel or resources to prepare applications for federal funding. Congress may consider providing supplemental funding for RD programs that provide technical assistance or training to help applicants assess needs and apply for RD funding. If Congress provides supplemental funding for

RD programs, it may also consider providing additional funding for program monitoring and evaluation to track the distribution of funds and assess the efficacy of projects in helping rural areas respond to and recover from the impacts of COVID-19.

In: COVID 19: Agricultural and Food Issues ISBN: 978-1-53618-561-4
Editor: Kristian Goldschmidt

Chapter 17

SUPPLEMENTAL APPROPRIATIONS FOR AGRICULTURE AND RELATED AGENCIES DUE TO COVID-19*

Jim Monke

INTRODUCTION

In March and April 2020, Congress passed and the President signed four supplemental appropriations acts in response to the COVID-19 pandemic (P.L. 116-123, P.L. 116-127, P.L. 116-136, and P.L. 116-139; Table 1). This In Focus summarizes over $36 billion of appropriations and policy changes in the jurisdiction of the Agriculture appropriations subcommittees. These supplemental appropriations acts are referred to here by order of passage.

---

* This is an edited, reformatted and augmented version of Congressional Research Service Publication No. IF11491, Updated May 8, 2020.

Food and Drug Administration (FDA)

For FDA to respond to the public health emergency of the COVID-19 pandemic, the first, third, and fourth supplemental appropriations acts provide a total of $163 million. The funds are to develop medical countermeasures and tests (e.g., drugs, vaccines), and monitoring of medical product supply chains.

Nutrition Assistance

Access to food has been a concern, particularly in light of school closures. Rising unemployment may also increase participation in the Supplemental Nutrition Assistance Program (SNAP) and other food assistance programs. The second and third supplemental appropriations acts provide a total of $26 billion for nutrition assistance and give the U.S. Department of Agriculture (USDA) certain flexibilities to increase program access and accommodate social distancing. The Congressional Budget Office (CBO) estimated that SNAP policies in the second law will increase mandatory spending by more than $21 billion over FY2020-FY2021, some of which the third act provides.

Agricultural Producers

On April 17, 2020, USDA announced a $19 billion Coronavirus Food Assistance Program (CFAP), composed of $16 billion of direct payments to farmers and $3 billion of commodity purchases.

The CFAP has three funding sources. The $16 billion of direct payments is derived from $9.5 billion in the CARES Act (the third supplemental) and $6.5 billion from the Commodity Credit Corporation (CCC) using the Secretary's general discretion to support agriculture. As of the date of this chapter, USDA has not announced how it will distribute the direct payments regarding commodities, payment rates, payment limits, and other terms.

The $3 billion commodity purchase program uses indefinite authority in the Families First Coronavirus Relief Act (the second supplemental) to make purchases for emergency distribution. On April 21, USDA announced plans to implement a Farmers to Families Food Box Program by purchasing about $100 million per month of fresh produce, $100 million per month of dairy products, and $100 million per month of meat products (chicken and pork). Products purchased from farmers and processors are to be assembled into household-ready boxes for expedited distribution. On May 4, 2020, USDA announced $470 million of additional purchases with Section 32 authority. USDA intends to buy $170 million of produce, $120 million of dairy products, $80 million of poultry, $70 million of fish, and $30 million of pork for distribution through food banks. The third supplemental appropriation also replenishes by July up to $14 billion of funding availability for the CCC, which operates with a $30 billion line of credit with the U.S. Treasury. The $14 billion is not new spending, and will reimburse CCC for past spending. This reimbursement could increase opportunities for USDA to use its executive authority to provide future support.

DISASTER DESIGNATION

For producers, the majority of USDA's existing disaster response programs cover natural disasters (physical and production). The losses from the pandemic are economic and do not trigger agricultural disaster assistance programs such as emergency loans, despite the President's declaration of a public health emergency under the Stafford Act. However, USDA has announced some loan flexibilities, including payment deferrals and moratoriums on foreclosures.

Rural Development

The third supplemental appropriation provides $146 million for USDA rural development programs, including $100 million for broadband grants, $25 million for rural telemedicine and distance learning, and $20.5 million to support rural business loans.

USDA Agency Operations

The third supplemental provides $141 million to six USDA agencies or offices, including the Animal and Plant Health Inspection Service (APHIS), Agricultural Marketing Service (AMS), Food Safety and Inspection Service (FSIS), Foreign Agricultural Service (FAS), Farm Service Agency (FSA), and Office of the Inspector General (OIG). These funds, except for OIG, were part of the Administration's request on March 17, 2020, for the third supplemental.

Most of the USDA request was to replace a combined $100 million of user fee revenues that are expected to decline for APHIS due to reduced air passenger traffic and for AMS because of reduced grading, inspections and audit services.

The $36 million combined amount for FSIS and FSA is to support temporary employees and adjustments to respond to COVID-19 workload demands. FAS received $4 million to repatriate staff and provide temporary housing.

Agricultural Businesses

In addition to the support for agricultural producers mentioned above, other provisions outside the jurisdiction of agriculture appropriations may provide loans and grants to agriculture-related businesses.

Table 1. Summary of supplemental appropriations for agriculture and related agencies due to COVID-19 jurisdiction of the House and Senate Agriculture Appropriations Subcommittees (dollars in millions)

Department/Agency/Program[a]	P.L. 116-123 3/6/2020	P.L. 116-127 3/18/2020	P.L. 116-136 3/27/2020	P.L. 116-139 4/24/2020
Department of Health and Human Services				
Food and Drug Administration	61		80	22
Department of Agriculture				
Food and Nutrition Service				
Supplemental Nutrition Assistance Program (SNAP)		—[b]		
SNAP reserve			15,510	
Northern Mariana Is., Puerto Rico, Am. Samoa		100	200	
Food Distribution on Indian Reservations			100	
Child Nutrition Programs			8,800	
The Emergency Food Assistance Program		400	450	
Special Supplemental Nutrition Program for Women, Infants, and Children		500		
Office of the Secretary (purchases and producer support)		—[c]	9,500	
Commodity Credit Corporation (CCC)			[14,000][d]	
Rural Development				
Rural e-Connectivity Pilot Program (ReConnect)			100	
Distance Learning and Telemedicine Program			25	
Business & Industry Loans			20.5	
Animal and Plant Health Inspection Service			55	
Agricultural Marketing Service			45	

Table 1. (Continued)

Department/Agency/Program[a]	P.L. 116-123 3/6/2020	P.L. 116-127 3/18/2020	P.L. 116-136 3/27/2020	P.L. 116-139 4/24/2020
Food Safety and Inspection Service			33	
Foreign Agricultural Service			4	
Farm Service Agency			3	
Office of Inspector General			0.75	
Total in the Jurisdiction of Agriculture Appropriations	61	1,000	34,926	22

Source: Compiled by CRS, using identified public laws and CBO score of H.R. 6201, Families First Coronavirus Response Act.

[a] All funding in this table is designated as emergency and does not count against budget caps. This table excludes other support that agriculture or rural entities may receive from other portions of these supplemental appropriations acts.

[b] CBO estimates that certain SNAP policy changes would increase mandatory spending by $21.24 billion in FY2020-FY2021.

[c] Includes authority for USDA to purchase commodities for emergency distribution; used for the $3 billion food purchase portion of CFAP.

[d] The $14 billion for CCC is not new spending and is not part of the $34.9 billion subtotal of the act. It is a reimbursement for prior obligations and increases the availability of funding (i.e., access to borrowing authority) for USDA to make future obligations.

INDEX

A

D

G

H

I

N

O

P

Q

R

S

T

U

V

W